Inverse Problems of Mathematical Physics

Related Titles of Interest

Modelling Non-Linear Wave Processes
Yu. A. Berezin

Inverse Sturm–Liouville Problems
B. M. Levitan

Electromagnetic Inverse Profiling: Theory and Numerical Implementation
A. G. Tijhuis

Group Theoretical Methods in Physics
Proceedings of the Third Yurmala Seminar, Yurmala, USSR, 1985

For further information please contact the Publisher

Inverse Problems of Mathematical Physics

V. G. ROMANOV

Siberian Branch, Computing Centre, USSR Academy of Sciences

Translated by L. Ya. Yuzina

VNU SCIENCE PRESS BV Utrecht, The Netherlands

VNU Science Press BV
P.O. Box 2093
3500 GB Utrecht
The Netherlands

First English edition published 1987

CIP-DATA KONINKLIJKE BIBLIOTHEEK, DEN HAAG
Inverse problems of mathematical physics/V. G. Romanov:
transl. [from the Russian] by L. Ya. Yuzina.—Utrecht:
VNU Science Press
With ref.
ISBN 90-6764-056-5 bound
SISO 517 UDC 53.01
Subject heading: mathematical physics.

Phototypesetting by Thomson Press (India) Limited, New Delhi
Printed in Great Britain by
Butler & Tanner Ltd, Frome and London

Contents

Foreword

The monograph surveys basic features of the comparatively newly emerged theory of inverse problems for equations of mathematical physics, i.e. of the problems aimed at defining the coefficients of a differential equation through some functionals of its solution. Problems of the kind arise in various fields of science when trying to describe internal characteristics of the substance where physico-chemical processes take place by the results of observing these processes within the available range of measurements.

The book offers in-depth coverage of inverse problems for second-order equations and for hyperbolic systems of first-order equations, including the kinematic problem of seismology, the Lamb dynamic problem for equations of the theory of elasticity, and the problem of electrodynamics.

The book is intended for specialists in applied mathematics, physics, and geophysics, as well as for university students of corresponding fields.

V. G. YAKHNO

NAUKA, Moscow 1984

Chapter 1
Introduction

1.1. Inverse problem concept: examples of formulating inverse problems

Mathematical physics commonly deals with problems of the following type: a differential equation, with its solution obeying certain additional conditions given; as a rule, the additional conditions single out one of the solutions of the differential equation from a whole set of possible solutions. There exists a differential classification based on their types, each class characterized by a typical formulation of problems. For instance, the Cauchy problem or the boundary problem for hyperbolic and parabolic equations, the Dirichlet and Newmann problems for equations of elliptical type; a characteristic feature of these typical problems being their correctness. Henceforth, correct boundary problems considered in mathematical physics will be referred to as direct ones.

Setting a direct problem implies the idea of setting a certain number of functions, one part of these functions determining the differential equation (for instance, linear equation coefficients), the other part determining the boundary conditions. A direct problem solution results in the fact that a new function, i.e. the boundary problem solution, is determined by a given set of functions, and a certain new operator, conditioned by the direct problem data, arises. Assume that some of the functions commonly set in direct problems are unknown (it is their determination that is of interest) and instead of them some additional information on the direct problem solution is given. Such problems will be referred to as inverse ones and will be the subject of this book.

Additional information on the direct problem solution (or on a solution of a series of direct problems) may be of various types. It can be, for instance, the solution proper given on a certain set of independent variables or some integral characteristics of the solution (solution functionals). If in an inverse problem the functions considered are members of the differential equation only, the task is to determine the differential equation. There exist some other types of inverse problems, for example, problems of determining initial and boundary conditions. Consider some examples of inverse problems.

Example 1. Let $q(x)$ be a continuous throughout the X-axis function. Consider the Cauchy problem for the function $u(x, y)$

$$\left[\frac{\partial}{\partial x} - \frac{\partial}{\partial y} - q(x)\right] u = 0, \quad (x, y) \in R^2 \tag{1.1}$$

$$u(x, 0) = \phi(x), \quad x \in R. \tag{1.2}$$

When functions q, ϕ are given, problems (1.1) and (1.2) are set correctly. For the solution of the problem to be classical (the solution is continuous together with its partial derivatives in the space R^2), it is sufficient to require continuous differentiability of the function $\phi(x)$. Let this condition be fulfilled. Now consider the problem of determining the function $q(x)$ on the basis of the following information on the solution of problems (1.1) and (1.2)

$$u(0, y) = \Psi(y), \quad y \in R. \tag{1.3}$$

Analysis of the problem does not present any difficulty. Indeed, the solution of problems (1.1) and (1.2) is set by

$$u(x, y) = \phi(x + y) \exp\left(\int_{x+y}^{x} q(\xi)\,\mathrm{d}\xi\right), \quad (x, y) \in R^2.$$

Condition (1.3) results in the equality

$$\psi(y) = \phi(y) \exp\left(\int_{y}^{0} q(\xi)\,\mathrm{d}\xi\right), \quad y \in R. \tag{1.4}$$

Equality (1.4) shows that uniqueness of the inverse problem solution depends greatly on the measure of the set of zeros of function $\phi(x)$. If the measure equals zero, the inverse problem has a unique solution, while in case of a positive measure there exists an unlimited number of solutions. Without going into the details of a general situation, let us assume that $\phi(x) \neq 0, x \in R$. In this case the inverse problem solution is unique and for its existence it is necessary and sufficient that the function $\psi(y)$ possesses the following properties: (1) $\psi(y)$ is continuously differentiable for $y \in R$, and (2) $\psi(y)/\phi(y) > 0$, $y \in R$, $\psi(0) = \phi(0)$.
In this case the inverse problem solution is set by the formula

$$q(x) = -\frac{\mathrm{d}}{\mathrm{d}x} \ln \frac{\psi(x)}{\phi(x)}, \quad x \in R. \tag{1.5}$$

Example 2. In connection with problems (1.1) and (1.2) let us consider the problem of determining continuous the function $q(x)$ on the basis of the following information on the solution:

$$u(x, h) = \psi(x), \quad x \in R, \quad h > 0. \tag{1.6}$$

The problem formulated is analogous to that of determining $q(x)$ by the equality

$$\psi(x) = \phi(x + h) \exp\left(\int_{x+h}^{x} q(\xi)\,\mathrm{d}\xi\right). \tag{1.7}$$

Let $\phi(x) \neq 0, x \in R$. For the solution to exist the following conditions are necessary: (1) $\psi(x)$ is continuously differentiable for $x \in R$; (2) $\psi(x)/\phi(x + h) > 0$, $x \in R$.

When the above conditions are fulfilled, (1.7) is equivalent to the equalities

$$q(x) - q(x+h) = f(x), \qquad f(x) \equiv \frac{\mathrm{d}}{\mathrm{d}x} \ln \frac{\psi(x)}{\phi(x+h)},$$

$$\int_0^h q(\xi)\,\mathrm{d}\xi = -\ln \frac{\psi(0)}{\phi(h)}. \tag{1.8}$$

Since at $\psi = 0$ there exist non-zero solutions for (1.8), i.e. periodic functions $q(x)$ with period h for which the integral over the length $[0, h]$ equals zero, the inverse problem has no unique solution. A unique solution requires some additional limitations as regards the class of functions $q(x)$, which allow one to eliminate nontrivial solutions of homogeneous equations (1.8). As a limitation of this kind, one can choose, for instance, the condition that $q(x)$ decrease at $x \to \infty$. Under the further supposition that the function $f(x)$ at $x \to \infty$ is characterized by a sufficiently great degree of decreasing, for instance, $f(x) = O(x^{-\alpha})$, $\alpha > 1$, a unique solution to the inverse problem exists and is governed by the formula

$$q(x) = \sum_{k=0}^{\infty} f(x + kh).$$

Example 3. The task is to determine the initial condition of a limited length of heated wire when the solution of the boundary problem

$$\left(\frac{\partial}{\partial t} - \frac{\partial^2}{\partial x^2}\right) u = 0, \quad 0 < x < \pi, \quad t > 0 \tag{1.9}$$

$$u(0, t) = u(\pi, t) = 0, \quad t > 0, \quad u(x, 0) = \phi(x), \quad 0 \leqslant x \leqslant \pi \tag{1.10}$$

is known at a fixed time $t = T$

$$u(x, T) = \psi(x), \quad 0 \leqslant x \leqslant \pi. \tag{1.11}$$

Under fairly general assumptions as regards the functions $\phi(x)$, the solution to problems (1.9) and (1.10) can be found by the Fourier method

$$u(x, t) = \sum_{n=1}^{\infty} \mathrm{e}^{-n^2 t} \phi_n \sin nx. \tag{1.12}$$

Here ϕ_n are the Fourier coefficients of the function $\phi(x)$ with respect to the function system $\sin nx, n = 1, 2, \ldots$. Setting $t = T$ in (1.12) we obtain

$$\psi(x) = \sum_{n=1}^{\infty} \mathrm{e}^{-n^2 T} \phi_n \sin nx, \quad x \in [0, \pi]. \tag{1.13}$$

Therefore

$$\phi_n = \mathrm{e}^{n^2 T} \psi_n, \quad n = 1, 2, \ldots$$

where ψ_n are the Fourier coefficients of the function $\psi(x)$. Since coefficients ϕ_n; $n = 1, 2, \ldots$ unambiguously determine any function $\phi(x) \in L^2[0, \pi]$, the inverse problem solution is unique within the class of functions $\phi(x) \in L^2[0, \pi]$. Note that in this case the limiting condition (1.10) is fulfilled only in the following sense

$$\lim_{t \to +0} \int_0^{\pi} [u(x,t) - \phi(x)]^2 \, dx = 0.$$

For the existence of the inverse problem solution it is necessary and sufficient to fulfil the condition

$$\sum_{n=1}^{\infty} \psi_n^2 \, e^{2n^2 T} < \infty.$$

In an applied sense, of greatest interest are the problems of finding variable coefficients of differential equations. Differential equations describe, as a rule, physical processes, while their coefficients pertain to physical characteristics of the medium wherein the processes take place. For instance, if the equation $u_{tt} = a^2 u_{xx}$ describes small oscillations of a string, then $a = \sqrt{(T/\rho)}$, where T is the string tension (constant within the approximation of small oscillations), ρ is its density. If density changes from point to point, a is the function of a string point, i.e. $a = a(x)$. It has been proved that density $\rho(x)$ of a semi-limited string can be found if one knows the time oscillation regime of a free end of the string (see Section 2.7). The system of equations of elasticity for an isotropic medium considers three parameters: density of the substance and the Lamé parameters characterizing elastic properties of substances; the Maxwell system takes into account the coefficients of dielectric and magnetic permeability and the coefficient of electric conductivity. It is often the case that these parameters are not constant but change from point to point, i.e. they are the coordinate functions. Since these parameters often elude direct measurement, the problem of determining properties of a substance is, essentially, an inverse one. As initial data for its solution, one often employs the characteristics of the physical process measured at the area boundary, which is often so in geophysical problems.

The principal task of geophysics is to study the internal structure of the Earth based on surface and subsurface observations. The available geophysical data convincingly prove that the physical parameters of the Earth are essentially the functions of three variables, changing considerably with depth. But since direct measurement of the physical parameters can be carried out only within an extremely narrow subsurface layer (suffice to say that the deepest geophysical borehole is only about 11 km deep), the only method of studying the Earth's internal structure is that based on solving inverse problems. Consider some of them here.

Earthquakes, being quite a regular phenomenon in various regions, cause seismic waves to propagate inside the Earth, which in turn, results in oscillation points on the Earth's surface. In a physical sense, the picture of seismic wave propagation over the Earth's surface resembles that of smooth water surface after having been disturbed by a stone. Along the Earth's surface, off the earthquake epicentre a seismic wave front moves that separates the points of the surface that have been disturbed by the source (i.e. by the earthquake hypocentre) from those still in a state of rest. As compared to the waves propagating on a water surface, there is only one distinction: in an elastic body, unlike liquid, there exist both longitudinal and transverse waves. The velocity of propagation of the longitudinal waves, which basically contribute to the wavefront on the Earth's surface, exceeds that of the transverse waves, while in a liquid medium there are no transverse waves at all. The existence of transverse waves results in the fact that along the Earth's surface, following the sharp boundary separating a disturbed medium from an undisturbed one, there moves the front of a transverse seismic wave. However, since this front propagates within the medium having been disturbed (as if dropped a second stone into the water shortly after the first one), this front cannot be always traced at all the points. To complete the picture, recall that besides longitudinal and transverse waves (often referred to as volume waves), there also exists on the surface of the Earth surface waves arising due to the ground–air boundary. A true kinematic picture of wave processes can only be formed when all the fronts propagating along the Earth's surface are taken into account.

Now consider the propagation of the front of a seismic wave along the Earth's surface. The front shape and the velocity of its points along the surface depend on the velocity of the seismic wave inside the Earth. For instance, for a spherically symmetrical model of the Earth, when the wave velocity is only depth-dependent, the fronts are represented by circumferences propagating from the epicentre with the velocity depending on the velocity of the seismic wave. In case when the velocity depends on all the three coordinates, the wave front can be represented by an intricate curve. At the beginning of this century geophysicists speculated on if it was possible to determine the velocity of seismic wave propagation inside the Earth judging by the picture of the propagation along the Earth's surface. Note, that this velocity depends on the elastic properties of a substance, for instance, velocities of the longitudinal v_p and transverse v_s waves are related to the Lame parameters λ, μ and density ρ through the following:

$$v_p = \sqrt{[(\lambda + 2\mu)/\rho]}, \quad v_s = \sqrt{(\mu/\rho)}.$$

Velocities v_p, v_s provide important information on the Earth's structure, that is why geophysicists seek ways to find v_p, v_s as functions of a point of the Earth's surface.

Now express the above thought mathematically. Let D be a domain of a three-dimensional space R^3 limited by a boundary S, wherein transfer of signals with the finite positive velocity $v(x)$, $x = (x_1, x_2, x_3)$ takes place. Let $\tau(x, x^0)$ be the time required for the signal to get from the point x^0 to the point x. The function $\tau(x, x^0)$ satisfies (see Chapter 3) the equation

$$|\nabla_x \tau(x, x^0)| = \frac{1}{v(x)} \tag{1.14}$$

and the condition

$$\tau(x, x^0) = O(|x - x^0|), \quad x \to x^0. \tag{1.15}$$

Here ∇_x is the gradient calculated with respect to the variable x. The equation $\tau(x, x^0) = t$ defines the wavefront from the source x^0 at the moment of time t. Therefore, the problem of finding $v(x)$ by the law of front motion along the surface

S can be set as follows: to find $v(x)$ for $x \in D$, if we know the function $\tau(x, x^0)$ for $x \in S$ and $x^0 \in M$ (M is a certain set belonging to $D \cup S = \bar{D}$), which is a solution to (1.14) under condition (1.15).

This problem is referred to as the inverse kinematic problem of seismic surveying. In solving the problem one often sets $M = S$, thus allowing a simple geometrical interpretation. Consider the characteristics of (1.14) and denote by $\Gamma(x, x^0)$ the characteristics passing through the points x, x^0 [assume that each pair of points x, x^0 in the domain D corresponds to a single $\Gamma(x, x^0)$]. Then the function $\tau(x, x^0)$ is calculated by

$$\tau(x, x^0) = \int_{\Gamma(x,x^0)} \frac{|\mathrm{d}x|}{v(x)}, \qquad |\mathrm{d}x| = \left(\sum_{k=1}^{3} \mathrm{d}x_k^2 \right)^{1/2}. \tag{1.16}$$

In this case

$$\mathrm{d}\tau = |\mathrm{d}x|/v(x). \tag{1.17}$$

Equation (1.17) sets the Riemann's metric in D, the curve $\Gamma(x, x^0)$ is the shortest in this metric (in other words geodesic), and $\tau(x, x^0)$ is its length. Hence, an inverse kinematic problem at $M = S$ can be formulated as follows: to find in D a metric of type (1.17) [i.e. to find function $v(x)$], if distances between the points of the boundary of the domain D in this metric are known.

The inverse kinematic problem was initially considered by the German geophysicists Herglotz and Wiehert at the beginning of the century. Under the supposition of a spherical symmetry of the Earth: $v = v(r)$, $r = |x|$, they showed that the function $v(r)$ could be found if the function $\tau(x, x^0)$ for a fixed point $x^0 \in S = \{x : |x| = 1\}$ and for an arbitrary one $x \in S$, is known. The obtained formula is of a local character: if x runs not along the whole boundary of a unit sphere S, but only through a spherical sector $|x - x^0| \leqslant \Delta$, then $v(r)$ is determined for $r \in [r_0, 1]$, $r_0 = r_0(\Delta)$. In this case $r_0(\Delta) \to 1$, if $\Delta \to 0$, and $r_0 \to 0$, if $\Delta \to 2$. Herglotz and Wiehert also found that the solution of the inverse kinematic problem is unique only when the condition

$$\frac{\mathrm{d}}{\mathrm{d}r}\left(\frac{r}{v(r)} \right) > 0$$

is met, which means monotonic decrease of the function $m(r) = r/v(r)$ with depth. An analogue to the above condition for a semi-space (one more popular geophysical model) is the condition of monotonic velocity increase with depth. If the condition of the function $m(r)$ monotonicity is not fulfilled, the inverse kinematic problem has no unique solution.

The solution of the inverse kinematic problem for the spherically symmetrical model of the Earth has played an important part in our understanding the internal structure of the planet. For the first time analysis of seismic data has resulted in graphic representation of distribution of longitudinal and transverse waves along the Earth's radius and in determining basic elements of the Earth's structure: crust, coating, and nucleus.

The spherically symmetrical model of the Earth is, of course, only a rough approximation of the real structure. Systematic deviations between seismic data and theoretical results derived on the basis of the above models have stimulated elaboration of more detailed variations of the inverse kinematic problem—two- and three-dimensional. The results of these studies are the subject of Chapter 3 which is devoted to the inverse kinematic seismic problem.

The drawbacks of the kinematic approach in studying the internal structure of the Earth, manifested in ambiguity of the solution of the inverse kinematic problem, make the investigators pay greater attention to the wave field on the surface. Available surface seismic stations record complete earthquake seismograms, i.e. they measure the vector of displacement of surface points $u(x, t, x^0)$ as the time function t. The function $u(x, t, x^0)$, as that of variables x, t, satisfies the dynamical system of equations of elasticity

$$\rho(x)\frac{\partial^2 u_i}{\partial t^2} = \operatorname{div} \sigma_i + F_i(x, t, x^0), \quad i = 1, 2, 3 \tag{1.18}$$

$$\sigma_{ij} = \mu(x)\left(\frac{\partial u_i}{\partial x_j} + \frac{\partial u_j}{\partial x_i}\right) + \lambda(x)\delta_{ij} \operatorname{div} u, \quad i, j = 1, 2, 3.$$

Here u_i is a projection of the displacement vector u on the x_i-axis; σ_i is the vector of stresses on an area with the normal parallel to the x_i-axis and the components σ_{i1}, σ_{i2}, σ_{i3}; δ_{ij} is the Kronecker symbol; F_i is a projection of the vector of external forces F onto the x_i-axis. System (1.18) should be supplemented with a boundary condition, which in case of a free surface S, is reduced to the fact that the vector of stresses on a plane with the direction n, i.e. with the direction of the normal to S, must equal zero

$$\sigma_n|_S = \sum_{i=1}^{3} \sigma_i|_S \cos(n, x_i) = 0. \tag{1.19}$$

The simplest case of earthquakes can be simulated by the function $F(x, t, x^0)$ of type

$$F(x, t, x^0) = \delta(x - x^0) f(t, x^0) \tag{1.20}$$

where $\delta(x - x^0)$ is the Dirac function concentrated at the point x^0; $f(t, x^0) = 0$ for $t < 0$. The function $f(t, x^0)$ for an inverse problem must be included in a number of unknown functions together with the parameters ρ, λ, and μ defining an elastic medium. An inverse problem can be formulated as follows: to find in domain D the medium density ρ, the Lamé elastic parameters λ and μ and the vector function of the source $f(t, x^0)$ provided we know the vector $u(x, t, x^0)$ for $x \in S$, $t \geqslant 0$, and $x^0 \in M \subset \bar{D}$. This problem is called the inverse dynamic problem of seismic surveying. Note, that the formulation of the problem given above is far from being complete. Even under the assumption (1.20), provided $x^0 \bar{\in} S$, coordinates of the point x^0 (an earthquake hypocentre) cannot be known in advance and, hence, must be included in the number of the required parameters [1]. But even formulated in such a simplified manner, the inverse dynamic problem of seismic surveying is still highly complex and can be solved only under heavy additional assumptions (see Chapter 4).

Associated with gravitational and magnetic surveying is one more inverse problem—the theory of potential. This is generally formulated as follows: outside a certain domain limited by a surface S, a potential (of bulky masses, ordinary layer or magnetic) is given, which is induced by a body lying inside S; the task is to find the shape and density of the body. This formulation of the problem is used, for instance, in ore geophysics for locating deposits of mineral resources by gravity anomalies.

Consider a simple mathematical model associated with the inverse problem of Newton's potential. In space R^3 inside a limited surface S let there be located a body with density $\sigma(x)$. Let D denote a medium of the function $\rho(x)$ [i.e. a set of points where $\rho(x) \neq 0$], and $\chi(D)$ denote a characteristic function of the point set D. Newton's potential $u(x)$, induced by this body, obeys the Poisson equation

$$\Delta u = -4\pi\rho(x)\chi(D). \tag{1.21}$$

The inverse problem is to find the functions $\rho(x)$, $\chi(D)$ using the known outside S solution to (1.21). Under fairly general assumptions on the functions $\rho(x)$, $\chi(D)$, a solution to (1.21) can be written as a Poisson formula

$$u(x) = \int_D \frac{\rho(\xi)\,d\xi}{|x-\xi|}, \quad \xi = (\xi_1, \xi_2, \xi_3). \tag{1.22}$$

Thus, the above problem is reduced to solving a first-order integral equation [at given $u(x)$ for x lying outside S] with respect to the function $\rho(x)$ and the function defining the boundary D. Note that this problem has no unique solution without additional assumptions since bodies having different density and different support of the density, can induce one and the same gravitational field outside S (for instance, a body with a small support and great density and a body with greater support and less density). Therefore, as regards density certain simplifications are usually made; the commonest being to consider it given. Under such a supposition the first theorem of uniqueness for an inverse problem of the Newton's potential was formulated by Novikov in 1938 [2]. Later the inverse problem of the potential theory has been investigated by Tikhonov, Ivanov, Sretensky, Lavrent'ev, Prilepko, Strakhov and their co-workers. At present the theory of the potential inverse problem has been substantially developed and numerical methods for solving such problems have also been elaborated.

One of most important applied problems is that of the inverse problem of electromagnetic surveying. Interaction of electromagnetic fields with media is described by a system of Maxwell equations

$$\operatorname{rot} H = \frac{\partial}{\partial t}(\varepsilon E) + \sigma E + j, \qquad \operatorname{rot} E = -\frac{\partial}{\partial t}(\mu H) \tag{1.23}$$

involving as coefficients the dielectric ε and magnetic μ medium permeabilities, as well as its electric conductivity σ. Electromagnetic surveying frequently measures electric and magnetic fields on the Earth's surface, which are induced by a certain source of electric oscillations. The task is to find the electromagnetic parameters ε, μ, and σ on the basis of such data. In practice, a determined medium is viewed as a system of homogeneous layers and under this assumption a direct problem is solved, followed by selecting the layer parameters to fit the calculation results to observational ones. Up to the 1950s electromagnetic surveying had been based on direct current, until in 1949 Tikhonov [3] proved uniqueness of definition of the function σ, which is only depth-dependent, within the class of piecewise analytical functions by alternating electromagnetic field measured on the Earth's surface.

The problem discussed is closely related, as regards its formulation and interpretation, to the so-called method of magnetotelluric probing of the Earth (MTP), when one measures variations in the electromagnetic field of the Earth induced by its rotation around the axis and by interaction of its electromagnetic field with solar radiation. Such measurements carried out at special observatories can also lead to certain conclusions as regards electromagnetic properties of ores composing the Earth's crust and coating.

Recall two more inverse problems: the Sturm–Liouville problem and the problem of scattering, associated with one and the same differential operator

$$l_q \equiv -\frac{d^2}{dx^2} + q(x).$$

The theory of common differential equations yields the Sturm–Liouville problem, referring to finding eigenfunctions and eigenvalues of the differential operator l_q under certain boundary conditions. In a regular case (length $[a, b]$ is finite, the function $q(x)$ on $[a, b]$ is continuous), the problem is to determine the λ values which allow a solution to the boundary problem

$$l_q y = \lambda y, \qquad y'(a) - hy(a) = 0, \qquad y'(b) + Hy(b) = 0 \tag{1.24}$$

other than zero, and corresponding to them solutions. It is well known that the eigenvalues of this problem form a denumerable set λ_n, $n = 1, 2, \ldots$, with the only point of concentration $+\infty$. The corresponding eigenfunctions $y_n(x)$ can be normalized by the condition

$$y_n(a) = 1, \qquad y_n'(a) = h. \tag{1.25}$$

The inverse Sturm–Liouville problem is set as follows: the spectral function $\rho(\lambda)$ of problem (1.24) is known, the task is to find $q(x)$. The function $\rho(\lambda)$ is nondecreasing, determined along the whole numerical axis of λ and defined by the condition $\rho(-\infty) = 0$. In the case in question it is a step function: it is constant for the λ value which is in between two neighbouring eigenvalues λ_{n-1}, λ_n and at the point λ_n it has a finite step numerically equal to $\| y_n \|^{-2}$, where $\| y_n \|$ is the norm of the function $y_n(x)$ obeying conditions (1.25), in $L_2[a, b]$

$$\| y_n(x) \|^2 = \int_a^b y_n^2(x)\, dx.$$

Thus, two sequences of numbers

$$\lambda_1, \lambda_2, \ldots, \lambda_n, \ldots \quad \text{and} \quad \| y_1 \|, \| y_2 \|, \ldots, \| y_n \|, \ldots$$

define function $\rho(\lambda)$ completely.

The spectral function $\rho(\lambda)$ also exists in a singular case, when the function $q(x)$

infinitely increases at $x \to b$, remaining continuous on a semi-open length $[a, b)$, or when $b \to +\infty$.

The first results on the Sturm–Liouville inverse problem were obtained by Ambartzumyan in 1929 and Borg in 1945, while its intensive theoretical exploration began only in the 1950s. Of fundamental importance were the papers by Marchenko, Krein, Gel'fand and Levitan. This book does not consider the theory of the problem; only in Sections 2.6 and 2.8 are there some references concerning relations between one-dimensional formulations of certain inverse problems for equations in partial derivatives and the problem in question. The theory of the Sturm–Liouville inverse problem is considered in detail in Levitan [4]; Levitan and Sargasyan [5]; and Marchenko [6].

An inverse problem of scattering for the differential operator l_q is commonly associated with the class of functions $q(x)$ fairly quickly decreasing at $x \to +\infty$. For simplicity assume that $q(x) \in C[0, \infty)$ and $q(x) = 0$ for $x \geqslant b > 0$, and consider the limit $(0, \infty)$ on solutions to the equation

$$l_q y = \lambda y, \quad \lambda \in (-\infty, \infty) \tag{1.26}$$

with the boundary condition

$$y'(0) - hy(0) = 0. \tag{1.27}$$

Limited and nonzero solutions to problems (1.26) and (1.27) are called eigenfunctions and the corresponding λ values eigenvalues. It can be easily deduced that the eigenvalues fill in the whole real positive axis, forming a continuous spectrum of the operator of the boundary condition. Indeed, denote by $y(x, \lambda)$ the solutions to (1.26) with the Cauchy data

$$y(0, \lambda) = 1, \qquad y'(0, \lambda) = h. \tag{1.28}$$

It is obvious, that at any $\lambda > 0$ a solution to problem (1.26), (1.28) for $x \geqslant b$ can have the form

$$y(x, \lambda) = C(\lambda)[e^{i\sqrt{\lambda}x} - S(\lambda)e^{-i\sqrt{\lambda}x}]$$

in which case $C(\lambda) \neq 0$ (otherwise, due to uniqueness of the Cauchy problem solution with the zero Cauchy data at $x = b$ we would have $y(x, \lambda) \equiv 0$, $x \in (0, \infty)$, which contradicts (1.28). Thus, for any $\lambda > 0$ there exists other than zero and $(0, \infty)$ a solution to problems (1.26) and (1.27), i.e. any $\lambda > 0$ is an eigenvalue of this problem. The situation is quite different at $\lambda \leqslant 0$, in which case at $x \geqslant b$ a solution to problems (1.26) and (1.28) can be written as a linear combination of two linearly independent solutions to the homogeneous equation $y'' + \lambda y = 0$ $(\lambda \leqslant 0)$, one of them infinitely growing at $x \to +\infty$. Only in the case when the coefficient at the infinitely growing function becomes zero do we get the eigenfunction of problems (1.26) and (1.27). Analysis has shown that for $\lambda < 0$ there exists only a finite number of eigenvalues $\lambda = \lambda_n$, $n = 1, 2, \ldots, N$, in which case the eigenfunctions for $x \geqslant b$ are as follows:

$$y_n(x) \equiv y(x, \lambda_n) = m_n e^{-\sqrt{(|\lambda_n|)}x}$$

and hence $y_n(x) \in L_2(0, \infty)$.

A set of values of $S(\lambda)$, λ_n, $\| y_n \|$, $n = 1, 2, \ldots, N$, is called the data of scattering of (1.26) and (1.27). An inverse scattering problem is formulated; data on scattering are given, the task is to find l_q.

Scattering data prove to define uniquely a spectral function of the differential operator l_q under condition (1.27), in which case the inverse scattering problem is reduced to the Sturm–Liouville inverse problem. Marchenko [6] suggested a method of finding function $q(x)$ directly from scattering data. A matrix variation of the problem in question is considered by Agranovitch and Marchenko [7]. A modern approach to the theory of the problem is detailed by Sabatier and Chadan [8]; the book by Lax and Phillips [9], devoted to an abstract variation of the inverse problem, refers to the same direction. Faddeev [10] gives a detailed survey of the results obtained in the inverse scattering problem.

For the equations in partial derivatives Alekseev and Megrabov [11–13 and 14–21] considered different variations of the inverse problem of scattering of plane waves in non-homogeneous laminar media. The problem of defining the shape of a domain using scattering data is the subject of [22–26].

In particular, [27–29] are devoted to inverse problems for transport equations, [30, 31] to those in cardiology, and [32–34] to those in seismic surveying.

There is also a whole number of special monographs [35–43] and papers [44–47] devoted to different ways of formulating inverse problems.

Hereafter, refer to an inverse problem as n-dimensional ($n \geqslant 1$), when at least one of the required functions depends on n variables, while the other functions depend on not more than n variables.

1.2. On correctness of direct and inverse problems of mathematical physics

The notion of correctness is usually considered in the theory of mathematical physics (see, for instance, [26, 48–50]). When dealing with inverse problems in mathematical physics it proves expedient to alter this notion to some extent. Therefore, consider the problem briefly.

The theory of differential equations states that a differential equation defines a whole set of solutions, which depend on a certain number of arbitrary constants or arbitrary functions. For the problem to have a definite physical sense, it is necessary to single out of the whole class of the differential equation solutions the solution describing a given physical process. It is usually realized by giving some additional information on the solution; for differential equations, for instance, it commonly means setting initial and boundary conditions. Here are some examples to illustrate the above.

The equation

$$u_{tt} = c^2 u_{xx} \tag{1.29}$$

describes the process of small oscillations of a string. In this case the function $u(x, t)$ is equal to the string displacement at a point x at a moment of time t from the equilibrium position coinciding with x-axis. The numerical coefficient c characterizes the velocity of signal transfer in the string (the speed of sound). In the case of a limited string $x \in [0, l]$, fixed at the points $x = 0$, $x = l$, the boundary conditions for (1.29) have the form

$$u(0, t) = u(l, t) = 0, \quad t \geqslant 0. \tag{1.30}$$

Setting boundary conditions, however, is not sufficient for singling out a unique solution to (1.29). This can only be completely defined if some additional initial conditions are used. In this case these are: the string displacement and the initial velocities of its points

$$u(x, 0) = \phi(x), \qquad u_t(x, 0) = \psi(x), \quad 0 \leqslant x \leqslant l. \tag{1.31}$$

In mathematical physics, setting conditions (1.30) and (1.31) is shown to define the solution to (1.29) uniquely.

In the case of the equation of heat conduction

$$u_t = \Delta u, \tag{1.32}$$

considered within a cylindrical domain $G = \Omega \times [0, \infty]$, where $\Omega \subset R^n$ is the domain limited with a closed surface S, to single out a unique solution it is sufficient to set a heat regime on the surface S, for instance

$$u(x, t) = 0, \quad x \in S, \quad t \geqslant 0 \tag{1.33}$$

and the initial distribution of temperature inside S

$$u(x, 0) = g(x), \quad x \in \Omega. \tag{1.34}$$

Thus, setting initial and boundary conditions is aimed at singling out a unique solution out of the whole class of solutions to the differential equation. But the number of such conditions should be minimal, otherwise they can contradict one another.

Recall one more important thing: when considering a given differential equation with set initial and boundary conditions, we can pose the task of finding its solutions belonging, generally speaking, to different functional domains, whose choice depends on the physical sense of the problem. For instance, we can consider the problem of finding the solution, obeying conditions (1.29)–(1.31), in the class of functions continuous together with the second-order derivatives in the domain $D = \{(x, t): 0 \leqslant x \leqslant l, t \geqslant 0\}$; or in the class of functions $W_2^2(D)$, having generalized second-order derivatives, summed with the square; or in the class of $W_2^1(D)$; or in the class of generalized functions. In other words, one may choose a functional space of solutions to a differential equation in a quite arbitrary way. But in this case the functions involved in the boundary and initial conditions cannot be arbitrary; they must ensure that the solution belongs to the chosen functional space. To do this they must belong to a certain special functional space corresponding to the solution domain. This becomes clear if one considers the problems for differential equations from the point of view of functional analysis. Choose a space U for the solutions of a differential equation. The differential equation together with some additional conditions defines the operator A, that relates any solution $u \in U$ to the set of functions involved in the additional (initial and boundary) conditions. For (1.29) these are the functions ϕ and ψ, for (1.32) it is the function g. Viewing this set of functions as the element f of the functional space F, one comes to the conclusion that solving a problem for a differential equation is equivalent to solving the operator equation

$$Au = f \tag{1.35}$$

under the condition that $u \in U$.

For a solution to this equation to exist, it is necessary and sufficient that f be an image of a certain element of $u \in U$, i.e. it must belong to a set of values of the operator A. Thus, a set of data of the problem is reasonably defined by setting the solution space U.

The conditions of existence and uniqueness of the solution to (1.35) guarantee the existence of the inverse operator A^{-1}

$$u = A^{-1}f \tag{1.36}$$

that solves the problem by fitting the solution $u \in U$ to the initial data of the problem, i.e. to the element f. One should, however, remember that the chief aim of solving mathematical problems is to describe certain physical processes mathematically. In this case the initial data are measured experimentally and since these measurements cannot be absolutely accurate they contain measurement errors. For a mathematical model to describe a real physical process, the problem should be supplemented with some additional requirements reflecting in a physical sense small variations in the solution under slight changes in the initial data or, as is put conventionally, stability of the solution to small perturbations in the initial data. Let us now express the above in mathematical terms.

Let u be the solution to (1.35), wherein the operator A functions from the normalized space U into the normalized space F. A solution to (1.35) is said to be stable to small variations in the right-hand side of $f \in F$, if for any $\varepsilon > 0$ there exists $\delta > 0$, so that for any element $\tilde{f} \in F$, for which

$$\| f - \tilde{f} \|_F < \delta$$

the inequality

$$\| u - \tilde{u} \|_U < \varepsilon.$$

holds. In this case $Au = f$, $A\tilde{u} = \tilde{f}$.

Stability of the nonlinear operator A depends, generally speaking, on the element f; at one set of elements the operator being stable, at the other unstable. In the case of the linear operator A, either stability or instability takes place for all the elements $f \in F$ at once.

A mathematical problem of solving (1.35) obeying the requirements of existence, uniqueness and stability of the solution towards small variations in the initial data is called a correct problem. Formulate the notion in detail.

Let A be the operator functioning from the normalized space U to the normalized space F. The problem of solving (1.35) is said to be formulated correctly in the spaces U, F if it obeys the following requirements: (1) the problem solution exists at any $f \in F$ and belongs to U; (2) the problem solution is unique in the space U; (3) the problem solution is stable at any element $f \in F$.

Note, that the problem can be correct within one pair of spaces and incorrect in the other. It is clear, for instance, that when broadening the space F the requirement of existing a solution at any $f \in F$ is violated. In case when the operator A is linear, for the correctness of problem (1.35) in a pair of the Banach spaces U, F it is necessary and sufficient that for the operator A there exists a limited inverse A^{-1}, functioning from F into U, the domain of determining the inverse operator coinciding with the space F (see [51, p. 507]).

Here is an example of a correct problem. Let us consider (1.29) in the semiplane $D = \{(x, t): -\infty < x < \infty, t > 0\}$. In this case it is sufficient to set the Cauchy data

$$u(x, 0) = \phi(x), \qquad u_t(x, 0) = \psi(x), \qquad -\infty < x < \infty. \tag{1.37}$$

The solution to problems (1.29) and (1.37) is unique and is set by the d'Alembert formula

$$u(x, t) = \frac{1}{2}[\phi(x - ct) + \phi(x + ct)] + \frac{1}{2c}\int_{x-ct}^{x+ct} \psi(\xi)\,\mathrm{d}\xi. \tag{1.38}$$

For any arbitrary $T \in (0, \infty)$ let us consider a characteristic triangle

$$D_T = \{(x, t): 0 \leqslant t \leqslant T - |x|\}.$$

Equation (1.38) yields that for $\phi \in C^2[-T, T]$, $\psi \in C^1[-T, T]$ the solution is $u \in C^2(D_T)$, besides, small variations in the functions ϕ and ψ in the norm of the corresponding spaces result in small changes in the solution in the norm $C^2(D_T)$. Thus, the problem is correct for $\phi \in C^2[-T, T]$, $\psi \in C^1[-T, T]$ and $u \in C^2(D_T)$ at any $T \in (0, \infty)$.

Note, that the same problem will be incorrect, if one considers ϕ and ψ to be continuous functions and preserve for the solution the same space $C^2(D_T)$, as the first and the third conditions of correctness are not fulfilled in this case.

Now consider an example of an incorrect problem, when no demand on smoothness of the initial information results in its correctness. Such a problem was first formulated by Hadamard [48] to emphasize the importance of the third condition of correctness.

Let $u = u(x, y)$ be a solution to the Laplace equation

$$\Delta u = 0 \tag{1.39}$$

in a semiband $G = \{(x, y): -\pi < x < \pi, y > 0\}$, obeying the conditions

$$u(-\pi, y) = u(\pi, y) = 0, \qquad u(x, 0) = 0, \qquad u_y(x, 0) = \mathrm{e}^{-\sqrt{n}} \sin nx. \tag{1.40}$$

It can be easily demonstrated that a solution to this problem is the function

$$u(x, y) = \frac{1}{n}\mathrm{e}^{-\sqrt{n}} \sin nx \sinh ny. \tag{1.41}$$

The solution of the given problem can be proved to be unique. At $n \to \infty$ the function $\mathrm{e}^{-\sqrt{n}} \sin nx$, which represents the data of problems (1.39) and (1.40), tends uniformly to zero together with the derivatives of all orders. Nevertheless, the problem solution, as is seen from (1.41), at any fixed $y > 0$ and sufficiently great n has the shape of a harmonic curve with any great amplitude. Hence, any small changes in the problem data in $C^2[-\pi, \pi]$ or $W_2^l[-\pi, \pi]$ at any finite l result in far from being small variations in the solution. Thus, problems (1.39) and (1.40) are incorrect due to instability.

The examples cited show that there generally exist two types of incorrect problems. There are problems that are incorrect in one set of spaces but can be made correct in another one. It is useful to refer to such problems as weakly incorrect ones. Also, there exist incorrect problems [as, for instance, (1.39) and (1.40)], which are incorrect in any functional spaces whose norm makes use of a finite number of derivatives. Problems of this kind will be called strongly incorrect. A great number of examples of incorrect problems discussed in appendices can be found [41, 52, 53].

For the problems which are not correct in a classical sense Tikhonov [54] suggested a new notion of correctness that is physically justified for many applied problems. Within his approach, Tikhonov considers a very compact set $m \subset U$, that is essentially narrower than the whole space U. Let the image of the set m when reflected with the operator A in the space F is the set R, i.e. $R = Am$.

Equation (1.35) is termed conventionally correct (or correct, by Tikhonov) if the following conditions are met: (1) it is *a priori* known that the solution to the problem exists and belongs to a certain given set m of the functional space U; (2) the solution is unique in the set m; (3) for any $\varepsilon > 0$ there exists such $\delta > 0$, that for any $f, \tilde{f} \in R = Am$ and those that $\| f - \tilde{f} \|_F < \delta$, the inequality $\| u - \tilde{u} \|_U < \varepsilon$ holds.

The set m is referred to as the set of the problem correctness.

Note some peculiarities of the above definition which make it different from the classical one. In the classical approach to the definition of correctness, the problem data f are assumed to belong to a certain metric space F. The first of the requirements to the correctness demands to prove that for any $f \in F$ the solution to (1.35) exists and belongs to U. The corresponding requirement postulates that the solution of (1.35) belongs to the given set m, without discussing the properties of the set $R = Am$ necessary for it. It is associated with the fact that for many incorrect problems the conditions of belonging $f \in Am$ cannot be verified practically.

The requirement of uniqueness of the solution to the problem on the set m coincides with the classical requirement of uniqueness of the solution to the problem in the space of solutions U.

Consider now the third requirement. Within the classical approach the solution to the problem exists at any $f \in F$ and belongs to the space U. It would be essential that any small change in the element f in the norm of the space F is accompanied with a small change in the solution in the norm of the space U. Tikhonov stated that correctness of the problem does not require numerous initial data at all, but postulates belonging of the solution to a certain set m. In this case any small change in the initial data in the norm of F can result in the solution to the problem not existing at all or, if it does, does not belong to m. Therefore, it would be reasonable to modify the requirement of continuous dependence of the solution on the problem data by demanding continuous dependence for such variations in f that retain the solution within the set m, which is the third requirement of correctness.

In connection with the definition of conventional correctness note also that two out of three requirements of correctness are modified in it. Of course, if in case of an unstable problem one can constructively describe a set of data, i.e. the set $R = Am$, then it is unreasonable to abandon the first requirement of correctness in its classical variation.

Many problems that are incorrect in the classical sense are conventionally correct. For instance, the equations (1.39) and (1.40) are conventionally correct (see [55]) in the set of functions $u(x, y)$, $x \in [-\pi, \pi]$, $y \in [0, 1]$, belonging at any fixed y to the space $L_2[-\pi, \pi]$ and obeying the additional condition

$$\int_{-\pi}^{\pi} u^2(x, 1)\,dx \leqslant M^2$$

where M is a given constant.

When the operator A is continuous and m is a compact set, the general theorems of functional analysis yield that fulfilment of the first and the second conditions of correctness implies that the third condition is also fulfilled. It is this that serves as the basis for defining conventional correctness of the problem. A problem unstable throughout the whole space F can prove stable on the set $R = Am$, a correct type of stability estimation depending in this case greatly on the set m. Within a new approach to the definition of correctness the focus point is uniqueness of the problem solution, i.e. establishing the theorem of uniqueness.

I shall now clarify why the Tikhonov notion of correctness proves fairly simple when considering numerous problems and especially in studying inverse problems for differential equations. Consider, for instance, the inverse problem on oscillations of a string. When the string density varies from point to point, the process of small oscillations of the string is described by (1.29), where $c = c(x)$. Let it be the problem of defining the function $c(x)$ by certain known functionals from the solution of the direct problem (1.29)–(1.31), by considering displacement of some points of the string at different moments of time, for instance (see Chapter 2). The physical formulation of the problem prompts that the solution should be sought in the class of functions that are positive and limited from above and below by certain constants induced by a possible set of materials used in designing the string. It is possible that some other properties of the string are known, that its density changes continuously or in a piecewise-continuous way, for instance. As a result, we can outline the set m, which the function $c(x)$ must belong to, beforehand.

Provided the data of the problem are taken from a particular physical experiment with a string whose $c(x)$ belongs to m, there is no doubt that there exists a real $c(x)$ corresponding to these measurements. The problem is how to find it and whether this information is sufficient to define it uniquely. But the point is that the data are measured with a certain error δ and, hence, instead of the element f we get $\tilde{f}$, in which case $\| f - \tilde{f} \| < \delta$. The element $\tilde{f}$ will generally not belong to the set Am. Consequently, returning to the physical experiment, we should say that there exists the function $c(x)$ and the theoretical data f calculated for it will differ from the experimental ones $\tilde{f}$ not greater than by δ.

Let the data of the physical experiment permit us to find $c(x)$ from the class $c(x) \in m$ and, besides, there exists a continuous dependence of the problem solution on the data on the set $R = Am$. Let there exist in the space F, a sphere $S(\tilde{f}, \delta)$ with the radius δ with the centre at the point $\tilde{f}$. It is known that there is at least one element $f \in R$ in the sphere and, hence, there exists its inverse image $c(x) \in m$. If in the sphere $S(\tilde{f}, \delta)$ there are also other elements from R, they are separated from one another not greater than by 2δ. Therefore, at small δ (i.e. at high accuracy of the experiment) due to continuous dependence of $c(x)$ on $f \in R$ their inverse images also differ slightly, which means that at any inverse images considered, the function $c(x)$ is close to the real one. This situation is typical of many problems considered in geophysics.

At present there are a great number of papers devoted to the investigation of various problems of mathematical physics as regards conventional correctness, with the view of creating stable methods of numerical solution of classically incorrect problems (see [37, 41, 52, 54, 56–75]).

Chapter 2

Inverse problems for the operator $L_q = \frac{\partial^2}{\partial t^2} - \frac{\partial^2}{\partial x^2} + q(x)$

Studying any theory usually begins with considering the commonest conclusions illustrating its essence and basic peculiarities. For the theory of inverse problems of mathematical physics, fairly pithy in both mathematical and applied sense, there are the inverse problems associated with the simplest equation of a hyperbolic type

$$L_q u \equiv \left(\frac{\partial^2}{\partial t^2} - \frac{\partial^2}{\partial x^2} + q(x) \right) u = F(x, t). \tag{2.1}$$

The task is usually to find the operator I_q, i.e. the function $q(x)$, in line with this or that information on the solutions to (2.1). The choice of this information determines the formulation of an inverse problem and the properties of its solution. With one set of data the problem can be correct, with another incorrect, but normally the formulation of an inverse problem is conditioned by its physical applications. As in the case of direct problems of mathematical physics, formulation of inverse problems can be, on the one hand, fairly diverse, and on the other, there exist typical formulations of the problems associated with most urgent applied problems. The problems discussed in this chapter are of the two kinds: one set of problems, for instance, those considered in Sections 2.1 and 2.2, serve to illustrate the method of investigation, the other problems are important in an applied sense.

2.1. Problems with nonfocused initial data

For (2.1) consider in the domain $D = \{(x, t): -\infty < x < \infty,\ t > 0\}$ the Cauchy problem with the following initial conditions

$$u|_{t=0} = \phi(x), \quad u_t|_{t=0} = \psi(x), \quad -\infty < x < \infty. \tag{2.2}$$

At fixed q, F, ϕ, and ψ the problem is correct, provided the functional spaces for the problem data and the solution space are chosen in an appropriate way. In particular, (2.1) and (2.2) are correct if q, ϕ, and ψ belong to the space of continuous functions $C(-\infty, \infty)$, and $F, u \in C(D)$. In this case the function $u(x, t)$ is a generalized solution to (2.1) and (2.2). I shall now impose somewhat more rigid requirements on the data of the problem, i.e. that $q(x) \in C(-\infty, \infty)$, $\phi \in C^2(-\infty, \infty)$, $\psi \in C^1(-\infty, \infty)$, $F, F_t \in C(D_T)$, $D_T = \{(x, t): -\infty < x < \infty, 0 < t \leqslant T\}$.

The above conditions, as we shall see, guarantee the existence of a classical solution to (2.1) and (2.2) in D_T, i.e. $u \in C^2(D_T)$.

The following inverse problem can be formulated for (2.1) and (2.2): to find $q(x) \in C(-\infty, \infty)$, if at fixed functions ϕ, ψ and F the solution to (2.1) and (2.2) is known in the set $x = x_0$, $t \geq 0$ together with its derivative with respect to x

$$u(x_0, t) = f_1(t), \qquad u_x(x_0, t) = f_2(t), \quad t \geq 0. \tag{2.3}$$

In other words, the task is to find $q(x)$ by known functions ϕ, ψ, F, f_1, and f_2. Indeed, if one finds $q(x)$, one will find the function $u(x, t)$ as a solution to (2.1) and (2.2) as well. That is why the inverse problem is often formulated as the problem of finding the functions q, u.

In chosen classes of functions q, ϕ, ψ, and F, the functions f_1, f_2, being traces of functions u, u_x on the straight line $x = x_0$, cannot be arbitrary; in particular, they must obey certain conditions of smoothness. Hence, the necessary conditions for the functions f_1, f_2 must be derived from the properties of solutions to the direct problems (2.1) and (2.2), typical of the whole class of differential operators L_q, $q \in C(-\infty, \infty)$. It is these properties that will now be considered.

Let $\Delta(x_0, t_0)$ denote a triangle on the plane x, t limited by the x-axis and by the characteristics of (2.1) drawn through the point (x_0, t_0) (Fig. 2.1).

Lemma 2.1. If for a $t_0 > 0$, $q(x) \in C[x_0 - t_0, x_0 + t_0]$, $\phi(x) \in C^2[x_0 - t_0, x_0 + t_0]$, $\psi(x) \in C^1[x_0 - t_0, x_0 + t_0]$, F, $F_t \in C(\Delta(x_0, t_0))$, *then in domain* $\Delta(x_0, t_0)$ *there exists a single classical solution to problems (2.1) and (2.2).*

To prove the lemma, reduce problems (2.1) and (2.2) to an integral equation with respect to the function $u(x, t)$. For this purpose we can make use of the d'Alembert formula

$$u_0(x, t) = \tfrac{1}{2}[\phi(x - t) + \phi(x + t)] + \frac{1}{2}\int_{x-t}^{x+t} \psi(\xi)\, d\xi + \frac{1}{2}\iint_{\Delta(x,t)} F(\xi, \tau)\, d\xi\, d\tau, \tag{2.4}$$

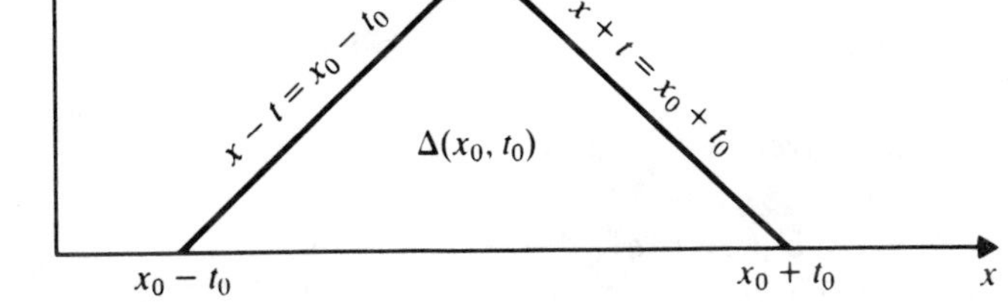

Figure 2.1.

which provides a solution to the following:

$$\left(\frac{\partial^2}{\partial t^2}-\frac{\partial^2}{\partial x^2}\right)u_0=F(x,t) \tag{2.5}$$

$$u_0\bigg|_{t=0}=\phi(x),\qquad \frac{\partial u_0}{\partial t}\bigg|_{t=0}=\psi(x). \tag{2.6}$$

Transferring the term containing qu to the right-hand part of the equality in (2.1) and using (2.4), we get an integral equation for the function $u(x,t)$

$$u(x,t)=u_0(x,t)-\frac{1}{2}\iint_{\Delta(x,t)} q(\xi)u(\xi,\tau)\,d\xi\,d\tau,\quad (x,t)\in\Delta(x_0,t_0). \tag{2.7}$$

I shall now demonstrate that (2.7) determines a single continuous solution within the domain $\Delta(x_0,t_0)$. For this purpose the method of successive approximations will be used, presenting $u(x,t)$ as a series

$$u(x,t)=\sum_{n=0}^{\infty} u_n(x,t) \tag{2.8}$$

where $u_n(x,t)$, $n\geqslant 1$, are obtained by

$$u_n(x,t)=-\frac{1}{2}\iint_{\Delta(x,t)} q(\xi)u_{n-1}(\xi,\tau)\,d\xi\,d\tau,$$

$$n\geqslant 1,\quad (x,t)\in\Delta(x_0,t_0). \tag{2.9}$$

It can be easily verified that under the lemma conditions $u_0(x,t)\in C^2(\Delta(x_0,t_0))$. Equation (2.9) demonstrates that all $u_n(x,t)\in C(\Delta(x_0,t_0))$. Denote

$$U_n(t)=\max_{x_0+(t-t_0)\leqslant x\leqslant x_0-(t-t_0)} |u_n(x,t)|,\quad 0\leqslant t\leqslant t_0$$

$$\|u\|_k=\sum_{|\alpha|\leqslant k}\max_{(x,t)\in\Delta(x_0,t_0)} |D^\alpha u(x,t)|,\quad k=0,1,2 \tag{2.10}$$

$$D^\alpha=\frac{\partial^{|\alpha|}}{\partial x^{\alpha_1}\partial t^{\alpha_2}},\quad \alpha=(\alpha_1,\alpha_2),\quad |\alpha|=\alpha_1+\alpha_2.$$

Then (2.9) yields the estimate

$$U_n(t)\leqslant\frac{1}{2}\int_0^t \max_{x_0-(t_0-t)\leqslant x\leqslant x_0+(t_0-t)}\left|\int_{\tau-t+x}^{-\tau+t+x} q(\xi)u_{n-1}(\xi,\tau)\,d\xi\right| d\tau$$

$$\leqslant \|q\|_0\int_0^t (t-\tau)U_{n-1}(\tau)\,d\tau,\quad n\geqslant 1, 0\leqslant t\leqslant t_0. \tag{2.11}$$

Applying it successively for $n=1,2,\ldots,$ we get

$$U_n(t)\leqslant(\|q\|_0)^n\frac{t^{2n}}{(2n)!}\|u_0\|_0,\quad n\geqslant 1. \tag{2.12}$$

Estimate (2.12) shows that series (2.8) converges uniformly in the domain $\Delta(x_0,t_0)$

since it is majorized in D_T by a convergent numerical series

$$\|u_0\|_0 \sum_{n=0}^{\infty} \frac{(\|q\|_0 t_0^2)^n}{(2n)!}$$

and hence, determines the continuous within the domain $\Delta(x_0, t_0)$ function $u(x, t)$ which is the solution to (2.7). This solution is unique, since the uniform equation corresponding to (2.7)

$$u(x,t) = -\frac{1}{2}\iint_{\Delta(x,t)} q(\xi)u(\xi,\tau)\,d\xi\,d\tau \qquad (2.13)$$

has only the zero solution in the class of continuous in $\Delta(x_0, t_0)$ functions. Indeed, if

$$\max_{x_0-(t_0-t)\leqslant x\leqslant x_0+(t_0-t)} |u(x,t)| = U(t)$$

then (2.13) yields

$$U(t) \leqslant \|q\|_0 \int_0^t (t-\tau)U(\tau)\,d\tau, \quad t\in[0,t_0]. \qquad (2.14)$$

It is known that the only solution to the integral inequality (2.14) is $U(t) = 0$ and, hence, $u(x, t) = 0$, $(x, t)\in\Delta(x_0, t_0)$.

Note that the obtained estimate (2.12) yields the following estimate of the solution to (2.1) and (2.2)

$$\|u\|_0 \leqslant \|u_0\|_0 \coth(\sqrt{(\|q\|_0)}t).$$

Thus, the existence of a unique solution to (2.7) has been proved. Let us now make sure that in $\Delta(x_0, t_0)$ this solution indeed has continuous derivatives up to the second order and determines the classical solution to (2.1) and (2.2). For this purpose write the double integral over the domain $\Delta(x, t)$ from (2.7) as an iterated one

$$u(x,t) = u_0(x,t) - \frac{1}{2}\int_{x-t}^{x+t} q(\xi)\left[\int_0^{t-|x-\xi|} u(\xi,\tau)\,d\tau\right]d\xi, \quad (x,t)\in\Delta(x_0,t_0). \qquad (2.7')$$

Since $u_0\in C^2(\Delta(x_0, t_0))$, the expression in the right-hand part of (2.7′) has first-order partial derivatives with respect to x, t. Consequently, the left-hand part of equality (2.7′), i.e. the function $u(x, t)$, also has first-order derivatives

$$u_t(x,t) = \frac{\partial}{\partial t}u_0(x,t) - \frac{1}{2}\int_{x-t}^{x+t} q(\xi)u(\xi, t-|x-\xi|)\,d\xi, \qquad (2.15)$$

$$u_x(x,t) = \frac{\partial}{\partial x}u_0(x,t) - \frac{1}{2}\int_{x-t}^{x+t} q(\xi)u(\xi, t-|x-\xi|)\operatorname{sign}(\xi-x)\,d\xi. \qquad (2.16)$$

The equalities obtained show that u_t, u_x are continuous functions in $\Delta(x_0, t_0)$. This fact indicates that the right-hand parts of the equalities have continuous in $\Delta(x_0, t_0)$ first-order derivatives with respect to x, t, and, hence, the function $u(x, t)$ has continuous in $\Delta(x_0, t_0)$ second-order derivatives. As the expressions for these derivatives will be needed later, write them as

$$u_{tt}(x,t)=\frac{\partial^2}{\partial t^2}u_0(x,t)-\tfrac{1}{2}[q(x+t)u(x+t,0)+q(x-t)u(x-t,0)]$$
$$-\frac{1}{2}\int_{x-t}^{x+t}q(\xi)u_t(\xi,t-|x-\xi|)\,d\xi \tag{2.17}$$

$$u_{xt}(x,t)=\frac{\partial^2}{\partial x\partial t}u_0(x,t)-\tfrac{1}{2}[q(x+t)u(x+t,0)-q(x-t)u(x-t,0)]$$
$$-\frac{1}{2}\int_{x-t}^{x+t}q(\xi)u_t(\xi,t-|x-\xi|)\,\mathrm{sign}\,(\xi-x)\,d\xi \tag{2.18}$$

$$u_{xx}(x,t)=\frac{\partial^2}{\partial x^2}u_0(x,t)-\tfrac{1}{2}[q(x+t)u(x+t,0)+q(x-t)u(x-t,0)]$$
$$+q(x)u(x,t)-\frac{1}{2}\int_{x-t}^{x+t}q(\xi)u_t(t-|\xi-x|)\,d\xi. \tag{2.19}$$

Equations (2.17)–(2.19) demonstrate that u_{tt}, u_{xt}, $u_{xx}\in C(\Delta(x_0,t_0))$. Thus, the lemma is proved.

Implication. When fulfilling the lemma conditions, the functions f_1, f_2, which are the data of the inverse problem, must satisfy the following requirements of smoothness

$$f_1(t)\in C^2[0,t_0], \qquad f_2(t)\in C^1[0,t_0]. \tag{2.20}$$

It is quite natural that the functions f_1, f_2 must satisfy not only the above requirements of smoothness but also some other conditions of agreement with the data of the direct problem. These conditions are obtained by comparing the values of the function $u(x,t)$ and its derivatives at the point $(x_0,0)$, calculated by the data given in (2.2) and (2.3). The conditions of agreement are as follows:

$$\phi(x_0)=f_1(0), \quad \psi(x_0)=f_1'(0), \quad \phi'(x_0)=f_2(0), \quad \psi'(x_0)=f_2'(0). \tag{2.21}$$

Conditions (2.20) and (2.21), which are necessary requirements for the existence of the inverse problem solution, are also sufficient for its existence in the small. The following theorem is valid.

Theorem 2.1. If at a $t_0>0$, the conditions of Lemma 2.1 for the functions ϕ, ψ, and F, conditions (2.20) and (2.21) for the functions f_1, f_2, and the condition

$$|\phi(x)|\geqslant\alpha>0, \quad x\in[x_0-t_0, \quad x_0+t_0] \tag{2.22}$$

are fulfilled, then at sufficiently small $h>0$ the solution to the inverse problem on the intercept $[x_0-h, x_0+h]$ exists, is unique and belongs to the class $C[x_0-h, x_0+h]$.

The smallness of h is conditioned by two factors: firstly, h cannot exceed t_0, secondly, by the method used to prove the theorem, which is based on the principle of contracted mapping. To use this method, it is necessary that inequality $h < h^*$, where h^* is determined by the data of the inverse problem, holds.

To prove the theorem, begin with setting $x = x_0$ in (2.17) and (2.18) and using data from (2.2) and (2.3). This results in two equalities

$$f_1''(t) = \frac{\partial^2}{\partial t^2} u_0(x_0, t) - \tfrac{1}{2}[q(x_0+t)\phi(x_0+t) + q(x_0-t)\phi(x_0-t)]$$

$$-\frac{1}{2}\int_{x_0-t}^{x_0+t} q(\xi)u_t(\xi, t-|x_0-\xi|)\,\mathrm{d}\xi \tag{2.23}$$

$$f_2'(t) = \frac{\partial^2}{\partial x \partial t} u_0(x,t)|_{x=x_0} - \tfrac{1}{2}[q(x_0+t)\phi(x_0+t) - q(x_0-t)\phi(x_0-t)]$$

$$-\frac{1}{2}\int_{x_0-t}^{x_0+t} q(\xi)u_t(\xi, t-|x_0-\xi|)\,\mathrm{sign}\,(\xi - x_0)\,\mathrm{d}\xi, \quad t\in[0, t_0] \tag{2.24}$$

from which we can find $q(x)$ for $x \geqslant x_0$ and $x \leqslant x_0$

$$q(x) = \frac{1}{\phi(x)}\left\{ -f_1''(x-x_0) - f_2'(x-x_0) + \left[\left(\frac{\partial^2}{\partial t^2} + \frac{\partial^2}{\partial x \partial t}\right)u_0(x,t)\right]_{x=x_0, t=x-x_0}\right.$$

$$\left. - \int_{x_0}^{x} q(\xi)u_t(\xi, x-\xi)\,\mathrm{d}\xi \right\}, \quad x_0 \leqslant x \leqslant x_0 + t_0$$

$$q(x) = \frac{1}{\phi(x)}\left\{ -f_1''(x_0-x) + f_2'(x_0-x) + \left[\left(\frac{\partial^2}{\partial t^2} - \frac{\partial^2}{\partial x \partial t}\right)u_0(x,t)\right]_{x=x_0, t=x_0-x}\right.$$

$$\left. - \int_{x}^{x_0} q(\xi)u_t(\xi, \xi-x)\,\mathrm{d}\xi \right\}, \quad x_0 - t_0 \leqslant x \leqslant x_0.$$

These two equalities can be united into one

$$q(x) = q_0(x) + \frac{1}{\phi(x)}\int_{x_0}^{x} q(\xi)u_t(\xi, |\xi - x|)\,\mathrm{d}\xi \cdot \mathrm{sign}\,(x_0 - x), \quad x\in[x_0 - t_0, x_0 + t_0]$$

$$q_0(x) = \frac{1}{\phi(x)}\left[-f_1''(|x-x_0|) - f_2'(|x-x_0|)\,\mathrm{sign}\,(x-x_0)\right.$$

$$\left. + \left(\frac{\partial^2}{\partial t^2} + \mathrm{sign}\,(x-x_0)\frac{\partial^2}{\partial x \partial t}\right)u_0(x,t)|_{x=x_0, t=|x-x_0|}\right]. \tag{2.25}$$

Despite the presence of a discontinuous multiplier $\mathrm{sign}\,(x - x_0)$, the function $q_0(x)$ is continuous on the intercept $[x_0 - t_0, x_0 + t_0]$, which results from fulfilling the last of the agreement conditions (2.21). Indeed, since

$$-f_2'(0) + \frac{\partial^2}{\partial x \partial t}u_0(x,t)|_{x=x_0, t=0} = -f_2'(0) + \psi'(x_0) = 0$$

the multiplier preceding sign $(x - x_0)$ turns to zero at $x = x_0$.

Consider in the domain (x_0, t_0) equalities (2.7), (2.15) and (2.25) which determine the system of nonlinear integral second-order equations with respect to the functions u, u_t, and q. This system of equations is of a small parameter, which is expressed through the measure of the integration domain in them. Due to the presence of this small parameter, one can apply to the system, in a sufficiently small domain, the principle of contracted mapping. Indeed, write the system of equations as an operator equation

$$g = Ag \tag{2.26}$$

where g is the vector function of two variables x, t with the components g_1, g_2, and g_3, in which case

$$g_1(x,t) = u(x,t), \quad g_2(x,t) = u_t(x,t), \quad g_3(x,t) \equiv g_3(x) = q(x)$$

and the operator A is determined in the set of functions $g \in C(\Delta(x_0, t_0))$ and, in line with equalities (2.7), (2.15), and (2.25), has the form $A = (A_1, A_2, A_3)$

$$\begin{aligned} A_1 g &= u_0(x,t) - \frac{1}{2}\iint_{\Delta(x,t)} g_3(\xi) g_1(\xi,\tau)\,d\xi\,d\tau, \\ A_2 g &= \frac{\partial}{\partial t}u_0(x,t) - \frac{1}{2}\int_{x-t}^{x+t} g_3(\xi) g_1(\xi, t-|x-\xi|)\,d\xi, \\ A_3 g &= q_0(x) + \frac{\operatorname{sign}(x-x_0)}{\phi(x)}\int_{x_0}^{x} g_3(\xi) g_2(\xi, |\xi - x|)\,d\xi. \end{aligned} \tag{2.27}$$

Denote

$$\|g\|(t_0) = \max_{1 \leqslant k \leqslant 3}\ \max_{(x,t)\in\Delta(x_0,t_0)} |g_k(x,t)|$$

$$g_0(x,t) = \left[u_0(x,t),\quad \frac{\partial}{\partial t}u_0(x,t),\quad q_0(x)\right]$$

and consider in the space $C(\Delta(x_0, h)$, $0 < h \leqslant t_0$ the set $m(h)$ of functions $g(x,t)$, which obey the inequality

$$\|g - g_0\|(h) \leqslant \|g_0\|(t_0). \tag{2.28}$$

It can be demonstrated that at sufficiently small h the operator A realizes contracted mapping of the set $m(h)$ onto itself. Indeed, for $g \in m(h)$ the inequality

$$\|g\|(h) \leqslant 2\|g_0\|(t_0)$$

is valid. On the other hand, by way of estimating the integrals in (2.27) one gets

$$\|Ag - g_0\|(h) \leqslant 4\|g_0\|^2(t_0)\max\left(\tfrac{1}{2}h^2, h, \frac{h}{\alpha}\right)$$

and hence for

$$h \leqslant h^* = \min\left(\frac{1}{\sqrt{[2\|g_0\|(t_0)]}}, \frac{\min(\alpha, 1)}{4\|g_0\|(t_0)}, t_0\right) \tag{2.29}$$

the operator A self-maps the set $m(h)$.

Now let $g^{(1)}, g^{(2)}$ be any two elements of the set $m(h)$, $h \leqslant h^*$. In this case, using the obvious inequalities

$$|g_k^{(1)}g_s^{(1)} - g_k^{(2)}g_s^{(2)}| \leqslant |g_k^{(1)} - g_k^{(2)}||g_s^{(1)}| + |g_k^{(2)}||g_s^{(1)} - g_s^{(2)}|$$
$$\leqslant 4\|g_0\|(t_0)\|g^{(1)} - g^{(2)}\|(h)$$

one gets

$$\|Ag^{(1)} - Ag^{(2)}\|(h) \leqslant 4\|g_0\|(t_0)\|g^{(1)} - g^{(2)}\|(h)$$
$$\times \max\left(\tfrac{1}{2}h^2, h, \frac{h}{\alpha}\right) \leqslant \frac{h}{h^*}\|g^{(1)} - g^{(2)}\|(h).$$

This indicates that at any $h < h^*$ the operator A realizes contracted mapping of the set $m(h)$ onto itself. Then, according to the Banach theorem (see, for instance, [76]), in the set $m(h)$ there exists only one fixed point of transformation, i.e. there exists only one solution to (2.26). Hence, solving the system of (2.7), (2.15), and (2.25) by the method of successive approximations, for instance (the Banach theorem demonstrates its convergence to the solution), we uniquely find the functions u, u_t, and q for $h \in (0, h^*)$ in the domain $\Delta(x_0, h)$, thus defining the solution to the inverse problem, i.e. the function $q(x)$ on the intercept $[x_0 - h, x_0 + h]$.

Note 1. The conditions imposed by the theorem on the data of the inverse problem are essential. Here are some examples to illustrate it.

Example 1. Let $\phi(x) = x^2$, $\psi(x) = 0$, $f(x,t) = 0$, $x_0 = 0$, $f_1(t) = 0$, $f_2(t) = 0$. One can easily prove that in this case the solution to the inverse problem exists: $q(x) = 2/x^2$, $u(x,t) = x^2$, although $q(x)$ is not a continuous function. Thus, violation of condition (2.22) at only one point results in violation of the theorem of existence.

Example 2. Let $\phi(x) = e^{2/3\sqrt{x^3}}$, $\psi(x) = 0$, $f(x,t) = 0$, $x_0 = 0$, $f_1(t) = 1$, $f_2(t) = 0$. The inverse problem solution exists: $q(x) = x + 1/2\sqrt{x}$ [in this case $u(x,t) = \phi(x)$], but is not a continuous function. This example illustrates the importance of conditions of smoothness for validity of the theorem. The function $\phi(x)$ is continuous here together with its first derivative, but its second derivative has a discontinuity of the second kind at $x = 0$.

Note 2. Condition (2.22) of the theorem can be replaced with the condition

$$\phi(x) \equiv 0, \quad |\psi(x)| \geqslant \alpha > 0, \quad x \in [x_0 - t_0, x_0 + t_0] \qquad (2.22')$$

combined with the requirement of greater smoothness of the functions ϕ, F, f_1, and f_2. Indeed, expressing $v = (\partial/\partial t)u$, conditions (2.1)–(2.3) yield in terms of this function the problem

$$L_q v = F_t(x,t) \qquad (2.1')$$

$$v|_{t=0} = \psi(x), \qquad v_t|_{t=0} = F(x,0) \qquad (2.2')$$

$$v|_{x=x_0} = f_1'(t), \qquad v_x|_{x=x_0} = f_2'(t) \qquad (2.3')$$

which differs from (2.1)–(2.3) only as far as the data is concerned, the function $\psi(x)$ playing the role here of the function $\phi(x)$. Therefore, Theorem 2.1 can be applied to system (2.1′)–(2.3′) by requiring greater smoothness from the functions ψ, F, f_1, and f_2.

In an analogous way, condition (2.22) in Theorem 2.1 can be replaced by the condition

$$\phi(x) = \psi(x) = F(x,0) = \frac{\partial}{\partial t} F(x,t)|_{t=0} = \cdots = \frac{\partial^{s-1}}{\partial t^{s-1}} F(x,t)|_{t=0} = 0$$

$$\frac{\partial^s}{\partial t^s} F(x,t)|_{t=0} \neq 0, \quad x \in [x_0 - t_0, x_0 + t_0] \tag{2.22''}$$

under an additional requirement of twice continuous differentiability of the functions $(\partial^s/\partial t^s)F(x,t)$, $f_1^{(s+2)}(t)$, $f_2^{(s+1)}(t)$ in the domain $\Delta(x_0, t_0)$.

Note 3. For sufficiently small t_0, $h^* = t_0$. Indeed, at $t_0 \to 0 \| g_0 \|(t_0)$ tends to a finite positive value, and thus h^*, determined by (2.29) for small values of t_0, coincides with t_0. There exists such a critical t_0^* that for $t_0 \leqslant t_0^*$ we always have $h^* = t_0$, and for $t_0 > t^*$ the inequality $h^* < t_0$ holds. This critical t_0^* is found as the root of the equation

$$t_0 = \min\left[\frac{1}{\sqrt{[2\|g_0\|(t_0)]}}, \frac{\min(1, \alpha(t_0))}{4\|g_0\|(t_0)}\right]$$

where

$$\alpha(t_0) = \min_{x_0 - t_0 \leqslant x \leqslant x_0 + t_0} |\phi(x)|.$$

At great values of $t_0 > t_0^*$ the parameter h^* can be considerably less than t_0. The question arises, if it is possible at any finite t_0 to extend the solution to the inverse problem, obtained on the intercept $[x_0 - h, x_0 + h]$, $h < h^*$, to a broader intercept $[x_0 - h_1, x_0 + h_1]$, $h_1 > h^*$, or to the whole intercept $[x_0 - t_0, x_0 + t_0]$. The first half of the question can be answered in the affirmative. Indeed, if one knows $q(x)$ on the intercept $[x_0 - h, x_0 + h]$, one can find the function $u(x, t)$ as a solution to the direct problem, making use of equivalency of the variables x, t in the domain $G_h(x_0, t) = \{(x,t): 0 \leqslant t \leqslant t_0 - |x - x_0|, |x - x_0| \leqslant h\}$, and, hence, one can calculate $u(x_0 \pm h, t)$, $u_x(x_0 \pm h, t)$. In this case the problem of finding $q(x)$ for $x > x_0 + h$ is reduced to the problem discussed above, where one has $x_0 + h$ instead of the point x_0, and $t_0 - h$ instead of t_0.

Using the above theorem, one can state that the solution to the direct problem exists and is unique on an intercept centred at the point $x_0 + h$, the same being true for the extension for $x < x_0 - h$. These considerations, however, do not give an answer to the question whether the solution can be extended to the whole intercept $[x_0 - t_0, x_0 + t_0]$. In all likelihood, the existence of the inverse problem solution as a whole (for any finite t_0) can be ensured only by some additional demands for the inverse problem data, which is not the subject of this book.

The answer to the question of uniqueness of the inverse problem solution as a whole is given in the following theorem.

Theorem 2.2. If under the conditions of Theorem 2.1 a solution to the inverse problem exists and belongs to the class $C[x_0 - t_0, x_0 + t_0]$, it is unique.

Suppose that there exist at least two solutions $q_1(x)$, $q_2(x)$ to the inverse problem in the class of continuous functions. Then according to Theorem 2.1, they coincide on a certain intercept $[x_0 - h, x_0 + h]$. Hence, continuity of $q_1(x)$, $q_2(x)$ yields that there is such a limited intercept $[x_1, x_2]$, containing the point x_0 inside, that

$$x_2 = \sup\{x : q_1(\xi) = q_2(\xi), x_0 \leqslant \xi \leqslant x \leqslant t_0 + x_0\}$$
$$x_1 = \inf\{x : q_1(\xi) = q_2(\xi), t_0 - x_0 \leqslant x \leqslant \xi \leqslant x_0\}.$$

Since $q_1(x) \not\equiv q_2(x)$ on $[x_0 - t_0, x_0 + t_0]$, then at least $x_2 < x_0 + t_0$ or $x_1 > x_0 - t_0$. Let, for the sake of definiteness, $x_2 < x_0 + t_0$. Then, as $q_1(x)$ and $q_2(x)$ coincide on the intercept $[x_0, x_2]$, the function $u(x, t)$, being a solution to the direct problem with the Cauchy data on the boundaries of the domain $x \geqslant x_0$, $t \geqslant 0$, is uniquely located in the semiband $x_0 \leqslant x \leqslant x_2$, $t \geqslant 0$. Consequently, the functions $q_1(x)$ and $q_2(x)$ have one and the same corresponding functions $u(x_2, t) = \tilde{f}_1(t)$, $u_x(x_2, t) = \tilde{f}_2(t)$, $t \in [0, t_0 - (x_2 - x_0)]$. But in this case, in line with Theorem 2.1, we get that $q_1 = q_2$ on a certain intercept $[x_2, x_2 + h]$, $h > 0$, which fact contradicts the definition of x_2 as a point belonging to the upper boundary of x values for which $q_1(\xi) \equiv q_2(\xi)$, $\xi \in [x_0, x]$. The theorem is proved.

Consider one further theorem characterizing the estimation of conventional stability of the inverse problem solution. Such an estimate can be obtained if we set a certain class of data $M(\alpha, K, x_0, t_0)$ for the functions ϕ, ψ, and F and a class $Q(M, x_0, t_0)$ for the functions $q(x)$. Assume that ϕ, ψ, and F belong to the class $M(\alpha, K, x_0, t_0)$ if they obey the inequalities

$$|\phi(x)| \geqslant \alpha > 0, \qquad x \in [x_0 - t_0, x_0 + t_0]$$
$$\|\phi\|_{C^2[x_0 - t_0, x_0 + t_0]} \leqslant K, \qquad \|\psi\|_{C^1[x_0 - t_0, x_0 + t_0]} \leqslant K$$
$$\|F\|_{C(\Delta(x_0, t_0))} \leqslant K, \qquad \|F_t\|_{C(\Delta(x_0, t_0))} \leqslant K$$

having universal for the whole class positive constants α and K. Analogously, $q \in Q(M, x_0, t_0)$ if

$$\|q\|_{C[x_0 - t_0, x_0 + t_0]} \leqslant M.$$

Theorem 2.3. Let $q(x)$, $\bar{q}(x)$ be two solutions to the inverse problem (2.1)–(2.3) with the data ϕ, ψ, F, f_1, f_2 and $\bar{\phi}, \bar{\psi}, \bar{F}, \bar{f}_1, \bar{f}_2$, respectively. Also, let $q, \bar{q} \in Q(M, x_0, t_0)$ and $\phi, \psi, F, \bar{\phi}, \bar{\psi}, \bar{F} \in M(\alpha, K, x_0, t_0)$. Then the estimate

$$\|q - \bar{q}\|_0 \leqslant C[\|\phi - \bar{\phi}\|_2 + \|\psi - \bar{\psi}\|_1 + \|F - \bar{F}\|_0 + \|F_t - \bar{F}_t\|_0 + \|f_1'' - \bar{f}_1''\|_0 + \|f_2' - \bar{f}_2'\|_0] \tag{2.30}$$

is valid; the constant C depending here only on the choice of classes $M(\alpha, K, x_0, t_0)$, $Q(M, x_0, t_0)$.

Let $u, \bar{u}$ denote the solutions to (2.1) and (2.2) corresponding to the functions q, $\bar{q}$. If the difference between two functions, whose only difference in notation is the overbar, is denoted by the same letter with a tilde ($\sim$), for instance, $\tilde{u} = u - \bar{u}$,

$\tilde{q} = q - \bar{q}$, etc., then equations (2.7), (2.15), and (2.25) give the following system of equalities

$$\tilde{u}(x,t) = \tilde{u}_0(x,t) - \frac{1}{2}\iint_{\Delta(x,t)} [q(\xi)\tilde{u}(\xi,\tau) + \tilde{q}(\xi)\bar{u}(\xi,\tau)]\,d\xi\,d\tau$$

$$\tilde{u}_t(x,t) = \frac{\partial}{\partial t}\tilde{u}_0(x,t) - \frac{1}{2}\int_{x-t}^{x+t} [q(\xi)\tilde{u}(\xi, t-|x-\xi|).$$

$$+ \tilde{q}(\xi)\bar{u}(\xi, t-|x-\xi|)]\,d\xi$$

$$\tilde{q}(x) = -\frac{\bar{q}(x)}{\phi(x)}\tilde{\phi}(x) + \frac{1}{\phi(x)}[-\tilde{f}_1''(|x-x_0|) - \tilde{f}_2'(|x-x_0|)\,\mathrm{sign}\,(x-x_0)]$$

$$+\left(\frac{\partial^2}{\partial t^2} + \mathrm{sign}\frac{\partial^2}{\partial x\,\partial t}\right)\tilde{u}_0(x,t)|_{x=x_0, t=|x-x_0|} + \frac{\mathrm{sign}\,(x-x_0)}{\phi(x)}$$

$$\times \int_{x_0}^{x} [q(\xi)\tilde{u}_t(\xi,|\xi-x|) + \tilde{q}(\xi)\bar{u}_t(\xi,|\xi-x|)]\,d\xi. \qquad (2.31)$$

Consider this system with respect to the functions $\tilde{u}, \tilde{u}_t$, and $\tilde{q}$. Note that the functions $\bar{u}, \tilde{u}_0, \bar{u}_t, q$, and $\bar{q}$ included in it can be estimated on the basis of the *a priori* information on the problem data. Indeed, there is the obvious estimate

$$\|\tilde{u}_0\|_2 \leqslant C_1[\|\tilde{\phi}\|_2 + \|\tilde{\psi}\|_1 + \|\tilde{F}\|_0 + \|\tilde{F}_t\|_0]. \qquad (2.32)$$

On the other hand, earlier we obtained the estimate for $\|u\|_0$, that affords

$$\|\bar{u}\|_0 \leqslant \|\bar{u}_0\|_0 \coth\sqrt{(M)}t_0 \leqslant 4C_1 K \coth\sqrt{(M)}t_0 \equiv C_2 K$$
$$\|\bar{u}_t\|_0 \leqslant \|\bar{u}_0\|_1[1 + \sqrt{(M)}\sinh\sqrt{(M)}t_0] \leqslant C_3 K. \qquad (2.33)$$

Here the constants C_2, C_3 are only t_0-dependent. Denote

$$U(t) = \max_{(\xi,\tau)\in\Delta(x_0,t)} |\tilde{u}(\xi,\tau)|$$
$$V(t) = \max_{(\xi,\tau)\in\Delta(x_0,t)} |\tilde{u}_\tau(\xi,\tau)|$$
$$Q(t) = \max(|\tilde{q}(x_0+t)|, |\tilde{q}(x_0-t)|).$$

Then from relations (2.31), making use of inequality (2.33) and the obvious inequality

$$|u(\xi,\tau)| \leqslant U(\tau + |\xi - x_0|), \qquad (\xi,\tau)\in\Delta(x_0,t)$$

one gets

$$U(t) \leqslant \|\tilde{u}_0\|_2 + \frac{1}{2}\iint_{\Delta(x_0,t)} [MU(\tau+|\xi-x_0|) + C_2K|q(\xi)|]\,d\xi\,d\tau$$

$$\leqslant \|\tilde{u}_0\|_2 + Mt_0\int_0^t U(\tau)\,d\tau + C_2Kt_0\int_0^t Q(\tau)\,d\tau$$

$$V(t) \leqslant \|\tilde{u}_0\|_2 + t_0 M U(t) + C_2 K \int_0^t Q(\tau)\,d\tau$$

$$Q(t) \leqslant \frac{1}{\alpha}[M\|\Phi\|_0 + \|\tilde{f}_1''\|_0 + \|\tilde{f}_2'\|_0 + \|\tilde{u}_0\|_2]$$

$$+ \frac{1}{\alpha}\left[C_3 K \int_0^t Q(\tau)\,d\tau + M t_0 V(t)\right].$$

To demonstrate that these inequalities result in estimate (2.30) let

$$W(t) = \max(U(t), Q(t)).$$

Then one has

$$U(t) \leqslant \|\tilde{u}_0\|_2 + (M + C_2 K) t_0 \int_0^t W(\tau)\,d\tau$$

$$V(t) \leqslant \|\tilde{u}_0\|_2 (1 + t_0 M) + [C_2 K + (M + C_2 K) t_0^2 M] \int_0^t W(\tau)\,d\tau$$

$$Q(t) \leqslant \frac{1}{\alpha}[M\|\Phi\|_0 + \|\tilde{f}_1''\|_0 + \|\tilde{f}_2'\|_0 + \|\tilde{u}_0\|_2$$

$$\times (1 + M t_0) M t_0] + \frac{1}{\alpha}\{C_3 K + M t_0 [C_2 K$$

$$+ M t_0^2 (M + C_2 K)]\} \int_0^t W(\tau)\,d\tau.$$

Hence

$$W(t) \leqslant W_0 + \lambda \int_0^t W(\tau)\,d\tau \tag{2.34}$$

where

$$W_0 = \max\left\{\|\tilde{u}_0\|_2, \frac{1}{\alpha}[M\|\Phi\|_0 + \|\tilde{f}_1''\|_0 + \|\tilde{f}_2'\|_0 + \|\tilde{u}_0\|_2 M t_0 (1 + M t_0)]\right\}$$

$$\lambda = \max\left\{(M + C_2 K) t_0, \frac{1}{\alpha}[C_3 K + M t_0 (C_2 K + M^2 t_0^2 + M C_2 K t_0^2)]\right\}.$$

As is known, inequality (2.34) affords $W(t) \leqslant W_0 e^{\lambda t}$. Taking inequality (2.32) into account, it yields estimate (2.30).

2.2. Some aspects associated with inverse problem for the equation $L_q u = F$

Consider some of the details associated with the inverse problem discussed in the previous section with the aim of achieving an insight into the basis whereon the closed system of integral equations arose and paying greater attention to possible modification of the method of the inverse problem solution.

What idea was laid in the basis of obtaining a closed system of integral equations? To find this system, we carried out a number of operations associated with differentiating integral equation (2.7) and, as a result, having used the additional information on the solution to the direct problem, we found the closed system of integral equations with respect to the functions u, u_t, and q. The question arises how could one guess that it is these operations that were necessary in order to get the closed system? To answer this question, let us see in what way the solution to the direct problem, i.e. the function $u(x,t)$, depends on the function $q(x)$. It has already been mentioned that the solution to (2.7) can be obtained for any continuous function $q(x)$ in any finite domain $\Delta(x_0, t_0)$ by using the successive approximations method (2.8) and (2.9), the $u(x,t)$ solution depending on the $q(x)$ nonlinearly. In series (2.8) only the function $u_1(x,t)$ depends on $q(x)$ linearly. Let us single out an explicitly linear part of the solution to (2.7) by substituting the unknown quantity $u(x,t)$ with the function

$$v(x,t) = u(x,t) - u_0(x,t). \tag{2.35}$$

If one expresses $u(x,t)$ from (2.35) through $v(x,t)$ and substitute it into (2.7), one will get the equation for $v(x,t)$

$$v(x,t) = -\frac{1}{2}\iint_{\Delta(x,t)} q(\xi)u_0(\xi,\tau)\,d\xi\,d\tau - \frac{1}{2}\iint_{\Delta(x,t)} q(\xi)v(\xi,\tau)\,d\xi\,d\tau. \tag{2.36}$$

The first of the integrals in the right-hand part of this equation expresses the linear with respect to q part of the solution, the second one expresses the nonlinear part. Solving (2.36) at a given function q, we find $v(x,t)$. In this case the nonlinear operator $v = A(q)$, that correlates any continuous function $q(u)$ with the function $v(x,t)$, is determined. Equation (2.36) yields the inequality

$$v(x,t) = -\frac{1}{2}\iint_{\Delta(x,t)} q(\xi)u_0(\xi,\tau)\,d\xi\,d\tau - \frac{1}{2}\iint_{\Delta(x,t)} q(\xi)A(q)\,d\xi\,d\tau.$$

To obtain from this equation an expression for $q(x)$, one needs information (2.3) on the direct problem solution. In the case in question it is reduced to the fact that functions $v(x_0,t)$ and $v_x(x_0,t)$ are known, which results in the equalities

$$f_1(t) - u_0(x_0,t) = -\frac{1}{2}\iint_{\Delta(x_0,t)} q(\xi)u_0(\xi,\tau)\,d\xi\,d\tau - \frac{1}{2}\iint_{\Delta(x_0,t)} q(\xi)A(q)\,d\xi\,d\tau$$

$$f_2(t) - \frac{\partial}{\partial x}u_0(x,t)|_{x=x_0} = -\frac{1}{2}\frac{\partial}{\partial x}\left[\iint_{\Delta(x,t)} q(\xi)u_0(\xi,\tau)\,d\xi\,d\tau + \iint_{\Delta(x,t)} q(\xi)A(q)\,d\xi\,d\tau\right]_{x=x_0}$$

which are a system of first-order integral nonlinear equations. To derive from it a system of second-order equations, which are far easier for investigation, it is sufficient to inverse two linear operators

$$B_1 q = -\frac{1}{2}\iint_{\Delta(x_0,t)} q(\xi)u_0(\xi,\tau)\,d\xi\,d\tau$$

$$B_2 q = -\frac{1}{2}\left[\frac{\partial}{\partial x}\iint_{\Delta(x,t)} q(\xi)u_0(\xi,\tau)\,d\xi\,d\tau\right]_{x=x_0}$$

$$= -\frac{1}{2}\int_{x_0-t}^{x_0+t} q(\xi)u_0(\xi, t-|x_0-\xi|)\,\mathrm{sign}\,(\xi - x_0)\,d\xi.$$

Taking into account inequality (2.22) for the function $\phi(x) = u_0(x, 0)$, the inversion of the first operator is carried out by way of double differentiating with respect to t, and of the second one—by differentiating it once, which was the reason for calculating the derivatives u_{tt} and u_{xt}.

By way of conclusion one may say that obtaining equations of the second kind for the inverse problem is associated with the inversion of the linear part of the direct problem operator using the variety of the inverse problem data that serve as an additional information.

Now consider a possible modification of the solution to the inverse problem. It appears that the function $q(x)$ can be found in the domain $x \geqslant x_0$ and $x \leqslant x_0$, since for the functions q, u one can obtain a closed system of integral equations in each of the domains $D_1 = \{(x,t): x_0 \leqslant x < \infty, t \geqslant 0\}$, $D_2 = \{(x,t): -\infty < x \leqslant x_0, t \geqslant 0\}$. Now see how it can be obtained, say, in the domain D_1.

Equation (2.25) for $x \geqslant x_0$ can be written as

$$q(x) = q_0(x) - \frac{1}{\phi(x)}\int_{x_0}^{x} q(\xi)v(\xi, x-\xi)\,d\xi, \quad x \geqslant x_0. \tag{2.37}$$

Here $v(x,t) = u_t(x,t)$. Relation (2.37) shows that the function $q(x)$ is associated for $x \geqslant x_0$ with those values of the function $v(x,t)$ which it takes on only in the domain D_1. At the same time function $v(x,t)$ in the domain D_1 satisfies the equation

$$L_q v = F_t(x,t), \quad (x,t)\in D_1 \tag{2.38}$$

and the boundary conditions

$$v|_{t=0} = \psi(x), \qquad v_t|_{t=0} = \phi_{xx}(x) - q(x)\phi(x) + F(x,0), \quad x \geqslant x_0 \tag{2.39}$$

$$v|_{x=x_0} = f_1'(t), \quad t \geqslant 0. \tag{2.40}$$

Now one can easily obtain an integral equation in the domain D_1 for the function $v(x,t)$. For this purpose first consider an auxiliary problem of the same type as (2.38)–(2.40) on finding in D_1 a solution to the problem

$$\left(\frac{\partial^2}{\partial t^2} - \frac{\partial^2}{\partial x^2}\right)w = \Phi(x,t), \quad (x,t)\in D_1 \tag{2.41}$$

$$w|_{t=0} = \psi(x), w_t|_{t=0} = h(x), \quad x \geqslant x_0 \tag{2.42}$$

$$w|_{x=x_0} = f_1'(t), \quad t \geqslant 0. \tag{2.43}$$

Write the wave operator as a product of first-order operators

$$\frac{\partial^2}{\partial t^2} - \frac{\partial^2}{\partial x^2} = \left(\frac{\partial}{\partial t} - \frac{\partial}{\partial x}\right)\left(\frac{\partial}{\partial t} + \frac{\partial}{\partial x}\right).$$

Integrating (2.41) along the characteristics of the differential operator $(\partial/\partial t) - (\partial/\partial x)$ from an arbitrary point $(x,t)\in D_1$ to the point of crossing with the straight line $t=0$, and using data (2.42), one obtains

$$\left(\frac{\partial}{\partial t}+\frac{\partial}{\partial x}\right)w(x,t)=\psi'(x+t)+h(x+t)+\int_x^{x+t}\Phi(\xi,x+t-\xi)\,d\xi,(x,t)\in D_1. \tag{2.44}$$

Now integrating the first-order equation (2.44) and using condition (2.43), we find the solution to (2.41)–(2.43) as

$$w(x,t)=f'_1(t-x+x_0)+\frac{1}{2}[\psi(x+t)-\psi(t-x+2x_0)]$$
$$+\frac{1}{2}\int_{t-x+2x_0}^{x+t}h(\xi)\,d\xi+\int_{t-x+x_0}^{t}d\tau\int_{\tau-t+x}^{2\tau-t+x}\Phi(\xi,2\tau-t+x-\xi)\,d\xi,$$
$$(x,t)\in D_1. \tag{2.45}$$

If (2.45) is now used for (2.38)–(2.40), we obtain in the domain D_1 the integral equation

$$v(x,t)=v_0(x,t)-\frac{1}{2}\int_{t-x+2x_0}^{x+t}q(\xi)\phi(\xi)\,d\xi$$
$$-\int_{t-x+x_0}^{t}d\tau\int_{\tau-t+x}^{2\tau-t+x}q(\xi)v(\xi,2\tau-t+x-\xi)\,d\xi,\quad (x,t)\in D_1 \tag{2.46}$$

in which the function $v_0(x,t)$ is equal to the right-hand part of (2.45), if we set in it that $h=\phi_{xx}-F(x,0)$, $\Phi=F_t(x,t)$.

The system of equalities (2.37) and (2.46) is closed in any triangle domain $\Delta'(x,t)=\{(x,t):0\leqslant t\leqslant t_0+x_0-x,\ x_0\leqslant x\leqslant x_0+t_0\}$ and the principle of contracted mapping can be applied to it at small t_0. All the three theorems: of existence, uniqueness, and solution stability formulated in Section 2.1 are also valid in this case, but obviously with certain corrections in the definitions, since in the domain $D_1 x\geqslant x_0$.

The above investigation shows, in particular, that the inverse problem (2.1)–(2.3) can be considered in a semi-limited, with respect to x, domain $x\geqslant x_0$. In this case one of the conditions (2.3) should be regarded as a boundary condition, and the other one as an additional information for defining $q(x)$ on the intercept $x\geqslant x_0$.

2.3. Problems with a focused source of disturbance

In applied problems, in particular, in problems of prospecting geophysics, the data are often the finite functions localized within a relatively small area of space. It is convenient to model such situations mathematically by regarding the data to the problems as the generalized functions whose supports are concentrated at some fixed points of the space. In this case the solution of the boundary problem is also a generalized function, and the fulfilment of the differential equalities and boundary conditions is realized in terms of the theory of generalized functions (see [77–79]). The inverse problems where the direct problem data are generalized functions are worth studying separately. In an applied sense, these problems are most interesting.

Consider in the domain $D=\{(x,t): -\infty<x<\infty,\ t>0\}$ the differential equation

$$L_q u = 0, \qquad (x,t)\in D \tag{2.47}$$

and the Cauchy data

$$u|_{t=0}=0, \qquad u_t|_{t=0}=\delta(x). \tag{2.48}$$

Formulate the inverse problem, fairly identical to (2.1)–(2.3); to find a continuous function $q(x)$, if the solution to (2.47) and (2.48) is known at $x=0$ together with the partial derivative with respect to the variable x

$$u(0,t)=f_1(t), \qquad u_\alpha(0,t)=f_2(t), \quad t>0. \tag{2.49}$$

Now study this problem according to the scheme suggested in Section 2.1, beginning with investigating the properties of the solution to (2.47) and (2.48). Using the basic property of the delta-function

$$\int_{-\infty}^{\infty} \psi(x)\delta(x)\,\mathrm{d}x = \psi(0)$$

for any continuous finite function $\psi(x)$ and the d'Alembert formula, one finds that the generalized solution to (2.47) and (2.48) is a piecewise continuous solution of the integral equation

$$u(x,t)=\tfrac{1}{2}\theta(t-|x|)-\frac{1}{2}\iint_{\Delta(x,t)} q(\xi)u(\xi,\tau)\,\mathrm{d}\xi\,\mathrm{d}\tau, \quad (x,t)\in D. \tag{2.50}$$

Here $\theta(t)$ is the Heavyside function

$$\theta(t)=\begin{cases}1, & t\geqslant 0\\ 0, & t<0.\end{cases}$$

From (2.50) one has

$$u(x,t)\equiv 0, \quad t<|x|, \quad (x,t)\in D. \tag{2.51}$$

Indeed, consider the domain $\Delta(x_1,t_1)$ with an arbitrary fixed point $(x_1,t_1)\in D, t_1<|x_1|$. Then for the points $(x,t)\in\Delta(x_1,t_1)$ (2.50) gives the uniform equation

$$u(x,t)=-\frac{1}{2}\iint_{\Delta(x,t)} q(\xi)u(\xi,\tau)\,\mathrm{d}\xi\,\mathrm{d}\tau, \quad (x,t)\in\Delta(x_1,t_1). \tag{2.52}$$

By denoting

$$\max_{x_1-t_1\leqslant x\leqslant x_1+t_1} |q(x)| = Q$$

$$\max_{x_1-(t_1-t)\leqslant x\leqslant x_1+(t_1-t)} |u(x,t)| = U(t)$$

from (2.52) one gets the inequality

$$U(t) \leqslant Qt_1 \int_0^t U(\tau)\,d\tau, \quad 0 \leqslant t \leqslant t_1.$$

The inequality denotes that $U(t) = 0$, $t \in [0, t_1]$, and, hence, $u(x, t) = 0$ in $\Delta(x_1, t_1)$. Due to the arbitrariness of the choice of the point (x_1, t_1) it is equivalent to (2.51).

Equality (2.51) demonstrates that the support of the function $u(x, t)$ is concentrated inside the angle formed by the characteristics outgoing from the point (0, 0). Denote

$$D' = \{(x, t): t > |x|\}$$

and henceforce consider the function $u(x, t)$ only in the domain D'. From (2.50) and (2.51) one finds

$$u(x, t) = \frac{1}{2} - \frac{1}{2} \iint_{\square(x,t)} q(\xi) u(\xi, \tau)\,d\xi\,d\tau, \quad (x, t) \in D'. \tag{2.53}$$

Through $\square(x, t)$ we denote here the domain having the form of a rectangle (Fig. 2.2) formed by the characteristics passing through the points (0, 0) and (x, t).

As far as (2.53) is concerned, the following lemma is valid.

Lemma 2.2. If for a point $(x_1, t_1) \in D'$ the function $q(x) \in C[(x_1 - t_1)/2, (x_1 + t_1)/2]$, then the solution to (2.53) exists and belongs to the class $C^2(\square(x_1, t_1))$.

The lemma will be proved as follows. First, using the method of successive approximations

$$u(x, t) = \sum_{n=0}^{\infty} u_n(x, t),\ u_0 = \frac{1}{2} \tag{2.54}$$

$$u_n(x, t) = -\frac{1}{2} \iint_{\square(x,t)} q(\xi) u_{n-1}(\xi, \tau)\,d\xi\,d\tau, \quad n \geqslant 1, (x, t) \in \square(x_1, t_1),$$

we show that the sum of the series provides a continuous solution to (2.53) and then, by differentiating equality (2.53) on this solution we verify the required smoothness. For the sum of the series (2.54) to define the continuous solution to (2.53), it is sufficient to demonstrate that all the $u_n(x, t)$, $n = 0, 1, 2, \ldots$, are continuous and the series uniformly converges in the domain $\square(x_1, t_1)$. The first of the required properties is explicit, the second one can be derived from the easily verified by induction inequalities

$$|u_n(x, t)| \leqslant \frac{1}{2^{n+1}} [\|q\|_0 (t_1 - |x_1|)]^n \frac{t^n}{n!}, \quad (x, t) \in \square(x_1, t_1), n = 0, 1, 2, \ldots.$$

Uniqueness of the continuous solution to (2.53) results from the fact that a uniform equation has only the zero solution. Demonstration of this fact is identical to that of equality (2.51).

Equation (2.53), in particular, shows that on the boundary of the domain D' the function $u(x, t)$ is constant and equals 1/2

$$u(x, |x|) = 1/2. \tag{2.55}$$

Now write the formulas defining in the domain D' the partial derivatives u_t, u_x, u_{tt}, u_{xt}, which will be useful for investigating the inverse problem. For this purpose rewrite equality (2.53), having substituted the double integral by the iterated one (see Fig. 2.2)

$$u(x,t)=\frac{1}{2}-\frac{1}{2}\int_{(x-t)/2}^{(x+t)/2}\int_{|\xi|}^{t-|x-\xi|} q(\xi)u(\xi,\tau)\,d\xi\,d\tau, \quad (x,t)\in D'. \tag{2.53'}$$

By differentiating this equality, we obtain

$$u_t=-\frac{1}{2}\int_{(x-t)/2}^{(x+t)/2} q(\xi)u(\xi,t-|x-\xi|)\,d\xi \tag{2.56}$$

$$u_x(x,t)=-\frac{1}{2}\int_{(x-t)/2}^{(x+t)/2} q(\xi)u(\xi,t-|x-\xi|)\,\mathrm{sign}\,(\xi-x)\,d\xi \tag{2.57}$$

$$u_{xt}(x,t)=-\frac{1}{8}\left[q\left(\frac{x+t}{2}\right)-q\left(\frac{x-t}{2}\right)\right] -\frac{1}{2}\int_{(x-t)/2}^{(x+t)/2} q(\xi)u_t(\xi,t-|x-\xi|)\,\mathrm{sign}\,(\xi-x)\,d\xi, \tag{2.58}$$

$$u_{tt}(x,t)=-\frac{1}{8}\left[q\left(\frac{x+t}{2}\right)+q\left(\frac{x-t}{2}\right)\right] -\frac{1}{2}\int_{(x-t)/2}^{(x+t)/2} q(\xi)u_t(\xi,t-|x-\xi|)\,d\xi. \tag{2.59}$$

Lemma 2.2 and equalities (2.55)–(2.58) result in the lemma characterizing the properties necessary for the data of the inverse problem.

Lemma 2.3. If $q(x)\in C[-T/2, T/2]$, $T>0$, then $f_1\in C^2[0,T]$, $f_2\in C^1[0,T]$ and

$$f_1(0)=1/2, \quad f'_1(0)=f_2(0)=f'_2(0)=0 \tag{2.60}$$

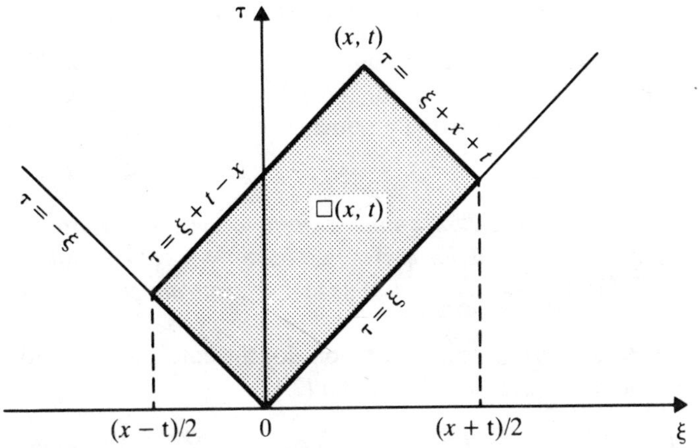

Figure 2.2.

in which case functions f_1 and f_2 and their derivatives at $t=0$ are viewed as the limits of the corresponding functions from the side of values $t>0$.

Equalities (2.58) and (2.59) written at $x=x_0$ make it possible to obtain the equation of the second kind with respect to the function $q(x)$

$$q(x)=q_0(x)-4\,\mathrm{sign}\,(x)\int_0^x q(\xi)u_t(\xi,2|x|-\xi)\,d\xi$$

$$q_0(x)=-4[f_1''(2|x|)+f_1'(2|x|)\,\mathrm{sign}\,x]. \tag{2.61}$$

This equation, as well as (2.53) and (2.56) define in D' the closed system of integral equations with respect to the three functions u, u_t, and q. Using the obtained system we can easily prove the three following theorems.

Theorem 2.4. If at a $t_0>0$ $f_1\in C^2[0,t_0]$, $f_2\in C^1[0,t_0]$ and conditions (2.60) are met, then there exists such $h\in(0,t_0/2)$ that the solution to the inverse problem (2.47)–(2.49) in the class of functions $q(x)\in C[-h,h]$ exists and is unique.

Theorem 2.5. If the conditions of Theorem 2.4 are fulfilled, the function $q(x)\in C[-t_0/2,t_0/2]$ is uniquely defined by information (2.49) for $t\in(0,t_0]$.

Theorem 2.6. Let $q(x)$, $\bar{q}(x)$ be two solutions to the inverse problem (2.47)–(2.49) with the data f_1, f_2 and $\bar{f}_1, \bar{f}_2$, respectively. Then the estimate

$$|q(x)-\bar{q}(x)|\leqslant C(\|f_1''-\bar{f}_1''\|_{C[0,t_0]}+\|f_2'-\bar{f}_2'\|_{C[0,t_0]}),\quad x\in[-t_0/2,t_0/2],$$

is valid, with C depending only on the norms in the metric $C[-t_0/2,t_0/2]$ of the functions $q,\bar{q}$ and on the parameter t_0.

These theorems are proved according to the scheme given in Section 2.1, therefore the proof is omitted.

Consider now the case when an additional information on the solution to the Cauchy problem (2.47) and (2.48) is set at the point $x_1\neq 0$

$$u(x_1,t)=f_1(t),\qquad u_x(x_1,t)=f_2(t),\quad t>0. \tag{2.62}$$

In this case the situation with the uniqueness of the inverse problem solution appears essentially different. Let, for the sake of definiteness, $x_1>0$. Then the following theorem, characterizing uniqueness of the inverse problem solution, is valid.

Theorem 2.7. The data (2.62) on the solution to the Cauchy problem (2.47) and (2.48) uniquely define the function $q(x)$ in the domain $x\geqslant x_1>0$, while in the domain $x\leqslant x_1$ they define it to the accuracy of its almost arbitrary setting on the intercept $[0,x_1]$.

First of all write the necessary conditions for the data (2.62), resulting from the analysis of the direct problem (2.47) and (2.48)

$$f_1(t)=f_2(t)\equiv 0,\quad t<x_1$$

$$f_1(t)\in C^2[x_1,t_0],\qquad f_2(t)\in C^1[x_1,t_0],\quad t_0>x_1,$$

if $q(x) \in C[(x_1 - t_0)/2, (x_1 + t_0)/2]$. Besides, from inequalities (2.55)–(2.57) we have

$$f_1(x_1+0) = 1/2, \qquad f_1'(x_1+0) = -f_2(x_1+0). \tag{2.63}$$

Setting $x = x_1$ in equalities (2.58) and (2.59) and using the data (2.62), we get

$$q\left(\frac{x_1+t}{2}\right) = -4[f_1''(t) + f_2'(t)] - 4\int_{x_1}^{(x_1+t)/2} q(\xi)u_t(\xi, t - \xi + x_1)\,d\xi$$

$$q\left(\frac{x_1-t}{2}\right) = -4[f_1''(t) - f_2'(t)] - 4\int_{(x_1-t)/2}^{x_1} q(\xi)u_t(\xi, t - x_1 + \xi)\,d\xi, \quad t \geqslant x_1. \tag{2.64}$$

The first of equalities (2.64) serves to define $q(x)$ in the domain $x \geqslant x_1$. Actually, with t changing from x_1 to t_0, the argument $(x_1 + t)/2$ of the function q varies within the limits from x_1 to $(x_1 + t_0)/2$, monotonically increasing with growing t. In this case the function $q(x)$ is expressed through the data of the inverse problem and in an integral way through itself and the function $u_t(x, t)$ for $x \geqslant x_1$.

The second of equalities (2.64) defines $q(x)$ for $x \leqslant 0$, since with growing of argument t from x_1 to t_0, the expression $(x_1 - t)/2$ decreases from 0 to $(x_1 - t_0)/2$. At the same time

$$[(x_1 - t)/2, x_1] \supset [0, x_1], \quad \forall t \in [x_1, t_0].$$

Therefore, the function $q(x)$ in the domain $x \leqslant 0$ is expressed in an integral way through its values in a broader domain $x \leqslant x_1$, which accounts for the fact that $q(x)$ cannot be defined uniquely in the domain $x \leqslant x_1$.

Let us make use of the fact that the inverse problem can be solved separately in the domains lying to the left and to the right from the straight line on which the additional information was set. The function $u_t(x, t) \equiv v$ in the domain D' satisfies the equation

$$L_q v = 0, \qquad (x, t) \in D' \tag{2.65}$$

and the condition on the boundary of the domain D'

$$v(x, |x| + 0) = -\frac{1}{4}\int_0^x q(\xi)\,d\xi \operatorname{sign} x \tag{2.66}$$

resulting from equalities (2.55) and (2.57). The additional condition

$$v(x_1, t) = f_1'(t), \quad t > 0 \tag{2.67}$$

enables one to obtain the integral equation for the function $v(x, t)$ separately in the domains

$$D_1(x_1) = \{(x, t): (x, t) \in D', x \geqslant x_1\}$$
$$D_2(x_1) = \{(x, t): (x, t) \in D', x \leqslant x_1\}.$$

For instance, in order to obtain the equation in the domain $D_1(x_1)$, it is sufficient, after substituting the wave operator $(\partial^2/\partial t^2) - (\partial^2/\partial x^2)$ with the product

$$\left(\frac{\partial}{\partial t} - \frac{\partial}{\partial x}\right)\left(\frac{\partial}{\partial t} + \frac{\partial}{\partial x}\right)$$

to integrate equality (2.65) first along the characteristics $\xi+\tau=x+t$ from the point $(x,t)\in D_1(x_1)$ to that of its crossing the characteristics $\tau-\xi=0$, followed by the integration of the resulting relation along another characteristics $\tau-\xi=t-x$ from the point (x,t) to the straight line $x=x_1$. As a result one has

$$v(x,t)=f_1'(t-x+x_1)-\frac{1}{4}\int_{x_1-(t-x)/2}^{(x+t)/2} q(\xi)\,d\xi-\int_{t-x+x_1}^{t} d\tau$$
$$\times\int_{\tau-t+x}^{(2\tau+x-t)/2} q(\xi)v(\xi,2\tau+x-t-\xi)\,d\xi,\quad (x,t)\in D_1(x_1). \qquad (2.68)$$

An analogous equation can be easily written for the domain $D_2(x_1)$.

Equation (2.68) together with the first of equalities (2.64) uniquely defines the continuous functions $q, v=u_t$ in the domain $D_1(x_1)$. Indeed, under the supposition that (q_1,v_1) and (q_2,v_2) are different pairs of functions satisfying (2.64) and (2.68), we obtain for their differences $q_1-q_2=\tilde{q}$, $v_1-v_2=\tilde{v}$ the following relations

$$\tilde{q}\left(\frac{x_1+t}{2}\right)=-4\int_0^{(x_1+t)/2}[\tilde{q}v_1+q_2\tilde{v}](\xi,t-\xi+x_1)\,d\xi$$
$$\tilde{v}(x,t)=-\frac{1}{4}\int_{x_1+(t-x)/2}^{(t+x)/2}\tilde{q}(\xi)\,d\xi-\int_{t-x+x_1}^{t} d\tau\int_{\tau-t+x}^{\tau+(x-t)/2}[\tilde{q}v_1+q_2\tilde{v}]$$
$$(\xi,2\tau+x-t-\xi)\,d\xi,\quad (x,t)\in D_1(x_1),$$

which define the system of the Volterra integral uniform equations of the second kind. The only possible solution to this system in the class of continuous functions is $\tilde{q}=\tilde{v}=0$, and, hence, $q_1=q_2$, $v_1=v_2$.

It is also the case in the domain $D_2(x_2)$, provided the function $q(x)$ is set on $[0,x_1]$. Then for the values $x\leqslant 0$ it can be uniquely defined from information (2.62). For the function $q(x)$ to be continuous, it is necessary when setting it on the intercept $[0,x_1]$ to fulfil the conditions of sewing

$$q(x_1)=-4[f_1''(x_1+0)+f_2'(x_1+0)]$$
$$q(0)=-4[f_1''(x_1+0)-f_2'(x_1-0)]+\int_0^{x_1} q(\xi)\int_0^{\xi} q(\xi_1)\,d\xi_1\,d\xi \qquad (2.69)$$

resulting from equalities (2.64) at $t=x_1$. It is equalities (2.69) that define the only requirement to setting $q(x)$ on $[0,x_1]$. No doubt that there exists a great number of the continuous on $[0,x_1]$ functions satisfying conditions (2.69), which fact is stressed in the formulation of Theorem 2.7 by 'to the accuracy of almost arbitrary setting'.

Note: Problems (2.47) and (2.48) are equivalent to the problem of finding a generalized solution to the nonuniform equation

$$L_q u=\delta(x,t),\quad t\geqslant 0 \qquad (2.70)$$

under the condition

$$u|_{t<0}\equiv 0. \qquad (2.71)$$

The validity of this conclusion can be easily derived from the d'Alembert formula. Therefore, the inverse problem (2.70), (2.71), and (2.49) is equivalent to (2.47)–(2.49) discussed above.

The case when a source of disturbance is focused at a fixed point of the space but distributed with respect to time can now be considered

$$L_q u = \delta(x)\theta(t)F(t), \quad t > 0. \tag{2.72}$$

In this case if $F(t) \in C^1(0, \infty)$, then the solution to (2.72) and (2.71) is a twice continuously differentiated function in closure of the domain D' and equals zero outside $\bar{D}'$. There is a simple relation between the solution u_F to (2.72) and (2.71) and the solution u_δ to (2.70) and (2.71)

$$u_F(x, t) = \int_0^\infty F(t_0)u_\delta(x, t - t_0)\,\mathrm{d}t_0 = \int_0^t F(t - \tau)u_\delta(x, \tau)\,\mathrm{d}\tau. \tag{2.73}$$

This formula enables one to find the function u_F using the known function u_δ. If the function $F(t)$ has the property that if there exists such t_0 that $F(t) \not\equiv 0$ for $t \in (0, t_0)$, then according to the results obtained by Titchmarsh [80], the opposite is valid: u_δ can be uniquely defined by u_F. For instance, if one knows $u_F|_{x=0}$ one can find $u_\delta|_{x=0}$, or if $(\partial/\partial x)u_F|_{x=0}$ is known one can find $(\partial/\partial x)u_\delta|_{x=0}$. Returning to the inverse problem (2.72), (2.71), and (2.49), it can be reduced to (2.70), (2.71), and (2.49) using the above statement and provided the function $F(t)$, possessing the necessary properties, is known. As has already been mentioned, (2.70), (2.71), and (2.49) are equivalent to problem (2.47)–(2.49) studied earlier.

2.4. Reducing the problem with a focused source of disturbance to a linear integral equation: necessary and sufficient conditions for the inverse problem solvability

Investigation of problem (2.47)–(2.49) can be carried out on the basis of a different approach, within which necessary and sufficient conditions of its solvability as a whole can be found, i.e. in any vicinity of the point x_0. This approach to studying inverse problems was suggested by Parijskij [81] and Blagoveshchenskij [82].

Let in condition (2.49) $f_2(t) = 0$. Denote $f_1(t)$ by $f(t)$. The function $u(x, t)$, as the solution to (2.47) and (2.48), is defined within the upper semi-plane $t \geqslant 0$. Extend its definition onto the whole plane x, t with the help of the odd extension $u(x, t) = -u(x, -t)$, $t \leqslant 0$. In this case the function $f(t)$ will also be extended in an odd way, and in line with condition (2.60), $f_1(+0) = 1/2$ will have at zero a discontinuity of the first kind $f_1(-0) = -1/2$. Thus, on the whole plane the function $u(x, t)$ extended in such a way satisfies the equation

$$L_q u = 0 \tag{2.74}$$

and the condition

$$u(x, t) \equiv 0, \quad |t| < |x|. \tag{2.75}$$

Suppose now that the solution to (2.47)–(2.49) does exist. Then the function $u(x, t)$ satisfies the additional conditions

$$u(0, t) = f(t), \qquad u_x(0, t) = 0 \tag{2.76}$$

in which case at $t \neq 0$ $f(t)$ is twice continuously differentiated.

Let the function w denote the solution of (2.74) with the Cauchy data on the axis t

$$L_q w = 0, \quad x > 0, -\infty < t < \infty \tag{2.77}$$

$$w|_{x=0} = \delta(t - t_0), \quad w_x|_{x=0} = 0. \tag{2.78}$$

Since the operator L_q is t-independent, then

$$w = w(x, t - t_0). \tag{2.79}$$

It is therefore sufficient to consider the partial value $t_0 = 0$.

The process of solution of (2.77) and (2.78) is in this case equivalent to that of the integral equation

$$w(x,t) = \frac{1}{2}[\delta(t-x) + \delta(t+x)] + \frac{1}{2}\iint_{\Delta'(x,t)} q(\xi)w(\xi,\tau)\,\mathrm{d}\xi\,\mathrm{d}\tau \tag{2.80}$$

where $\Delta'(x,t)$ denotes a triangle outlined by the characteristics passing through the point (x,t) and by the axis t

$$\Delta'(x,t) = \{(\xi,\tau): 0 < \xi \leqslant x, t - x + \xi < \tau < t + x - \xi\}.$$

According to (2.80), one can easily demonstrate that

$$w(x,t) \equiv 0, \quad 0 < x < |t| \tag{2.81}$$

and, therefore, the actual domain of integration in (2.80) for the points $(x,t) \in D''$

$$D'' = \{(x,t): x \geqslant |t|\}$$

is in fact a rectangle $\square'(x,t)$ outlined by the characteristics outgoing from the points $(0,0)$, (x,t) (Fig. 2.3).

Introduce the following notations

$$\tilde{w}(x,t) = w(x,t) - \tfrac{1}{2}[\delta(t-x) + \delta(t+x)]. \tag{2.82}$$

The function $\tilde{w}(x,t)$ is a common piecewise continuous function satisfying the equation

$$\tilde{w}(x,t) = \frac{1}{4}\theta(x - |t|)\left[\int_0^{(x+t)/2} q(\xi)\,\mathrm{d}\xi + \int_0^{(x-t)/2} q(\xi)\,\mathrm{d}\xi\right]$$

$$+ \frac{1}{2}\iint_{\square'(x,t)} q(\xi)\tilde{w}(\xi,\tau)\,\mathrm{d}\xi\,\mathrm{d}\tau, \quad x > 0, \tag{2.83}$$

resulting from (2.80).

Equation (2.83) shows that the function $\tilde{w}(x,t)$ equals zero outside D'', is even with respect to the variable t, has continuous inside D'' first derivatives by its arguments and besides

$$\tilde{w}(x,t) = \frac{1}{4}\int_0^x q(\xi)\,\mathrm{d}\xi, \quad x > 0. \tag{2.84}$$

Using the function $\tilde{w}(x,t)$ the solution to (2.74) under conditions (2.76) can be

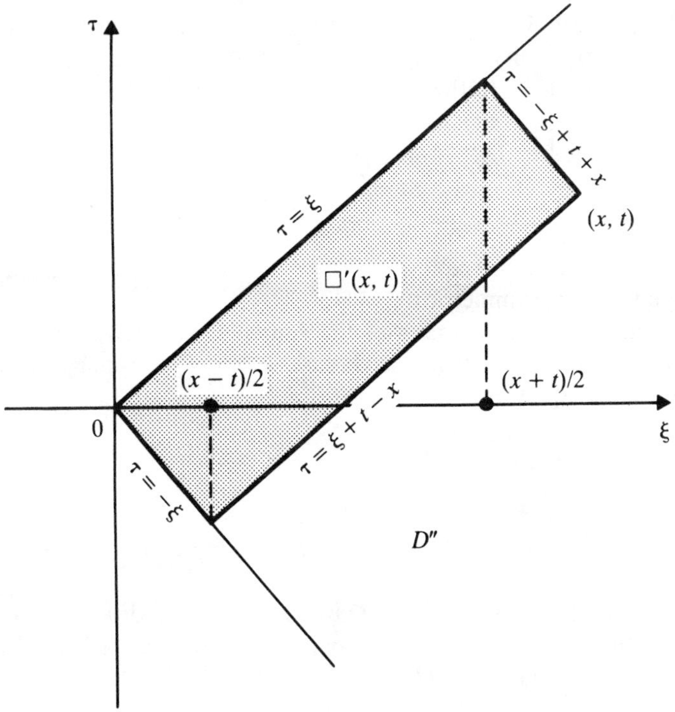

Figure 2.3.

written as

$$u(x,t) = \int_{-\infty}^{\infty} f(t_0)w(x, t-t_0)\,dt_0 = \int_{-\infty}^{\infty} f(t-\tau)w(x,\tau)\,d\tau. \tag{2.85}$$

Indeed, from (2.77) and (2.78) one has

$$L_q u = \int_{-\infty}^{\infty} f(t_0)L_q w(x, t-t_0)\,dt_0 = 0$$

$$u|_{x=0} = \int_{-\infty}^{\infty} f(t_0)\delta(t-t_0)\,dt_0 = f(t), \quad u_x|_{x=0} = 0.$$

Using definitions (2.82) and identities (2.81) formula (2.85) can be rewritten in the following way

$$u(x,t) = \tfrac{1}{2}[f(t-x) + f(t+x)] + \int_{-x}^{x} f(t-\tau)\tilde{w}(x,\tau)\,d\tau. \tag{2.86}$$

If the points (x, t) are now considered, at which identity (2.76) is valid, then

$$\tfrac{1}{2}[f(t-x) + f(t+x)] + \int_{-x}^{x} f(t-\tau)\tilde{w}(x,\tau)\,d\tau = 0, \quad x > |t|. \tag{2.87}$$

At any fixed $x > 0$ relation (2.87) can be viewed as an integral equation of the first kind with respect to $\tilde{w}(x,t)$, $t\in(-x,x)$. Since at $t=\tau$ the kernel of this equation has a finite discontinuity: $f(+0)=1/2$, $f(-0)=-1/2$, then its differentiation with respect to t results in the Fredholm equation of the second kind

$$\tilde{w}(x,t)+\int_{-x}^{x} f'(t-\tau)\tilde{w}(x,\tau)\,d\tau=-\tfrac{1}{2}[f'(t-x)+f'(t+x)],$$

$$x>0, \quad t\in(-x,x). \qquad (2.88)$$

Note that the kernel of this equation $f'(t-\tau)$ is continuous and, due to evenness of the function $f'(t)$, symmetrical.

Taking into account evenness of the function $\tilde{w}(x,t)$ with respect to the argument t, (2.88) can be replaced with a more convenient method for investigating the equation

$$\tilde{w}(x,t)+\int_{0}^{x}[f'(t-\tau)+f'(t+\tau)]\tilde{w}(x,\tau)\,d\tau$$

$$=-\tfrac{1}{2}[f'(t-x)+f'(t+x)], \quad x>0, \quad t\in[0,x). \qquad (2.89)$$

Equation (2.89) is equivalent to (2.87) under the additional condition of evenness of the function $\tilde{w}(x,t)$ with respect to the argument t. Indeed, any even with respect to t the solution to (2.87) evidently satisfies (2.89). On the other hand, any solution to (2.89), extended in an even way with respect to t for the values $t\in(-x,0)$, satisfies (2.88), obtained by differentiating (2.87) with respect to t, and the condition

$$\int_{-x}^{x} f(-\tau)\tilde{w}(x,\tau)\,d\tau=0$$

coinciding with equality (2.87) at $t=0$. Therefore, any solution to (2.89), when extended in an even way with respect to t, is the solution to (2.87). Thus, the equivalence of (2.87) and (2.89) is proved.

Assume that for any $x>0$ (2.88) is uniquely solvable in the class of continuous functions. In this case its solution defines the continuous in the domain D'' function $\tilde{w}(x,t)$ (with its even extension to the part of the domain D'' where $t<0$ accounted for). Since the right-hand part of (2.89) is a continuously differentiable in D'' function, while its kernel is piecewisely continuously differentiable, the solution to (2.89) will be continuously differentiable in D''. Equation (2.89), in particular, gives $\tilde{w}_t(x,+0)=0$. Therefore, the even extension to the domain $t<0$ is carried out with continuity of partial derivatives preserved.

From equality (2.84)

$$q(x)=4\frac{d}{dx}\tilde{w}(x,x), \quad x>0. \qquad (2.90)$$

In the same way one can get for the negative values of x the equation that differs from (2.89) only in that the upper limit x is replaced with $|x|$. Thus, we have $\tilde{w}(-x,t)=\tilde{w}(x,t)$ and, as a result, $q(-x)=q(x)$.

Thus the following algorithm for solving inverse problems arises: (2.89) is solved for every x and then $q(x)$ is found by formula (2.90).

In accordance with the Fredholm theory of equations, for small values of x (2.89) is uniquely solvable. While in this case the following statement proves to be valid.

Lemma 2.4. If a solution to the inverse problem exists, then (2.89) is uniquely solvable for any $x > 0$.

As is already known (see [83, 84]) for (2.89) to be uniquely solvable, it is necessary and sufficient that the uniform equation

$$\tilde{w}(x,t) + \int_0^x [f'(\tau+t) + f'(t-\tau)]\tilde{w}(x,\tau)\,d\tau = 0, \quad t\in[0,x), \tag{2.91}$$

have only the trivial solution $\tilde{w} \equiv 0$. Assume that there exists such $x_0 > 0$ that at $x = x_0$(2.91) has the solution $\tilde{w}(x_0, t) \not\equiv 0, t\in[0, x_0)$. Extend the function $\tilde{w}(x_0, t)$ in an even way onto the intercept $(-x_0, x_0)$ and with the zero outside the interval $(-x_0, x_0)$. Then from (2.91)

$$\int_{-\infty}^{\infty} f(t-\tau)\tilde{w}(x_0,\tau)\,d\tau = 0, \quad t\in(-x_0, x_0).$$

Consider the function

$$\Phi(x,t) = \int_{-\infty}^{\infty} u(x, t-\tau)\tilde{w}(x_0,\tau)\,d\tau. \tag{2.92}$$

It is obvious that

$$L_q\Phi = 0, \quad 0 < x < x_0$$
$$\Phi|_{x=0} = 0, \quad t\in(-x_0, x_0), \quad \Phi_x|_{x=0} = 0. \tag{2.93}$$

Hence, $\Phi(x,t) = 0$ for $(x,t)\in\Delta'(x_0, 0)$. With the obtained result used, (2.92) gives the equality

$$\int_{-\infty}^{\infty} u(x, t-\tau)\tilde{w}(x_0,\tau)\,d\tau = 0, \quad (x,t)\in\Delta'(x_0,0). \tag{2.94}$$

Differentiating this equality with respect to t and setting $t = 0$

$$\int_{-\infty}^{\infty} u_t(x, -\tau)\tilde{w}(x_0,\tau)\,d\tau = 0, \quad x\in[0, x_0]. \tag{2.95}$$

Taking into account the evenness of the functions $u_t(x,t)$ and $\tilde{w}(x_0,t)$, equality (2.95) can be rewritten as follows:

$$\int_0^{\infty} u_t(x,\tau)\tilde{w}(x_0,\tau)\,d\tau = 0, \quad x\in[0, x_0]. \tag{2.95'}$$

As the solution to the Cauchy problem (2.47) and (2.48), in the domain $x > 0, t > 0$ function $u(x,t)$ has the following form

$$u(x,t) = \theta(t-x)[\tfrac{1}{2} + \tilde{u}(x,t)]$$

where $\tilde{u}(x,t)$ is a continuous in domain $t \geqslant x \geqslant 0$ function together with its second-order derivatives, in which case $\tilde{u}(x,x) = 0$. Therefore

$$u_t(x,t) = \tfrac{1}{2}\delta(t-x) + \theta(t-x)\tilde{u}_t(x,t), \quad x > 0, t > 0.$$

Thus equality (2.95′) is transformed into

$$\tfrac{1}{2}\tilde{w}(x_0,x) + \int_x^{x_0} \tilde{u}_t(x,t)\tilde{w}(x_0,t)\,\mathrm{d}t = 0, \quad x \in [0,x_0]. \tag{2.96}$$

As regards the function $\tilde{w}(x_0,t)$, equality (2.96) is a uniform integral Volterra equation of the second kind with the continuous kernel and its only possible solution is $\tilde{w}(x_0,t) = 0$, $t \in [0,x_0]$. Thus one comes to the conclusion that the assumption on the existence of a nontrivial solution to (2.91) is erroneous, which fact proves the lemma. An implication of Lemma 2.4 is the theorem of the uniqueness of the inverse problem solution. The lemma inverse to Lemma 2.4 is valid.

Lemma 2.5. If for a $T > 0$ $f(t) \in C^2(0,T]$, $f(+0) = 1/2$, $f'(+0) = 0$ and (2.89), where $f(t)$ is extended onto the values $t \in [-T,0)$ in an odd way, is uniquely solvable for $x \in (0, T/2]$, a solution to the inverse problem exists on the intercept $[-T/2, T/2]$.

Let $f(t) \in C^3(0,T]$. Denote the part of domain D″, where $x \leqslant T/2$, through $D''(T)$. The solution to (2.89) defines (under even extension with respect to t for the values $t < 0$) in the domain $D''(T)$ the twice continuously differentiable function $\tilde{w}(x,t)$.

Assume

$$q(x) = 4\frac{\mathrm{d}}{\mathrm{d}x}\tilde{w}(x,x-0), \quad 0 < x \leqslant T/2, \quad q(-x) = q(x)$$

$$w(x,t) = \tfrac{1}{2}[\delta(t-x) + \delta(t+x)] + \theta(x-|t|)\tilde{w}(x,t), \quad 0 < x \leqslant T/2$$

$$\omega(x,t) = L_q w(x,t), \quad 0 \leqslant x \leqslant T/2, -\infty < t < \infty. \tag{2.97}$$

Now verify if the function $\omega(x,t)$ is a piecewise continuous function in the band $0 \leqslant x \leqslant T/2$. Indeed

$$\omega(x,t) = \tfrac{1}{2}L_q[\delta(t-x) + \delta(t+x)] + L_q[\theta(x-|t|)\tilde{w}(x,t)]$$

$$= \tfrac{1}{2}q(x)[\delta(t-x) + \delta(t+x)] - [\delta(t-x) + \delta(t+x)]2\frac{\mathrm{d}}{\mathrm{d}x}$$

$$\times\, \tilde{w}(x,x-0) + \theta(x-|t|)L_q\tilde{w} = \theta(x-|t|)L_q\tilde{w}.$$

Then, the function $w(x,t)$ in the domain $D''(T)$ satisfies the equation

$$\int_{-\infty}^{\infty} f(t-\tau)w(x,\tau)\,\mathrm{d}\tau = 0, \quad (x,t) \in D''(T) \tag{2.98}$$

or

$$\int_{-\infty}^{\infty} f(t_0)w(x,t-t_0)\,\mathrm{d}t_0 = 0, \quad (x,t) \in D''(T).$$

Applying to this equation the operator L_q, it is found that $\omega(x, t)$ is the solution to the equation

$$\int_{-\infty}^{\infty} f(t_0)\omega(x, t-t_0)\,\mathrm{d}t_0 = 0, \quad (x, t)\in D''(T)$$

that, accounting for the fact that $\omega(x, t) = 0$ for $x < |t|$, can be rewritten as

$$\int_{x}^{-x} f(t-\tau)\omega(x, \tau)\,\mathrm{d}\tau = 0, \quad (x, t)\in D''(T).$$

Differentiating this equation with respect to t and using the conditions

$$\omega(x, -t) = \omega(x, t), \qquad f(+0) = 1/2, \qquad f(-0) = -1/2,$$

one obtains

$$\omega(x, t) + \int_{0}^{x} [f'(t-\tau) + f'(t+\tau)]\omega(x, \tau)\,\mathrm{d}\tau = 0, \quad (x, t)\in D''(T). \tag{2.99}$$

Due to unique solvability of equation (2.89) equality (2.99) gives

$$\omega(x, t) \equiv 0, \quad (x, t)\in D''(T).$$

Therefore, in the band $0 \leqslant x \leqslant T/2$ $\omega = 0$. Thus, the function $w(x, t)$ is the solution to the equation

$$L_q w = 0.$$

Since $f'(0) = 0$, from (2.89) we have $\tilde{w}(0, 0) = 0$. By using the definition of the function $w(x, t)$ [see (2.97)], one obtains

$$w(0, t) = \delta(t), \quad w_x|_{x=0} = 0, -\infty < t < \infty$$

and hence, in the band $0 \leqslant x \leqslant T/2$ $w(x, t)$ is the solution to the Cauchy problem (2.77) and (2.78) at $t_0 = 0$.

In this case in the band $0 \leqslant x \leqslant T/2$ the function $u(x, t)$, which is defined by equality (2.85), satisfies conditions (2.76) for $t\in(-T, T)$ and, in line with equality (2.98), condition (2.75). Under these conditions its even extension with respect to t is also the solution to (2.74) and satisfies conditions (2.75) and (2.76). In this case one has

$$u(x, 0) = 0, \quad u_t|_{t=0} = \delta(x). \tag{2.100}$$

Thus, the solution to the inverse problem exists and $q(x)\in C^1[-T/2, T/2]$.

Now let the function $f\in C^2[0, T]$. Consider such a sequence of the functions $f_n(t)\in C^3[0, T]$ that $f_n(t)\to f(t)$ in the norm of the space $C^2[0, T]$. The sequence f_n corresponds to that of the functions $\tilde{w}_n(x, t)$, $q_n(x)$, $u_n(x, t)$. No doubt that at $n\to\infty$ the sequence $\tilde{w}_n\in C^2(D''(T))$ converges in the norm of the space $C^1(D''(T))$ to a certain finite function $\tilde{w}(x, t)\in C^1(D''(T))$; $q_n(x)$ and $u_n(x, t)$ converge uniformly to $q(x)\in C[-T/2, T/2]$ and to a certain piecewise continuous function $u(x, t)$, respectively. In this case for limited functions equalities (2.74), (2.76), and (2.100), viewed as generalized functions, are valid.

Lemmas 2.4 and 2.5 result in the following theorem.

Theorem 2.8. For an inverse problem to be uniquely solved on the intercept $[-T/2, T/2]$ in the class of continuous functions, it is necessary and sufficient that the function $f(t)$ satisfies the following conditions: (1) $f(t)\in C^2[0,T]$, $f'(+0)=1/2$, $f'(0)=0$; and (2) the integral equation (2.89) in which $f(-t)=-f(t)$, $t\in(0,T]$, is uniquely solvable at any $x\in(0,T/2)$.

The second condition of the theorem can be replaced with the equivalent condition of the positive definiteness of the operator A_x

$$A_x\phi = \phi(t) + \int_0^x [f'(t-\tau) + f'(t+\tau)]\phi(\tau)\,d\tau$$

for $x\in[0,T/2]$, i.e. with the condition

$$(A_x\phi, \phi) > 0, \quad x\in[0,T/2], \quad \phi\in L_2[0,x]$$

where the symbol (ϕ,ψ) denotes the scalar product of the functions ϕ,ψ in the space $L_2[0,x]$.

2.5. Inverse problems for differential equations in a limited domain

Consider the differential equation

$$L_q u = 0 \tag{2.101}$$

in the limited with respect to x domain

$$D = \{(x,t): 0 < x < l, t > 0\}$$

with the boundary conditions

$$(u_x - hu)_{x=0} = g_1(t), \qquad (u_x + Hu)_{x=l} = g_2(t), \quad t \geq 0 \tag{2.102}$$

and the initial data

$$u|_{t=0} = \phi(x), \qquad u_t|_{t=0} = \psi(x), \quad 0 \leq x \leq l. \tag{2.103}$$

Here h and H are real and finite numbers. Under the assumption $q(x)\in C(0,1)$ study the problem of finding L_q from the condition

$$u(0,t) = f(t), \quad t \geq 0. \tag{2.104}$$

Leaving aside the case of the regular functions g_1, g_2, ϕ, ψ, consider here two most interesting examples in the applied sense, when the data of the direct problem (2.101) and (2.102) are singular generalized functions, i.e.

$$g_1 = g_2 = \phi \equiv 0, \qquad \psi(x) = \delta(x); \tag{2.105}$$

$$g_2 = \phi = \psi = 0, \qquad g_1(t) = -\delta(t). \tag{2.106}$$

Note that in the first example the solution to problem (2.101)–(2.103) is the limit (in the sense of generalized functions) of the solutions to problem (2.101)–(2.103), corresponding to $\psi(x) = \delta(x - x_0)$ at $x_0 \to +0$, while in the second example it is the limit of the solutions corresponding to $g_1(t) = \delta(t - t_0)$ at $t_0 \to +0$. Later we shall see that these two solutions coincide.

Prior to analysing the inverse problem, consider one of the possible methods for solving the direct problem (2.101)–(2.103), which is based on reducing this problem to an integral equation inside the domain D.

Firstly we shall examine an additional problem for the equation

$$u_{tt} - u_{xx} = F(x,t), \quad (x,t) \in D, \tag{2.107}$$

under the conditions (2.102) and (2.103).

In the part of the domain D, limited by the intercept $(0, l)$ of the x-axis and by the characteristics outgoing from the terminal ends of this intercept, the solution to (2.107) under the conditions (2.102) and (2.103) can be obtained with the d'Alembert formula

$$u(x,t) = \tfrac{1}{2}[\phi(x-t) + \phi(x+t)] + \frac{1}{2}\int_{x-t}^{x+t} \psi(\xi)\,d\xi + \frac{1}{2}\iint_{\Delta(x,t)} F(\xi,\tau)\,d\xi\,d\tau$$

$$(x,t) \in D_0 = \left\{(x,t): 0 \leqslant x \leqslant l, 0 \leqslant \tau \leqslant \frac{l}{2} - \left|x - \frac{l}{2}\right|\right\}. \tag{2.108}$$

In the domain $D \backslash D_0$ the solution to the problem depends on the boundary conditions. Now we can demonstrate that, using (2.107) and initial data (2.103), we can first find u, u_x on the right and left boundaries of the domain D, and then obtain the solution in the inside of the domain as well. To find u, u_x at $x = 0$, $x = l$ consider a bundle of characteristics of the operator $(\partial/\partial t) + (\partial/\partial x)$, passing through the intercept $(0, l)$ of the x-axis and separating the intercept $(0, l)$ on the right boundary of the domain D. By writing the wave operator as the product

$$\left(\frac{\partial}{\partial t} + \frac{\partial}{\partial x}\right)\left(\frac{\partial}{\partial t} - \frac{\partial}{\partial x}\right)$$

integrating equality (2.107) along the intercept of the fixed characteristics of the bundle within D, and using data (2.103), one finds

$$(u_t - u_x)_{x=l} = \psi(l-t) + \phi'(l-t) + \int_0^t F(\tau - t + x, \tau)\,d\tau, \quad t \in (0, l).$$

By joining this equality and the boundary condition (2.102) at $x = l$, one obtains

$$(u_t + Hu)_{x=l} = g_2(t) + \psi(l-t) + \phi'(l-t)$$
$$+ \int_0^t F(\tau - t + x, \tau)\,d\tau, \quad t \in (0, l).$$

Hence

$$u(l,t) = \phi(l-t) + \int_0^t e^{H(\tau - t)}\left[g_2(\tau) + \psi(l-\tau)\right.$$
$$\left. + H\phi(l-\tau) + \int_0^\tau F(\tau_1 - \tau + x, \tau_1)\,d\tau_1\right]d\tau \equiv f_2(t)$$

$$u_x(l,t) = g_2(t) - Hf_2(t) \equiv \chi(t), \quad t \in (0, l). \tag{2.109}$$

Now, using data (2.103) and the values of u, u_x obtained on the intercept $t \in (0, l)$,

$x=l$ one can calculate the u, u_x values at $x=0$, $t\in(0,2l)$. For this purpose draw through the intercept $(0,2l)$ of the axis $x=0$ a bundle of characteristics of the differential operator $(\partial/\partial t)-(\partial/\partial x)$, and write the wave operator as

$$\left(\frac{\partial}{\partial t}-\frac{\partial}{\partial x}\right)\left(\frac{\partial}{\partial t}+\frac{\partial}{\partial x}\right).$$

Integrating (2.107) along the fixed characteristics of the bundle from one point of the domain D boundary to another, at the points $x=0$, $t\in(0,2l)$ a combination of the functions u_t+u_x can be found, which, together with the boundary condition (2.102) at $x=0$, makes it possible to calculate u, u_x, $x=0$, $t\in(0,2l)$. Using the resulting values one can calculate u, u_x at $x=l$, $t\in(l,3l)$, then u, u_x at $x=0$, $t\in(2l,4l)$ and so on. Thus, for all $t>0$ one can find

$$u(0,t)=f_1(t), \qquad u_x(0,t)=\chi_1(t)$$
$$u(l,t)=f_2(t), \qquad u_x(l,t)=\chi_2(t).$$

To obtain the solution $u(x,t)$ in the domain D, it is now sufficient to integrate (2.107) first along the characteristics of one class, for instance, the characteristics of the operator $(\partial/\partial t)-(\partial/\partial x)$, and then along the characteristics of another class. In this case the following are found in succession

$$u_t+u_x=G(x,t)$$

$$G(x,t)=\begin{cases} f_2'(x+t-l)+\chi_2(x+t-l)+\displaystyle\int_{t+x-l}^{t} F(x+t-\tau,\tau)\,\mathrm{d}\tau, & x+t>l,\\ \psi(x+l)+\phi'(x+t)+\displaystyle\int_0^t F(x+t-\tau,\tau)\,\mathrm{d}\tau, & x+t\leqslant l,\end{cases}$$

$$u(x,t)=\begin{cases} f_1(t-x)+\displaystyle\int_{t-x}^{t} G(\tau-t+x,\tau)\,\mathrm{d}\tau, & t-x>0,\\ \phi(x-t)+\displaystyle\int_0^t G(\tau-t+x,\tau)\,\mathrm{d}\tau, & t-x\leqslant 0.\end{cases}$$

If the above algorithm for problem (2.101)–(2.103) is now used, setting $F=-qu$, then for the function $u(x,t)$ we get an integral equation whose domain of integration contains the intercept $(0,t)$ with respect to the variable τ. Therefore, the integral equation possesses the property of the Volterra type equations: it is uniquely solvable at any t and the solution can be found by employing the standard method of successive approximations.

Now examine in more detail problem (2.101)–(2.103) under conditions (2.105) or (2.106) with the particular aim of obtaining the necessary properties of the function $f(t)=u(0,t)$. It will be demonstrated later that setting the function $f(t)$ for $t\in(0,2l)$ uniquely defines the function $q(x)$ on $(0,l)$. Therefore, let us restrict ourselves to studying the solution to the direct problem (2.101)–(2.103) in the domain $D_1\cup D_2$

$$D_1=\{(x,t):0<x<l,0<t<x\},$$
$$D_2=\{(x,t):0<x<l,x<t<2l-x\}.$$

First of all under conditions (2.105) and (2.106) show that

$$u(x,t) \equiv 0, \quad (x,t) \in D_1. \tag{2.110}$$

Indeed, in the domain $D_0 \subset D_1$ the d'Alembert formula (2.108) gives the uniform integral equation

$$u(x,t) = -\frac{1}{2}\iint_{\Delta(x,t)} q(\xi)u(\xi,\tau)\,d\xi\,d\tau, \quad (x,t)\in D_0.$$

Hence, $u(x,t)=0$, $(x,t)\in D_0$. Using formula (2.109), we find

$$u(l,t) = -\int_0^t e^{H(\tau-t)} \int_{l-\tau/2}^{l} q(\xi)u(\xi,\tau-l+\xi)\,d\xi\,d\tau. \tag{2.111}$$

Integrating (2.101) along the characteristics $dx/dt=1$

$$\left(\frac{\partial}{\partial t}-\frac{\partial}{\partial x}\right)u = -\int_{(l+x-t)/2}^{x} q(\xi)u(\xi,\xi+t-x)\,d\xi, \quad (x,t)\in D_1\backslash D_0.$$

Then, using (2.111), the equation for $u(x,t)$ in the domain $D_1\backslash D_0$ is found

$$u(x,t) = -\int_0^{t+x-l} e^{H(\tau-t-x+l)} \int_{l-\tau/2}^{l} q(\xi)u(\xi,\tau-l+\xi)\,d\xi\,d\tau$$
$$-\int_{t+x-l}^{t}\int_{(l+t+x-2\tau)/2}^{t+x-\tau} q(\xi)u(\xi,\xi+2\tau-t-x)\,d\xi\,d\tau, \quad (x,t)\in D_1\backslash D_0. \tag{2.212}$$

which is a uniform equation of Volterra type. Hence

$$u(x,t)\equiv 0, \quad (x,t)\in D_1\backslash D_0$$

and (2.110) is determined.

Now demonstrate that in the domain D_2 for the data (2.105), (2.106) the function $u(x,t)$ satisfies one and the same equation

$$u(x,t) = e^{h(x-t)} - \int_0^{t-x} e^{h(\tau-t+x)} \int_0^{\tau/2} q(\xi)u(\xi,\tau-\xi)\,d\xi\,d\tau$$
$$-\int_{t-x}^{t}\int_{\tau-t+x}^{(2\tau-t+x)/2} q(\xi)u(\xi,2\tau-t+x-\xi)\,d\xi\,d\tau, \quad (x,t)\in D_2. \tag{2.113}$$

which indicates that under conditions (2.105) or (2.106) the solutions to problems (2.101)–(2.103) coincide.

Let us begin with data (2.105). As has already been mentioned, in this case the solution to problem (2.101)–(2.103) is the limit, at $x_0\to+0$, of the solutions $u(x,t,x_0)$ to (2.101) under conditions (2.102) with the Cauchy data

$$u|_{t=0}=0, \qquad u_t|_{t=0}=\delta(x-x_0).$$

Now consider four domains D'_k, $k=1,2,3,4$, that are formed if, from the point $(x_0,0)$, we let the characteristics inside the domain D up to their crossing with the right and left D boundaries, and if from these points of crossing, in their turn, we

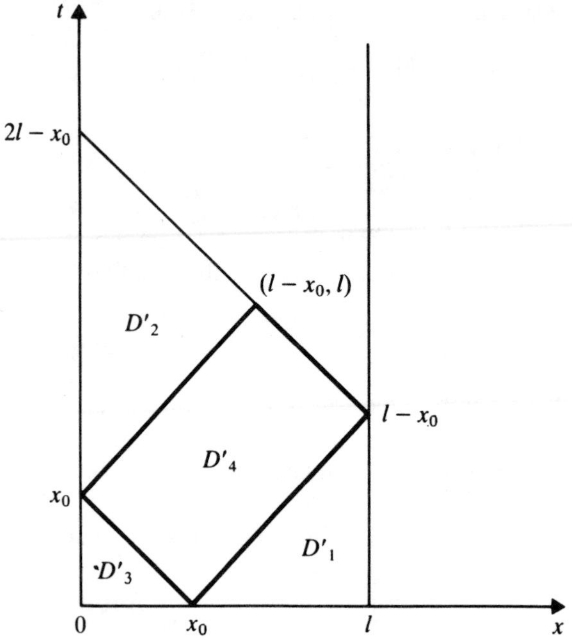

Figure 2.4.

let the characteristics inside the domain D (Fig. 2.4)

$$D_1' = \{(x,t): x_0 < x < l, 0 < \tau < \xi - x_0\}$$
$$D_2' = \{(x,t): 0 < x < l - x_0, x_0 + \xi < \tau < 2l - x_0 - \xi\}$$
$$D_3' = \{(x,t): 0 < x < x_0, 0 < \tau < x_0 - \xi\}$$
$$D_4' = \{(x,t): 0 < x < l, |\xi - x_0| < \tau < l - |\xi - x_0 + l|\}.$$

At $x_0 \to +0$ the measure of the domains D_3', D_4' tends to zero, while the domains D_1', D_2' contract to the domains D_1, D_2, respectively.

The solution $u(x, t, x_0)$ will become zero in the domains D_1' and D_3'. Integrating (2.101) along the characteristics $dx/dt = -1$, one obtains

$$(u_t + u_x)_{x=0} = \delta(t - x_0) - \theta(t - x_0) \int_0^{(t+x_0)/2} q(\xi)u(\xi, t - \xi)\, d\xi, \quad t \in (0, 2l - x_0).$$

From this equality and the boundary condition at $x = 0$ it follows that

$$u|_{x=0} = \theta(t - x_0)\left[e^{h(x_0 - t)} - \int_{x_0}^{t} e^{h(\tau - t)} \int_0^{(\tau + x_0)/2} q(\xi)u(\xi, \tau - \xi)\, d\xi\, d\tau \right],$$
$$t \in (0, 2l - x_0).$$

Using this formula, by integrating (2.101) under conditions (2.102) and (2.103) with respect to the characteristics $dx/dt = -1$ and $dx/dt = 1$, one obtains the integral equation in the domains D'_4, D'_2

$$u(x,t,x_0)=\frac{1}{2}-\int_{(t-x+x_0)/2}^{t}\int_{\tau-t+x}^{(2\tau-t+x+x_0)/2} q(\xi)u(\xi,2\tau-t+x-\xi,x_0)\,d\xi\,d\tau,$$
$$(x,t)\in D'_4$$

$$u(x,t,x_0)=e^{h(x_0+x-t)}-\int_{x_0}^{t-x} e^{h(\tau-t+x)}\int_{0}^{(\tau+x_0)/2} q(\xi)u(\xi,\tau-\xi,x_0)\,d\xi\,d\tau$$
$$-\int_{t-x}^{t}\int_{\tau-t+x}^{(2\tau-t+x+x_0)/2} q(\xi)u(\xi,2\tau-t+x-\xi,x_0)\,d\xi\,d\tau,\quad (x,t)\in D'_2.$$

Going over to the limit at $x_0\to+0$ results here in (2.113).

Data (2.106) can be treated in an analogous way. Introduce the parameter t_0, $g_1(t)$ substituted for $-\delta(t-t_0)$ and examine the limit of the solutions $u(x,t,t_0)$ of the corresponding problems (2.101)–(2.103). Begin with defining the domains

$$D'_1=\{(x,t):0<x<l,0<t<t_0+x\}$$
$$D'_2=\{(x,t):0<x<l,t_0+x<t<t_0+2l-x\}.$$

It is obvious that $D'_1\to D_1$, $D'_2\to D_2$ at $t_0\to 0$. In the domain D'_1 the function $u(x,t,t_0)=0$, and, in line with this fact

$$(u_t+u_x)_{x=0}=-\theta(t-t_0)\int_{0}^{(t+t_0)/2} q(\xi)u(\xi,t-\xi,t_0)\,d\xi,\quad t\in(0,t_0+2l).$$

Combined with the boundary condition $(u_x-hu)_{x=0}=-\delta(t-t_0)$ the above equality makes it possible to find $u(x,t,t_0)$ at $x=0$

$$u|_{x=0}=\theta(t-t_0)\left[e^{h(t_0-t)}-\int_{t_0}^{t} e^{h(\tau-t)}\int_{0}^{(\tau+t_0)/2} q(\xi)u(\xi,\tau-\xi,t_0)\,d\xi\,d\tau\right],$$
$$t\in(0,t_0+2l).$$

Using the resulting equality we obtain the integral equation for $u(x,t,t_0)$

$$u(x,t,t_0)=e^{h(t_0+x-t)}-\int_{t_0}^{t-x} e^{h(\tau-t+x)}\int_{0}^{(\tau+t_0)/2} q(\xi)u(\xi,\tau-\xi,t_0)\,d\xi\,d\tau$$
$$-\int_{t-x}^{t}\int_{\tau-t+x}^{(2\tau-t+x+t_0)/2} q(\xi)u(\xi,2\tau-t+x-\xi,t_0)\,d\xi\,d\tau,\quad (x,t)\in D'_2.$$

Going over to the limit at $t_0\to 0$ results here in (2.113).

Now examine (2.113); first of all, note that

$$u(x,x+0)=1, x\in(0,l).$$

Equation (2.113) is that of Volterra type and therefore its solution is unique in the class of functions belonging to the space $C(D_2)$ and can be obtained by the method of successive approximations, the existence of the solution in this class resulting from $q(x)$ belonging to the class $C(0,l)$. Moreover, differentiation of (2.113) easily proves that its solution belongs to the class $C^2(D_2)$. In this case

$$u_t(0,+0)=-h.$$

From here we can derive the necessary conditions for the function $f(t)$: (1) $f(t) \in C^2(0, l)$; (2) $f(+0) = 1$, $f'(+0) = -h$.

We can readily prove that these conditions are also sufficient for the solvability of the inverse problem in the small. For this purpose the equations for the derivatives u_t, u_{tt} can be written, set $x = 0$ in the second of them and use condition (2.104). In this case a system of nonlinear integral equations is obtained for $u, u_t, q(x)$, that is uniquely solvable at sufficiently small T in the domain

$$D_2(T) = \{(x,t): 0 < x < T/2, x < t < T - x\}.$$

However, a more convenient way of studying the inverse problem is that given in Section 2.4. Extend the solution to problem (2.101)–(2.103) into the domain $D^- = \{(x,t): 0 < x < l, t \leqslant 0\}$ in an odd way and introduce into consideration the function $w(x,t)$ as the solution to the problem

$$L_q w = 0, \quad (x,t) \in D \cup D^-, \quad w|_{x=0} = \delta(t), \quad w_x|_{x=0} = h\delta(t). \tag{2.114}$$

For the function $w(x,t)$ the following representation is valid

$$w(x,t) = \tfrac{1}{2}[\delta(t-x) + \delta(t+x)] + \theta(x - |t|)\tilde{w}(x,t), \tag{2.115}$$

where $\tilde{w}(x,t)$ is the even, with respect to t, function, continuous in the domain

$$D'' = \{(x,t): 0 \leqslant x \leqslant l, -x \leqslant t \leqslant x\}$$

together with the first-order derivatives. The function $\tilde{w}$ is the solution to the integral equation analogous to (2.83)

$$\tilde{w}(x,t) = \tfrac{1}{2}h + \frac{1}{4}\int_0^{(x+t)/2} q(\xi)\,d\xi + \frac{1}{4}\int_0^{(x-t)/2} q(\xi)\,d\xi$$
$$+ \frac{1}{2}\iint_{\square(x,t)} q(\xi)\tilde{w}(\xi,\tau)\,d\xi\,d\tau, \quad (x,t) \in D''. \tag{2.116}$$

This equation, in particular, gives

$$\tilde{w}(x,x) = \tfrac{1}{2}h + \frac{1}{4}\int_0^x q(\xi)\,d\xi, \quad x \in (0, l). \tag{2.117}$$

The solution to (2.101) satisfying the conditions

$$u(0,t) = f(t), \quad t \in (-2l, 2l), \quad f(-t) = -f(t),$$
$$(u_x + hu)_{x=0} = 0, \quad t \in (-\infty, \infty)$$

is uniquely determined in the domain

$$Q = \{(x,t): 0 < x < l, -2l + x < t < 2l - x\}.$$

Using the introduced function $w(x,t)$ this solution can be written as follows:

$$u(x,t) = \int_{-\infty}^{\infty} f(t-\tau)w(x,\tau)\,d\tau = \tfrac{1}{2}[f(t-x) + f(t+x)]$$
$$+ \int_{-x}^{x} f(t-\tau)\tilde{w}(x,\tau)\,d\tau, \quad (x,t) \in Q.$$

In this case using the conditions $u(x,t)=0$, $(x,t)\in D''$ for the function $\tilde{w}(x,t)$ results in (2.89). Having found the solution to this equation for $(x,t)\in D''$, from (2.117) we get

$$q(x)=4\frac{d}{dx}\tilde{w}(x,x), \quad x\in(0,l).$$

Equation (2.117) demonstrates that when setting the inverse problem, we could have considered unknown also the parameter h, which is included in the boundary condition (2.102). This parameter can be obtained by

$$h=2\tilde{w}(0,0)=-f'(+0).$$

If we continue examining the problem in the way it is done in Section 2.4, a theorem analogous to Theorem 2.8 will be obtained.

Theorem 2.9. For the inverse problem (2.101)–(2.104) to be uniquely solvable in the class of functions $q(x)\in C(0,l)$, it is necessary and sufficient that the function $f(t)$ satisfies the following conditions: (1) $f(x)\in C^2(0,2l)$, $f(+0)=1$, $f'(+0)=-h$; and (2) the integral equation (2.89), where $f(-t)=-f(t)$, $t\in(0,2l)$, is uniquely solvable at any $x\in(0,l)$.

Note that at $l\to\infty$ we come to setting the inverse problem for (2.101) in the semi-limited domain $x\in(0,\infty)$, $t>0$. In this case one can set $l=\infty$ in (2.101)–(2.104) and to preserve in (2.102) only the condition corresponding to $x=0$.

2.6. Relationship with the Sturm–Liouville problem

Until now the inverse problems for the operator L_q for the case when the solution to the direct problem itself only has been considered, at a fixed point x_0 and as a function of time t, serves as an information for obtaining the solution to the inverse problem. There are, however, some other ways of setting, when a number of functionals is set from the solution to the direct problem. For linear differential operators one can use the spectral characteristics of the operator as such functionals.

In a regular case, the essence of the known inverse Sturm–Liouville problem is in the fact that on $[a,b]$ the structure of the differential operator l_q is set

$$l_q y:\begin{cases}(-d^2/dx^2+q(x))y\\ (y'-hy)_{x=a}=0, \quad (y'+Hy)_{x=b}=0\end{cases} \tag{2.118}$$

where a, b, h, and H are the finite numbers; $q(x)\in C[a,b]$. The task is to find $q(x)$, provided we know the eigenvalues λ_n of this differential operator, i.e. such λ at which there exist the nontrivial solutions to the operator equation

$$l_q y=\lambda y \tag{2.119}$$

and the corresponding to them norms $\|y_n\|$ of the eigenfunctions $y_n(x)=y(x,\lambda_n)$ $(y(a,\lambda_n)=1,\ y'(a,\lambda_n)=h)$ are known

$$\|y_n\|^2=\int_a^b y_n^2(x)\,dx. \tag{2.120}$$

The spectral theory of the operator l_q affords (see [5, 6]) that other than zero solutions to (2.119), under the supposition on the finiteness of the intercept $[a, b]$ and on $q(x)$ belonging to the class $C[a, b]$, can exist only for the discrete $\lambda = \lambda_n$ which are real and have a point of concentration at $+\infty$. At n the asymptotic formula for λ_n is valid and corresponding to them eigenfunctions $y_n(x)$, normed by the conditions $y_n(a) = 1$, $y_n'(a) = h$

$$\lambda_n = \frac{\pi}{b-a} n + O\left(\frac{1}{n}\right), \qquad y_n(x) = \cos[\lambda_n(x-a)] + O\left(\frac{1}{n}\right).$$

It is also known that all $y_n(x)$ are orthogonal between one another and form a complete system in $L_2[a, b]$, which means that for any function $f(x) \in L_2[a, b]$ the Fourier series by the system of functions $y_n(x)$, $n = 1, 2, \ldots$

$$\sum_{n=1}^{\infty} f_n y_n(x), \qquad f_n = (f, y_n)/\|y_n\|^2$$

converges by the norm $L_2[a, b]$ to $f(x)$. This fact is reflected in the Parseval's equality

$$\int_a^b f^2(x)\,\mathrm{d}x = \sum_{n=1}^{\infty} f_n^2 \|y_n\|^2 = \sum_{n=1}^{\infty} \frac{(f, y_n)^2}{\|y_n\|^2}. \tag{2.121}$$

Usually the Sturm–Liouville inverse problem is set in the terms of the spectral function $\rho(\lambda)$ of the operator l_q. To clarify the meaning of this statement in the case considered, let us examine the function $y(x, \lambda)$ which is the solution to the Cauchy problem

$$-y'' + q(x)y = \lambda y, \quad x \in [a, b], \qquad y(a, \lambda) = 1, \qquad y'(a, \lambda) = h \tag{2.122}$$

denoting for an arbitrary function $g(x) \in L_2[a, b]$

$$\tilde{g}(\lambda) = \int_a^b g(x) y(x, \lambda)\,\mathrm{d}x, \quad \lambda \in (-\infty, \infty). \tag{2.123}$$

The function $\tilde{g}(\lambda)$ is called the image of the Fourier function $g(x)$. Since $y(x, \lambda_n) = y_n(x)$, then the function $g(x)$ can be obtained by the values $\tilde{g}(\lambda_n)$ using the Fourier series

$$g(x) = \sum_{n=1}^{\infty} \frac{\tilde{g}(\lambda_n)}{\|y_n\|^2} y(x, \lambda_n). \tag{2.124}$$

This formula can be easily transformed to a new one which is in a certain sense analogous to (2.123)

$$g(x) = \int_{-\infty}^{\infty} \tilde{g}(\lambda) y(x, \lambda)\,\mathrm{d}\rho(\lambda) \tag{2.125}$$

by introducing the function $\rho(\lambda)$ as a piecewise continuous one

$$\rho(\lambda) = \sum_{\lambda_n < \lambda} \frac{1}{\|y_n\|^2} \theta(\lambda - \lambda_n). \tag{2.126}$$

The diagram of the function is shown in Fig. 2.5.

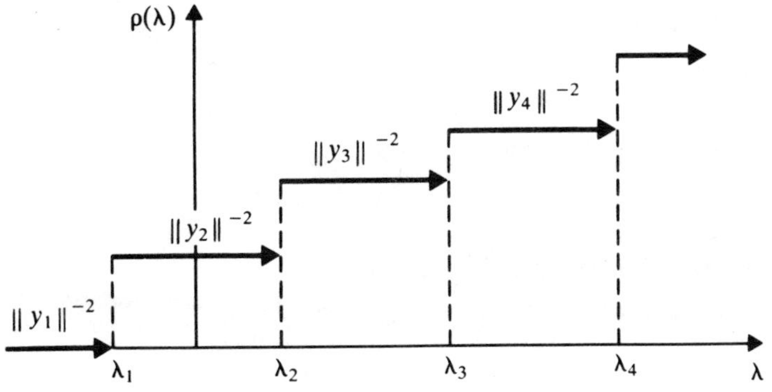

Figure 2.5.

The summation $\sum_{\lambda_n<\lambda}$ in (2.126) denotes the summation over all those n for which $\lambda_n < \lambda$. The function $\rho(\lambda)$ has finite discontinuities at the points where $\lambda = \lambda_n$, the quantity of the step being equal to $\| y_n \|^{-2}$. Thus, the spectral function includes the information on both the eigenvalues and the norms of the eigenfunctions. Hence, the Sturm–Liouville inverse problem can be formulated as follows: the task is to find the operator l_q when its spectral function is given. This way of setting proves more convenient since it can be directly related to the singular case of a semi-infinite intercept.

Demonstrate that this problem and the problem discussed in the preceding section are closely related. Examine the equation

$$L_q u = 0, \quad (x,t)\in D = \{(x,t): a < x < b, t > 0\} \tag{2.127}$$

in the conditions

$$u(x,0) = 0, \qquad u_t(x,0) = \delta(x-a) \tag{2.128}$$

$$(u_x - hu)_{x=a} = 0, \qquad (u_x + Hu)_{x=b} = 0. \tag{2.129}$$

Let

$$u(a,t) = f(t), \quad t\in(0, 2(b-a)) \tag{2.130}$$

and the task is to find $q(x)\in C(a,b)$.

Investigation of this problem in Section 2.5 was based on its reduction to an integral equation. There is, however, another approach based on the Fourier method of separation of variables. Write the solution to the direct problem (2.127)–(2.129) in the following form

$$u(x,t) = \sum_{n=1}^{\infty} T_n(t) y_n(x)$$

where $y_n(x)$ are the eigenfunctions of the operator $l_q(y_n(a) = 1,\ y_n'(a) = h)$. Then, according to (2.127) and (2.128), the function $T_n(t)$ satisfies the equation

$$T_n'' + \lambda_n T_n = 0$$

and the conditions

$$T_n(0) = 0, \qquad T_n'(0) = \| y_n \|^{-2}.$$

Hence

$$T_n(t) = \frac{1}{\sqrt{(\lambda_n)}\| y_n \|^2} \sin \sqrt{(\lambda_n)}t.$$

Therefore, the solution to (2.127)–(2.129) is as follows:

$$u(x, t) = \sum_{n=1}^{\infty} \frac{\sin \sqrt{(\lambda_n)}t}{\sqrt{(\lambda_n)}\| y_n \|^2} y_n(x). \tag{2.131}$$

By using the spectral function $\rho(\lambda)$, this formula can be rewritten as

$$u(x, t) = \int_{-\infty}^{\infty} \frac{\sin \sqrt{(\lambda)}t}{\sqrt{(\lambda)}} y(x, \lambda)\, \mathrm{d}\rho(\lambda). \tag{2.132}$$

By setting here $x = 0$, we can relate the spectral function $\rho(\lambda)$ and the function $f(t)$

$$f(t) = \int_{-\infty}^{\infty} \frac{\sin \sqrt{(\lambda)}}{\sqrt{(\lambda)}} \mathrm{d}\rho(\lambda). \tag{2.133}$$

Using (2.133) and knowing the function $\rho(\lambda)$, we can find $f(t)$ for $t \in (0, 2(b-a))$, solve the inverse problem (2.127)–(2.130) and, hence, find the operator l_q. In an analogous way, if $f(t)$ is known for $t \in (0, 2(b-a))$, we uniquely obtain L_q and, hence, the operator l_q. Therefore, one can uniquely find $\rho(\lambda)$ by $f(t)$, which demonstrates that there exists a one-to-one correspondence between the functions $f(t)$, $t \in (0, 2(b-a))$ and functions $\rho(\lambda)$.

In a singular case [$b = \infty$ and $q(x) \in C[a, l]$ for any finite $l > a$], depending on the behaviour of the function $q(x)$ at $x \to \infty$ there may be two possibilities: (1) a finite point and (2) a finite circle (see [5]).

In this case there exists at least one spectral function $\rho(\lambda)$ (in the case of a finite point only one), which does not decrease monotonically on $(-\infty, \infty)$, $\rho(-\infty) = 0$, and such that for any finite function $g(x) \in L_2(a, \infty)$ the equalities (2.123) and (2.125) are valid and the Parseval's equality

$$\int_a^{\infty} g^2(x)\, \mathrm{d}x = \int_{-\infty}^{\infty} \tilde{g}^2(\lambda)\, \mathrm{d}\rho(\lambda).$$

Finiteness of $g(x)$ here means that $g(x) = 0$ outside the finite intercept $[a, l]$.

The inverse Sturm–Liouville problem is posed here in the same way: to determine the operator l_q by $\rho(\lambda)$. Uniqueness of the solution to the Sturm–Liouville inverse problem on $[a, \infty]$ results from the Marchenko theorem, which demonstrates that when the spectral functions of two differential operators l_q[corresponding to different $q(x)$ and h] coincide to the accuracy of an arbitrary multiplier, these differential operators also coincide. It can be easily proved that in this case (2.133) remains valid. Problems associated with determining the function $q(x)$ by the function $\rho(\lambda)$ and with outlining necessary and sufficient conditions on $\rho(\lambda)$ for the inverse problem solution to exist have already been considered [85, 86].

2.7. One-dimensional inverse problems for second-order linear hyperbolic equations

Some ways of setting one-dimensional inverse problems for sufficiently common hyperbolic equations can be reduced to the problem discussed in the preceding sections. Consider the differential equation

$$u_{tt} = a_{00}(z)u_{zz} + 2\sum_{j=1}^{n} a_{0j}(z)u_{zy_j} + \sum_{k,j=1}^{n} a_{kj}(z)u_{y_k y_j} + b_0(z)u_z$$
$$+ \sum_{j=1}^{n} b_j(z)u_{y_j} + c(z)u \qquad (2.134)$$

with respect to the function $u(y,z,t)$, $y=(y_1,\dots,y_n)$, $n \geqslant 0$, in the domain

$$D = \{(y,z,t): y \in R^n, z>0, t>0\}.$$

Assume that the coefficients $a_{kj}=a_{jk}$, $k,j=0,1,\dots,n$, are such that equation (2.134) is that of a hyperbolic type and $a_{kj} \in C^2(0,\infty)$, $b_j \in C^1(0,\infty)$, $c \in C(0,\infty)$, with the following way of the inverse problem setting suggested. The task is to find the coefficients of the right-hand part of (2.134) provided that the solution to (2.134) of the boundary problem is known

$$u|_{t=0} = 0, \qquad u_t|_{t=0} = \delta(y,z) \qquad (2.135)$$

$$(u_z - hu)_{z=0} = 0 \qquad (2.136)$$

at $z=0$ is the given function

$$u|_{z=0} = f(y,t), y \in R^n, \quad t>0. \qquad (2.137)$$

Note the following peculiarity of the above setting: in order to define the finite number of coefficients depending on a single variable only, the function $n+1$ of the variable is set. Despite the fact that the above problem, as will be shown later, has no unique solution (only some combinations of the coefficients can be uniquely found), information contained in data (2.137) nonetheless proves to be 'too abundant'. The supposed paradox is accounted for by the fact that small portion of the information on the function $f(y,t)$ is sufficient for finding the combinations of the coefficients mentioned above, the rest portion appears excessive. No doubt, this fact points out to a certain defect in the way of setting the inverse problem. In a mathematical sense this defect reveals itself in difficulties of finding for the function $f(y,t)$ the necessary and sufficient conditions it must satisfy, which is of course connected with the theorem of the inverse problem existence.

Now investigate problem (2.134)–(2.137) using the approach of Blagoveshchenskij [87]. First of all demonstrate that the solution to the direct problem (2.134)–(2.136) at $z=0$ depends only on some combinations of the coefficients contained in (2.134). Now go over from the function $u(y,z,t)$ to its exponential

Fourier image with respect to the variables $y_1, \dots, y_n$

$$\tilde{u}(\lambda, z, t) = \int_{R^n} u(y, z, t) \exp [i(\lambda, y)] \, dy, \quad \lambda \in R^n.$$

Here $\lambda = (\lambda_1, \dots, \lambda_n)$ is the parameter of the Fourier transform; (λ, y) is the scalar product of λ and y; $dy = dy_1, dy_2, \dots, dy_n$. The Fourier transform of the function $u(y, z, t)$ exists at any finite t since the function $u(y, z, t)$ is a sum of a certain singular generalized finite-order function and a regular function [see Section 4.1 for the structure of solving problems of the type (2.134)–(2.136)], the support of the function $u(y, z, t)$ being finite.

At any fixed λ the function $\tilde{u}(\lambda, z, t)$ satisfies the differential equation

$$\tilde{u}_{tt} = a_{00}(z) u_{zz} + \left[b_0(z) - 2i \sum_{j=1}^{n} \lambda_j a_{0j}(z) \right] \tilde{u}_z$$
$$+ \left[c(z) - i \sum_{j=1}^{n} \lambda_j b_j(z) - \sum_{k,j=1}^{n} \lambda_j \lambda_k a_{kj}(z) \right] \tilde{u} \tag{2.138}$$

and the conditions

$$\tilde{u}(\lambda, z, 0) = 0, \qquad \tilde{u}_t(\lambda, z, 0) = \delta(z) \tag{2.139}$$
$$(\tilde{u}_z - h\tilde{u})_{z=0} = 0. \tag{2.140}$$

Now transform (2.138) by introducing a new independent variable

$$x = \int_0^z \frac{ds}{\sqrt{(a_{00}(s))}} \tag{2.141}$$

and the new function

$$v(\lambda, x, t) = \tilde{u}(\lambda, g(x), t) / S(\lambda, x). \tag{2.142}$$

Here $g(x)$ is the function interrelating x and z from equality (2.141); $S(\lambda, x)$ will be determined later. The function v satisfies the equation resulting from (2.138)

$$v_{tt} = v_{xx} + \left\{ \frac{2S_x(\lambda, x)}{S(\lambda, x)} - \frac{1}{\sqrt{(a_{00}(z))}} \left[\tfrac{1}{2} a'_{00}(z) - b_0(z) \right. \right.$$
$$\left. \left. + 2i \sum_{j=1}^{n} \lambda_j a_{0j}(z) \right]_{z=g(x)} \right\} v_x + \left\{ \frac{S_{xx}(\lambda, x)}{S(\lambda, x)} + \frac{S_x(\lambda, x)}{S(\lambda, x)} \frac{1}{\sqrt{(a_{00}(z))}} \right.$$
$$\times \left[\tfrac{1}{2} a'_{00}(z) - b_0(z) + 2i \sum_{j=1}^{n} \lambda_j a_{0j}(z) \right]_{z=g(x)}$$
$$\left. + \left[c(z) - i \sum_{j=1}^{n} \lambda_j b_j(z) - \sum_{k,j=1}^{n} \lambda_k \lambda_j a_{kj}(z) \right]_{z=g(x)} \right\} v.$$

Now choose the function $S(\lambda, x)$ from the conditions

$$\frac{S_x(\lambda, x)}{S(\lambda, x)} = \frac{1}{\sqrt{(a_{00}(z))}} \left[\tfrac{1}{4}(a'_{00}(z) - 2b_0(z)) + i \sum_{j=1}^{n} \lambda_j a_{0j}(z) \right]_{z=g(x)}, \qquad S(\lambda, 0) = 1.$$

Then one has

$$S(\lambda, x) = \sqrt[4]{\left[\frac{a_{00}(g(x))}{a_{00}(0)}\right]} \exp\left\{\int_0^{g(x)} \left[-\tfrac{1}{2}b_0(z) + i\sum_{j=1}^{n} \lambda_j a_{0j}(z)\right]\frac{dz}{a_{00}(z)}\right\}.$$

In this case the equation for the function v assumes the form

$$v_{tt} = v_{xx} + q(x, \lambda)v \tag{2.143}$$

where $q(x, \lambda)$ is the second-order polynomial with respect to λ

$$q(x, \lambda) = q_0(x) + i\sum_{j=1}^{n} \lambda_j q_j(x) + \sum_{k,j=1}^{n} \lambda_k \lambda_j q_{kj}(x)$$

$$q_0(x) = \left[c(z) + \frac{1}{4}\left(\frac{a'_{00}(z) - 2b_0(z)}{\sqrt{(a_{00}(z))}}\right)' \sqrt{(a_{00}(z))} - \frac{1}{16}\frac{(a'_{00}(z) - 2b_0(z))^2}{a_{00}(z)}\right]_{z=g(x)}$$

$$q_j(x) = \left[\left(\frac{a_{0j}(z)}{\sqrt{(a_{00}(z))}}\right)' \sqrt{(a_{00}(z))} - \frac{a_{0j}(z)}{2a_{00}(z)}(a_{00}(z) - 2b_0(z)) - b_j(z)\right]_{z=g(x)}$$

$$q_{kj} = q_{jk}(x) = \left[-a_{kj}(z) + \frac{a_{0k}(z)a_{0j}(z)}{a_{00}(z)}\right]_{z=g(x)}.$$

In terms of the function v conditions (2.139) and (2.140) take on the form

$$v|_{t=0} = 0, \qquad v_t|_{t=0} = [a_{00}(0)]^{-1/2}\delta(x) \tag{2.144}$$

$$(v_x - A(\lambda)v)_{x=0} = 0. \tag{2.145}$$

Here

$$A(\lambda) = h[a_{00}(0)]^{1/2} - S_x(\lambda, 0)$$

$$\equiv [a_{00}(0)]^{-1/2}\left[ha_{00}(0) - \tfrac{1}{4}(a'_{00}(0) - 2b_0(0)) - i\sum_{j=1}^{n} \lambda_j a_{0j}(0)\right]. \tag{2.146}$$

Equations (2.143)–(2.146) demonstrate that the function $v(\lambda, x, t)$ is uniquely determined by setting the functions $q_0(x)$, $q_j(x)$, and $q_{kj}(x)$ and a number of constants: $a_{00}(0)$, $a'_{00}(0) - 2b_0(0)$, and $a_{0j}(0)$. It results from (2.142) that the function $\tilde{u}(\lambda, 0, t)$ and hence the function $u(y, 0, t)$ are also determined by the functions and constants previously mentioned. Therefore, using information (2.137) one can find in the inverse problem only the functions q_0, q_j, and q_{kj} and constants $a_{00}(0)$, $a'_{00}(0) - 2b_0(0)$, and $a_{0j}(0)$. Now prove that they can be found uniquely.

For the function v (2.137) has the form

$$v(\lambda, 0, t) = \tilde{f}(\lambda, t)| \tag{2.147}$$

where $\tilde{f}(\lambda, t)$ is the Fourier image of the function $f(y, t)$.

At a fixed value of the parameter λ (2.143)–(2.147) is fairly identical to the problem discussed in Section 2.5 at $l=\infty$. It, in particular, results in the necessary conditions for the function $\tilde{f}(\lambda, t)$: (1) $\tilde{f}(\lambda, +0)=[a_{00}(0)]^{-1/2}>0$, $\tilde{f}_t(\lambda, +0)=-[a_{00}(0)]^{-1/2}A(\lambda)$; $\tilde{f}(\lambda, t)\in C^2(0, T)$ at any finite fixed λ and any $T>0$.

The first of these conditions allows one to find $a_{00}(0)$ and $A(\lambda)$, and hence the constants $a'_{00}(0)-2b_0(0)$, a_{0j}, $j=1,2,\dots,n$. Thus, the initial and boundary conditions, that the function v depends on, are fairly well defined. By solving the inverse problem (2.143)–(2.147) at any fixed λ by the scheme suggested in Section 2.5 we can obtain $q(x,\lambda)$ and, hence, $q_0(x)$, $q_j(x)$, and $q_{kj}(x)$. It can be realized, for instance, by calculating the derivatives of the function $q(x,\lambda)$ with respect to λ at $\lambda=0$

$$q_0(x)=q(x,0), \qquad q_j(x)=-i\frac{\partial}{\partial\lambda}q(x,\lambda)|_{\lambda=0}$$

$$q_{kj}(x)=\frac{1}{2}\frac{\partial^2}{\partial\lambda_k\partial\lambda_j}q(x,\lambda)|_{\lambda=0}.$$

These results allow one to formulate the following uniqueness theorem.

Theorem 2.10. Setting information (2.137) on the solution of problem (2.134)–(2.136) uniquely defines the functions

$$q_0(x), \qquad q_j(x), \quad j=1,2,\dots,n, q_{kj}(x)=q_{jk}(x), k, j=1,2,\dots,n$$

and the constants

$$a_{00}(0), \qquad a'_{00}(0)-2b_0(0), \qquad a_{0j}(0), \quad j=1,2,\dots,n$$

in which case the variable x is related to the initial coordinate z through equality (2.141).

Now consider some principal difficulties arising when attempting to investigate the problem of existing a solution to (2.134)–(2.137). The facts discussed above demonstrate that in order to find q_0, q_j, and q_{kj} it is sufficient to calculate at $\lambda=0$ the function $q(x,\lambda)$ and its derivatives with respect to λ up to the second order. These derivatives are uniquely defined by setting at $\lambda=0$ the function $\tilde{f}(\lambda,t)$ and its first- and second-order derivatives with respect to λ. Also note that having found q_0, q_j, and q_{kj}, we can find $q(x,\lambda)$ and hence the function $v(\lambda,0,t)$ for any λ. Therefore, the function $\tilde{f}(\lambda,t)$ is uniquely defined by its values and derivatives up to the second order with respect to λ at $\lambda=0$, which indicates that the conditions of the existence of the inverse problem solution impose fairly rigid restrictions on the function $\tilde{f}(\lambda,t)$, their analytical structure, however, being, obscure. It would, of course, be quite easy to formulate the theorem of existence in terms of the functions $\tilde{f}(0,t)$, $\tilde{f}_{\lambda_j}(0,t)$, $\tilde{f}_{\lambda_k\lambda_j}(0,t)$ but this theorem is, in essence, associated with another way of setting the information, different from (2.137). Setting the above derivatives of the function $\tilde{f}(\lambda,t)$ is equivalent to setting the moments of the function $f(y,t)$ up to the second order with respect to the variable y. Indeed

$$\tilde{f}(0,t)=\int_{R^n} f(y,t)\,dy, \qquad \tilde{f}_{\lambda_j}(0,t)=i\int_{R^n} y_j f(y,t)\,dy$$

$$\tilde{f}_{\lambda_k \lambda_j}(0,t) = -\int_{R^n} y_k y_j f(y,t)\,\mathrm{d}y.$$

For the inverse problem solution to exist, the function $f(y,t)$ must be defined by setting the above given moments.

No doubt, not all the coefficients in (2.134) can be found by the functions $q_0(x)$, $q_j(x)$, and $q_{kj}(x)$. The degree of ambiguity of the inverse problem solution here is such that we can arbitrarily set $(n+2)$ coefficients. In particular, if $c(z) = b_j(z) = 0$, $j = 0, 1, \ldots, n$, one can uniquely determine $a_{kj}(z)$, $k, j = 0, 1, \ldots, n$ by the functions q_0, q_j, and q_{kj}. Indeed, in this case for the function $\tilde{a}_{00}(x) = \ln a_{00}(g(x))$ one gets the equation

$$\frac{\mathrm{d}^2}{\mathrm{d}x^2}\tilde{a}_{00}(x) - \frac{1}{4}\left[\frac{\mathrm{d}}{\mathrm{d}x}\tilde{a}_{00}(x)\right]^2 = 4q_0(x)$$

with the known Cauchy data

$$\tilde{a}_{00}(0) = -2\ln \tilde{f}(0,+0), \qquad \tilde{a}'(0) = 4[(h + \tilde{f}_t(0,+0))/\tilde{f}(0,+0)].$$

Due to the uniqueness theory of the Cauchy problem the solution for common differential equations, $\tilde{a}_{00}(x)$ can be uniquely defined. Then, substituting $z = g(x)$ into (2.141) and differentiating the resulting identity with respect to x

$$1 = \frac{1}{\sqrt{[a_{00}(g(x))]}}\frac{\mathrm{d}g}{\mathrm{d}x}, \qquad g(x) = \int_0^x \exp\left[\tfrac{1}{2}\tilde{a}_{00}(s)\right]\mathrm{d}s.$$

Thus a correlation has been established between the variables x and z and hence the inverse correspondence $x = g^{-1}(z)$ can be considered known. In this case

$$a_{00}(z) = \exp \tilde{a}_{00}(g^{-1}(z))$$

$$a_{0j}(z) = a_{00}(z)\left[-i\tilde{f}_{t\lambda_j}(0,+0) + \int_0^z \frac{q_j(g^{-1}(s))}{a_{00}(s)}\,\mathrm{d}s\right], \quad j = 1, 2, \ldots, n$$

$$a_{kj}(z) = -q_{kj}(g^{-1}(z)) + a_{0k}(z)a_{0j}(z)/a_{00}(z), \quad k, j = 1, 2, \ldots, n.$$

A remark to setting the problem; we have proved that in order to determine the functions $q_0(x)$, $q_j(x)$, and $q_{kj}(x)$ it is sufficient to know the moments of the function $f(y,t)$ up to second order with respect to y. It would also be expected that the sought functions can be obtained provided the following information on the function $f(y,t)$ is available

$$f(0,t), \left.\frac{\partial f}{\partial y_j}\right|_{y=0}, \left.\frac{\partial^2 f}{\partial y_k \partial y_j}\right|_{y=0}, \quad k, j = 1, 2, \ldots, n.$$

A new way of setting the inverse problem arising here proves more interesting, since the data of the inverse problem in this case are in accordance with the number of the functions being determined as far as the dimension is concerned. This problem is much more difficult for investigation and only one particular case, when (2.134) has the form

$$u_{tt} = a^2(z)\Delta u$$

where Δ is the Laplace operator with respect to the variables y, z, has been as yet examined [88–93].

2.8. Problem of determining the operator L_q in a second-order hyperbolic equation

Consider the differential equation

$$L_q u - Mu = 0 \tag{2.148}$$

where M is a given linear uniformly elliptical second-order operator with respect to the variables $z_1, \ldots, z_n$ with the x-independent coefficients. Let (2.148) hold in the domain

$$D = \{(x, z, t): x > 0, z \in D_0, t > 0\}$$

where D_0 is a simply connected domain of the space R^n with the smooth boundary Γ_0. Define for (2.148) the initial and boundary conditions

$$u|_{t=0} = 0, \qquad u_t|_{t=0} = \delta(x, z) \tag{2.149}$$

$$(u_z - hu)_{x=0} = 0, \qquad mu|_{\Gamma_0} = 0. \tag{2.150}$$

Here m is a certain linear differential first-order operator (for instance, $mu \equiv u$ or $mu \equiv du/dn + h_1 u$; n is a co-normal to Γ_0).

For (2.148) examine the problem of defining L_q by the solution $u(x, z, t)$ to (2.148)–(2.150) at a fixed point of the space of the variables x, z

$$u(0, z^0, t) = f(t), \quad t > 0, z^0 \in D_0. \tag{2.151}$$

I shall demonstrate that under certain conditions this problem is reduced to the Sturm–Liouville inverse problem. For this purpose express the solution to (2.148)–(2.150) through its generalized Fourier transform by the system of functions $y(x, \lambda)$ of the operator $l_q y$ [see (2.122)–(2.125)]

$$u(x, z, t) = \int_{-\infty}^{\infty} \tilde{u}(\lambda, z, t) y(x, \lambda)\, d\rho(\lambda). \tag{2.152}$$

Then for the Fourier image $\tilde{u}(\lambda, z, t)$ one obtains

$$\tilde{u}_{tt} = M\tilde{u} - \lambda \tilde{u}, \quad z \in D_0, t > 0 \tag{2.153}$$

and the conditions

$$\tilde{u}|_{t=0} = 0, \qquad \tilde{u}_t|_{t=0} = \delta(z) \tag{2.154}$$

$$m\tilde{u}|_{\Gamma_0} = 0. \tag{2.155}$$

By solving (2.153)–(2.155) $\tilde{u}(\lambda, z, t)$ is obtained. Setting in (2.152) $x = 0$ and $z = z^0$ and using the condition $y(0, \lambda) = 1$, we obtain the equation interrelating the function $f(t)$ and the spectral function of the operator l_q

$$f(t) = \int_{-\infty}^{\infty} \tilde{u}(\lambda, z^0, t)\, d\rho(\lambda), \quad t > 0. \tag{2.156}$$

The kernel of (2.156) is known. Therefore, if (2.156) has a unique solution with respect to the function $\rho(\lambda)$ $[\rho(-\infty)=0]$, then, by finding it with $f(t)$, we come to the problem of obtaining $q(x)$ by the spectral function. Thus, of principal importance here is the problem of uniqueness of the solution to (2.156). In the case when the operator M has constant coefficients and the domain D_0 coincides with the whole space, (2.156) corresponds to known integral transformations. Condition (2.155) in this case is reduced to the condition of limitedness of the solution at $|z|\to\infty$. For instance, for $n=1$ and $Mu\equiv u_{zz}$, the solution to problem (2.153)–(2.155) is as follows:

$$\tilde{u}(\lambda, z, t)=\tfrac{1}{2}\theta(t-|z|)J_0(\sqrt{[\lambda(t^2-z^2)]}).$$

Equation (2.156) is in this case a Bessel transform and is uniquely solvable.

Chapter 3

Inverse kinematic problems in seismology

Physical interpretation of the inverse kinematic problem of seismology has been discussed in detail in Section 1.1. This chapter considers the modern state of the theory of the inverse kinematic problem, with the preceding consideration of the direct kinematic problem: of constructing rays and fronts, their differential properties, and possible peculiarities of behaviour. When studying kinematics of the wave process I will limit myself to the class of smooth media (continuous together with a certain number of derivatives). It is in these classes of media that the inverse kinematic problem has been studied in most detail.

3.1. Iconical equations, rays, and fronts

The important role of characteristic surfaces for the equations describing wave processes in mathematical physics is well known. For the wave equation

$$u_{tt} = v^2(x)\Delta u, \quad v(x) > 0 \tag{3.1}$$

describing the physical process of propagating acoustic oscillations, in the case when the equation of the characteristic surface is given in the form solved with respect to the variable $t{:}t = \tau(x)$, the function $\tau(x)$ satisfies the first-order differential equation

$$|\nabla_x \tau|^2 = n^2(x), \qquad n(x) = 1/v(x). \tag{3.2}$$

This equation is termed the iconical equation. Among all the possible solutions to (3.2), an important role belongs to those to which the characteristic surfaces $t = \tau(x, x^0)$, having a conic point at an arbitrary fixed point x^0, correspond to. Such surfaces are referred to as characteristic conoids. At $v(x) \equiv v_0$ a characteristic conoid is an ordinary cone $t = |x - x^0|/v_0$.

As far as the solutions to (3.2) are concerned, the point x^0 is a parameter. To find the function $\tau(x, x^0)$ (3.2) must be integrated under the condition

$$\tau(x, x^0) = O(|x - x^0|). \tag{3.3}$$

For (3.2) condition (3.3) replaces the Cauchy data corresponding to the degenerate case when the surface carrying the Cauchy data contracts to the point x^0.

A special role of the characteristic conoids manifests itself when considering the following physical experiment. Assume that a medium in the state of rest, at time $t = 0$ is disturbed by a source of oscillations that starts at a certain point of space x^0. In this case, outside the characteristic conoid $t = \tau(x, x^0)$, the solution to (3.1) is $u=0$, and inside the conoid it is generally not zero. Thus, from the physical point of view $\tau(x,x^0)$ is the time it takes the disturbance from the source of oscillations to get to the point x. If a fixed moment $t=t_0$ is considered and the domain in the space of the variables x, t, which is included inside a part of the conoid $0\leqslant\tau(x,x^0)\leqslant t\leqslant t_0$ is projected onto the space of x, we then get a set of points in the space of x that 'have felt' the action of the source by the moment t. The boundary of this set is called the front of the acoustic wave from the point source. In the case when the projections of the cross-sections of the surface $t=\tau(x,x^0)$ by the planes $t=t_1$, $0\leqslant t_1\leqslant t_0$ onto the space of x are successively inserted into one another, the front of the wave at the moment t_0 coincides with the projection of the cross-section of the characteristic conoid by the plane $t=t_0$, i.e. is determined by the equation $\tau(x,x^0)=t_0$. In a homogeneous medium $v(x)=v_0$ the wave front is a sphere of the radius v_0t centred at the point x^0.

As far as the system of equations of the theory of elasticity is concerned, the situation is analogous to (3.1). In this case for any point x^0 there exist (see [94, 95]) two characteristic conoids inserted one into the other. The equation for each of the conoids can be obtained by integrating the iconical equation (3.2) under the condition (3.3), provided we successively replace $v(x)$ with $v_p(x)$ or $v_s(x)$, which are the longitudinal and transverse velocities of the seismic waves. Since for a resilient substance $v_p>v_s$, the conoid corresponding to the longitudinal waves will contain inside the conoid corresponding to the transverse waves. Later the problem on constructing characteristic conoids for (3.2) will be considered. In this case it makes no difference which velocity is meant under $v(x)$, from the physical point of view it may be thought of as the velocity of signal transfer in the medium.

Throughout the chapter I set $x\in R^3$, the transition to space R^n presenting in this case no difficulty.

As is already known [83, 96] the method of constructing characteristic conoids $t=\tau(x,x^0)$ consists of constructing separate lines, called bicharacteristics, that lie on the conoid and jointly form it. Viewed formally, the procedure of obtaining the solution is as follows: consider the vector

$$p=\nabla_x\tau(x,x^0),\qquad p=(p_1,p_2,p_3) \tag{3.4}$$

and from the iconical equation $|p|^2=n^2(x)$ by differentiating with respect to x_k, we get the equation

$$pp_{x_k}=nn_{x_k},\quad k=1,2,3. \tag{3.5}$$

Then, using the equality $(p_i)_{x_k}=(p_k)_{x_i}$, resulting from (3.4), we can rewrite (3.5) as follows:

$$p\nabla_x p_k=nn_{x_k},\quad k=1,2,3. \tag{3.6}$$

Thus, each component of the vector p satisfies the first-order quasilinear equation. Along the curves satisfying the equality

$$\mathrm{d}x/\mathrm{d}t=p/n^2(x)$$

equation (3.6), after being divided into $n^2(x)$, can be written in the following way

$$\frac{\mathrm{d}p_k}{\mathrm{d}t}=\frac{\partial}{\partial x_k}\ln n(x),\quad k=1,2,3$$

and $\tau(x, x^0)$ satisfies the equation

$$\frac{d\tau}{dt} = \nabla_x \tau \frac{dx}{dt} = \frac{|p|^2}{n^2(x)} = 1.$$

By choosing the parameter t in such a way that $\tau = 0$ at $t = 0$, we get $\tau = t$, i.e. the parameter t is the time of the signal transfer from the point x^0 to the point x. The system of equalities

$$\frac{dx}{dt} = \frac{p}{n^2(x)}, \qquad \frac{dp}{dt} = \nabla_x \ln n(x) \tag{3.7}$$

is that of ordinary differential equations with respect to the pair of unknown vector-functions x, p.

Assume that

$$x|_{t=0} = x^0, \qquad p|_{t=0} = p^0 \equiv n(x^0)\nu^0 \tag{3.8}$$

where ν^0 is an arbitrary unit vector. By way of solving (3.7) and (3.8) find x, p as functions of t and parameters x^0, p^0

$$x = h(t, x^0, p^0), \qquad p = \psi(t, x^0, p^0). \tag{3.9}$$

The first of equalities (3.9) determines a two-parametric set of bicharacteristics, forming a characteristic conoid, in the space of the variables x, t at a fixed x^0, projection of the bicharacteristic onto the space of x is called a ray. The equality $x = h(t, x^0, p^0)$ can be considered as a parametric formulation of this ray. The first of the relations (3.7) demonstrates that a tangent to the ray coincides with the direction of the vector p and due to (3.4) the rays are orthogonal to the surfaces $\tau(x, x^0) = t$.

As is known from the variational calculation [96] the rays are the extremals of the functional

$$\tau(L) = \int_{L(x,x^0)} n(x)\, ds \tag{3.10}$$

where $L(x, x^0)$ is an arbitrary smooth curve connecting a pair of points x, x^0; ds is an element of its length. The product $n(x)\, ds$ determines the elementary time required by the signal to pass the length as long as ds with the velocity $1/n(x)$. As a whole, the integral over the curve $L(x, x^0)$ has, therefore, a physical sense of the time spent by the signal to pass along the curve $L(x, x^0)$ from the point x^0 to the point x. Those curves on which this integral reaches extremum (local, generally speaking) are termed the extremals of the functional $\tau(L)$. As is demonstrated in the variational calculation, the functional extremals satisfy system (3.7). Thus, the rays introduced as the projections of the bicharacteristics of the iconical equation onto the space of x are, in fact, the extremals of the functional $\tau(L)$.

Now consider the differential properties of the solutions to (3.7) and (3.8). Let D be an open domain of the space of x with a smooth boundary S, and let $\bar{D} = D \cup S$.

Lemma 3.1. Let $n(x) \in C^{s+1}(\bar{D})$, $s \geq 1$, $n(x) \geq n_0 > 0$, $x \in \bar{D}$. *Then for any point*

$x^0 \in D$ and any unit vector v^0 a solution to (3.7) and (3.8) exists and is unique for all t at which the point $h(t, x^0, p^0) \in \bar{D}$; functions $h(t, x^0, p^0)$, $\psi(t, x^0, p^0)$, $h_{tt}(t, x^0, p^0)$, $\psi_t(t, x^0, p^0)$ are continuous with respect to the series of the arguments together with the derivatives up to the sth order.

Indeed, under the lemma restrictions the conditions of the theorem of existence and of uniqueness of the solution to the Cauchy problem (3.7) and (3.8) hold for system (3.6). Hence, a solution to (3.7) and (3.8) exists on a certain intercept $0 \leqslant t \leqslant \delta$. Since the right-hand part in relations (3.7) is t-independent, this solution can be continued until the point x belongs to the domain $\bar{D}$. This solution is continuously dependent on the Cauchy data, i.e. on x^0, p^0, and moreover it has as many derivatives with respect to the arguments t, x^0, and p^0 as many times the right-hand part in equalities (3.7) is differentiated (see [97, p. 79]). The lemma is thus proved.

Lemma 3.2. Under the suppositions of Lemma 3.1 a solution to (3.7) and (3.8) can be presented in the following way:

$$x = f(\zeta, x^0), \qquad p = \frac{1}{t} n^2(f(\zeta, x^0))\zeta \frac{\partial}{\partial \zeta} f(\zeta, x^0), \qquad \zeta = p^0 t / n^2(x^0) \tag{3.11}$$

where the vector-function $f(\zeta, x^0)$, $f = (f_1, f_2, f_3)$ has s continuous derivatives over its arguments; $\partial f / \partial \zeta$ is the Jacobi's matrix

$$\frac{\partial f}{\partial \zeta} = \left\| \begin{matrix} \frac{\partial f_1}{\partial \zeta_1} & \frac{\partial f_2}{\partial \zeta_1} & \frac{\partial f_3}{\partial \zeta_1} \\ \frac{\partial f_1}{\partial \zeta_2} & \frac{\partial f_2}{\partial \zeta_2} & \frac{\partial f_3}{\partial \zeta_2} \\ \frac{\partial f_1}{\partial \zeta_3} & \frac{\partial f_2}{\partial \zeta_3} & \frac{\partial f_3}{\partial \zeta_3} \end{matrix} \right\|, \qquad \zeta = (\zeta_1, \zeta_2, \zeta_3).$$

Proof of this lemma can be found in [83, p. 267] and is based on the remark that follows. For a solution to (3.7) and (3.8) the relation

$$|p|^2 = n^2(x) \tag{3.12}$$

holds. Indeed, along the rays the following equality is valid

$$\frac{\mathrm{d}}{\mathrm{d}t}|p|^2 = 2p\frac{\mathrm{d}p}{\mathrm{d}t} = 2[\nabla \ln n(x)]n^2\frac{\mathrm{d}x}{\mathrm{d}t} = \frac{\mathrm{d}}{\mathrm{d}t}n^2(x)$$

which affords that along every ray

$$|p|^2 - n^2(x) = c$$

but due to (3.8) $c = 0$, that results in (3.12). Using this equality system (3.7) can be rewritten as follows:

$$\frac{\mathrm{d}x}{\mathrm{d}t} = \frac{p}{n^2(x)}, \qquad \frac{\mathrm{d}p}{\mathrm{d}t} = \frac{|p|^2}{n^2(x)} \nabla \ln n(x). \tag{3.13}$$

System (3.13) possesses a very interesting property: if we substitute $\bar{t}\alpha$ for t and $\bar{p}/\alpha$

for p, then the system will have with the variables x, p, and t in the same form as the initial one, while in the Cauchy data (3.8) for $\bar{p}$ it will have p^0/α instead of p^0. Hence

$$x = h\left(\bar{t}, x^0, \frac{p_0}{\alpha}\right) = h\left(t\alpha, x^0, \frac{p^0}{\alpha}\right) = h\left(n^2(x^0), x^0, \frac{p^0 t}{n^2(x)}\right).$$

Denoting $h(n^2(x^0), x^0, \zeta) = f(\zeta, x^0)$, one obtains the first of equalities (3.11). The second equality is obtained by using the equality resulting from (3.7)

$$p = n^2(x)\frac{dx}{dt}.$$

Differential properties of the function $f(\zeta, x^0)$ result from the corresponding properties of the function $h(t, x^0, p^0)$.

The point $\zeta = 0$ is conformed to by the value of the parameter $t = 0$. Write out the first terms of the function $f(\zeta, x^0)$ extension into a Taylor series in the vicinity of this point. Using the Cauchy data and those of equality (3.7) one finds

$$x = x^0 + \left.\frac{dx}{dt}\right|_{t=0} t + O(t^2) = x^0 + \frac{p^0}{n^2(x^0)}t + O(t^2).$$

Hence

$$f(\zeta, x^0) = x^0 + \zeta + O(|\zeta|^2). \tag{3.14}$$

At a fixed point x^0 and $|\zeta| = t/n(x^0)$ the equality $x = f(\zeta, x^0)$ is a parametric formulation of the surface $\tau(x, x^0) = t$.

Lemma 3.3. Under the suppositions of Lemma 3.1 in a small vicinity of the point $\zeta = 0$, the function $f(\zeta, x^0)$ has a unique inverse function

$$\zeta = g(x, x^0)$$

s times continuously differentiated with respect to the arguments x, x^0, in which case $g(x^0, x^0) = 0$.

Indeed, since from (3.14)

$$\left.\frac{\partial f}{\partial \zeta}\right|_{\zeta=0} = E \tag{3.15}$$

where E is a unit matrix, then $|\partial f/\partial\zeta|_{\zeta=0} = 1$ and the conditions of the theorem of uniqueness and existence of the inverse function are met. Smoothness of this inverse function coincides with that of the function $f(\zeta, x^0)$. Since $f(0, x^0) = x^0$, we get $g(x^0, x^0) = 0$.

Therefore, if x^0 is fixed and a set of points $x = f(\zeta, x^0)$ is considered for $|\zeta| \leqslant \delta$ [denoting it through $\Delta(x^0)$], then at a sufficiently small δ there exists a one-to-one correspondence between the points x and variable ζ. Because $\zeta = p^0 t/n^2(x^0)$ it means that each pair of the points x, x^0 corresponds to a single ray $\Gamma(x, x^0)$ which connects the points and is inside $\Delta(x^0)$. At the point x^0 the ray $\Gamma(x, x^0)$ has the direction $v^0 = p^0/n(x^0)$. The equations for determining v^0 and the time of the signal run $\tau(x, x^0)$ between the points x, x^0 are as follows:

$$t = \tau(x, x^0) = n(x^0)|g(x, x^0)| \tag{3.16}$$
$$v^0 = v^0(x, x^0) = g(x, x^0)/|g(x, x^0)|.$$

It should be remembered that for $x \in \Delta(x^0)$ there may exist the rays connecting the points x, x^0, but not entirely belonging to the vicinity of $\Delta(x^0)$. Generally even closely located points x, x^0 can be connected, but by not a single ray. It can be demonstrated (see [98, 99]) that ambiguity is absent provided the function $n(x)$ is close to a constant in the metric $C^2(\bar{D})$.

From (3.14)

$$g(x, x^0) = x - x^0 + O(|x - x^0|^2). \tag{3.17}$$

Lemma 3.4. Under the suppositions of Lemma 3.1 the functions $\tau(x, x^0)$, $v^0(x, x^0)$ at $x \neq x^0$ have continuous partial derivatives up to the sth order, and in the vicinity of x^0 the following estimations are valid for them

$$|D^\alpha \tau| \leqslant C/|x - x^0|^{|\alpha| - 1}, \qquad |D^\alpha v^0| \leqslant C/|x - x^0|^\alpha \tag{3.18}$$

Here $\alpha = (\alpha_1, \ldots, \alpha_6)$ is a multiindex

$$|\alpha| = \sum_{k=1}^{6} \alpha_k$$

and $D^\alpha = \partial^{|\alpha|}/(\partial x_1)^{\alpha_1} \ldots (\partial x_3^0)^{\alpha_6}$.

The validity of the lemma is obvious and results from smoothness of the function $g(x, x^0)$ and the fact that at the point x^0 it has zero of the first kind.

3.2. Boundary rays; waveguides: a sufficient condition for the absence of waveguides and boundary rays—ray regularity

So far the case when the point x^0 is an internal point of the domain D has been considered. As has been demonstrated, in this case one can always find such a vicinity of x^0 that any point x from this vicinity can be connected to the point x^0 with a single ray lying within this vicinity. The situation is essentially different in the case when x^0 is a boundary point of the domain D. Under such a restriction even any as much as desired close to x^0 point $x \in \bar{D}$ cannot always be connected by the curve which is a solution to (3.7) and (3.8) at any v^0. Here is an example to confirm the above statement. Let the function $n(x)$ be determined throughout the space R^3. Choose an arbitrary point x^0 and construct such its vicinity that inside it any point x is connected with x^0 by a single ray. Now choose a fixed point x from this vicinity and construct a domain D in such a way that its boundary S passes through x^0, crosses the ray $\Gamma(x, x^0)$ at least at one point $\bar{x}$ and contains the point x inside (Fig. 3.1).

In this case the only ray $\Gamma(x, x^0)$ connecting the points x, x^0 does not belong to the domain $\bar{D}$ and hence if we consider not the whole space but only the domain D, then a pair of points x and x^0 cannot be connected by the ray which is the solution to system (3.7). Nonetheless it appears that there exists such a curve where the functional $\tau(L)$, $L \subset D$, introduced in Section 3.1, reaches its minimum.

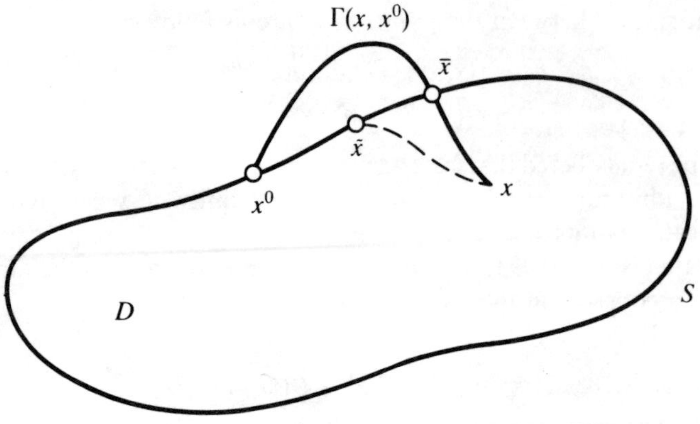

Figure 3.1.

It would be natural also to term this curve a ray. In particular, if the domain D is plane and the part of its boundary, lying between the points $x^0, \bar{x}$, is drawn in such a way that the rays connecting the point x^0 with the points on this boundary are outside D, then the ray interconnecting the point x^0 and point x consists of the part of the boundary S between x^0 and a certain point $\tilde{x}$ and the ray $\Gamma(x, \tilde{x})$, the ray $\Gamma(x, \tilde{x})$ touching boundary S at the point $\tilde{x}$. But if $x \in S$, then $x = \bar{x} = \tilde{x}$ and the ray $\Gamma(x, x^0)$ entirely belongs to S. Such rays are called boundary rays and to find them a system of Euler differential equations resulting from the functional $\tau(L)$ can be written. For this purpose it is sufficient to consider the curves $L \subset S$.

In physical processes associated with propagation of disturbances, for instance elastic and sound oscillations, one can encounter the phenomenon of a waveguide distribution of oscillations, which is related to a special behaviour of rays. Here is an example: consider a semi-space $x_3 \geqslant 0$ of the space R^3, and let the velocity of disturbance propagation in this semi-space depend only on coordinate x_3. In this case, as is seen from (3.7) coordinates p_1 and p_2 of the vector p remain unchanged along the ray. Since the vector p is colinear to the unit vector of the tangent, it demonstrates that the ray $\Gamma(x, x^0)$ connecting the points x, x^0 is plane. It lies on the plane that is parallel to axis x_3 and passes through the points x, x^0. Due to axial symmetry the ray picture will be the same on all the planes passing through x^0 and parallel to axis x_3, therefore, henceforce consider that the point x^0 lies on axis x_3, i.e. $x^0 = (0, 0, x_3^0)$, and the point x—on the plane x_1, x_3, and that $x_1 > 0$. In this case along the ray $\Gamma(x, x^0)$ we shall have $x_2 = 0$ and $p_2 = 0$. For the sake of simplicity denote: $x_1 = r$, $x_3 = z$, $p_1 = p$, $p_3 = q$. Then system (3.7) can be rewritten as follows:

$$\frac{\mathrm{d}r}{\mathrm{d}t} = \frac{p}{n^2(z)}, \qquad \frac{\mathrm{d}p}{\mathrm{d}t} = 0, \qquad \frac{\mathrm{d}z}{\mathrm{d}t} = \frac{q}{n^2(z)}, \qquad \frac{\mathrm{d}q}{\mathrm{d}t} = \frac{n'(z)}{n(z)}. \tag{3.19}$$

Let α denote an angle between the tangent v to the ray $\Gamma(x, x^0)$ at the point x and the z-axis. Then from (3.19) it is found that along the ray

$$p = n(z) \sin \alpha = \text{const.} \tag{3.20}$$

It is the first integral of system (3.19). Taking into account that along the ray the relation $p^2 + q^2 = n^2(z)$ holds, one has

$$q = \pm\sqrt{(n^2(z) - p^2)}. \tag{3.21}$$

Note that a plus sign is used in this equation if the angle α is acute, and a minus if the angle α is obtuse. An equation for the ray $\Gamma(x, x^0)$ in an explicit form can be written if coordinate r of the point on the ray is considered a function of z (generally speaking, ambiguous). From system (3.19)

$$\frac{dr}{dz} = \frac{p}{q} = \frac{p}{\pm\sqrt{(n^2(z) - p^2)}}.$$

From this formula for the ray outgoing from the point $x^0 = (0, 0, z^0)$ in the direction $v^0 = (\sin\alpha_0, 0, \cos\alpha_0)$ the following equation arises

$$r = \pm\int_{z_0}^{z} \frac{p\,dz}{\sqrt{(n^2(z) - p^2)}}, \qquad p = n(z_0)\sin\alpha_0. \tag{3.22}$$

The rule for choosing the sign is here the same as above. Note, however, that in any case $r \geqslant 0$. Formula (3.22) is generally only valid in the vicinity of the point x^0 since r is not necessarily a unique function of z. It is certainly valid, however, for those points x on the ray at which $p \neq n(z)$. Using (3.22) or even a simpler formula (3.20) one can easily analyse ray behaviour depending on the type of the function $n(z)$.

Now consider the following dependence of the velocity of signal transfer in the semi-space on z. Let $x(z)$ be a twice continuously differentiated function (Fig. 3.2). In the section $[0, z_1]$ the function $v(z)$ monotonically increases, then in the section $[z_1, \bar{z}]$ decreases and in the section $[\bar{z}, \infty]$ again monotonically increases, and at the point z_2 its value coincides with that of $v(z_1)$. Thus, the section $[z_1, z_2]$ is, as is described in geophysics, a zone of a decreased velocity. In this case the ray

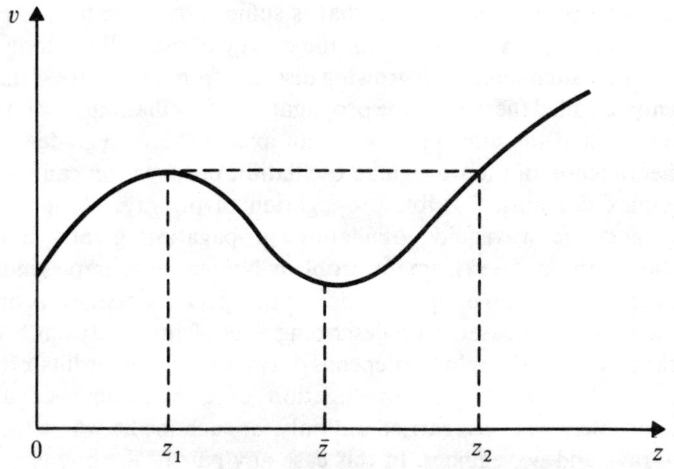

Figure 3.2.

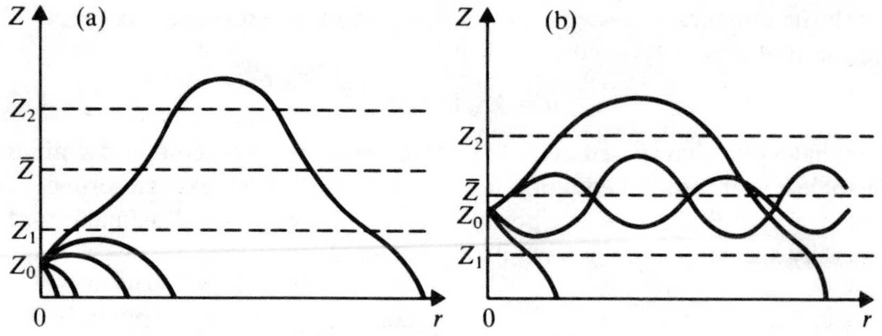

Figure 3.3.

behaviour is essentially dependent on the position of the point x^0 (Fig. 3.3a and b).

When $z_0 \in [0, z_1]$ each of the rays $\Gamma(x, x^0)$ goes beyond the boundary $z = 0$ at a certain point. The rays corresponding to the values $\alpha_0 \in (0, \pi/2)$, first reach the point with coordinate $z = \zeta$, where $n(\zeta) = n(z_0) \sin \alpha_0$, then make a turn at this point and appear on the plane $z = 0$. For any $\zeta \in [z_0, z_1) \cup (z_2, \infty)$ a ray can be found that has a turning point at $z = \zeta$, and it is the ray with the parameter $p = n(\zeta)$. On the other hand, $p = n(z_0) \sin \alpha_0$. Hence, $\alpha_0 = \arcsin n(\zeta)/n(z_0)$. It can be easily understood from equality (3.20) that there can be no rays outgoing from x^0 and turning in the layer $[z_1, z_2]$.

When $z_0 \in (z_1, z_2)$ alongside with the rays outgoing onto the boundary of the domain, there exist the rays that are entirely contained in the layer $z_1 < z < z_2$. These are all possible rays outgoing from the point x^0 at the angle

$$\alpha^0 \in [\arcsin(n(z_1)/n(z_0)), \pi - \arcsin(n(z_1)/n(z_0))].$$

Such rays have a sinusoidal behaviour, oscillating around axis $z = \bar{z}$. Oscillation energy is known to be transferred along the rays. In the layer (z_1, z_2) all the rays that have an angle of departure α_0, that is sufficiently close to $\pi/2$, propagate within the layer only. In line with this, the energy of oscillation along the layer (z_1, z_2) is weakly attenuated with growing distance from the source. This physical phenomenon is called the waveguide propagation of oscillations, the layer (z_1, z_2) is a waveguide and the intercept $z = \bar{z}$ is an axis of the waveguide.

The phenomenon of the waveguide oscillation propagation can also arise in more complex domains. Possible propagation of the rays along the domain boundary and the waveguide oscillation propagation greatly hamper the investigation of the inverse kinematic problem. Numerous examples indicate that in this case there can be no unique solution to the inverse kinematic problem. The problem arising in this case, i.e. on describing a set of ambiguity has been solved only for the case when the velocity depends only on a single coordinate (for details see Section 3.3). To date the investigation of a multidimensional inverse kinematic problem has been carried out only for such media where there are no boundary rays and waveguides. In this case any pair of the points x, x^0 of the domain D can be connected by a ray $\Gamma(x, x^0)$ (perhaps, not by a single one) and besides, any ray $\Gamma(x, x^0)$, when continued, crosses the boundary of the domain D. Now obtain a sufficient condition for such a ray behaviour.

Assume that the boundary S of the domain D is convex with respect to a set of rays, provided: (1) any two points x, x^0 can be connected with a ray belonging to D; (2) any ray, connecting a pair of different points belonging to S, does not touch S.

Consider a simply connected domain D having a boundary S (not necessarily limited) and let $F(x)$ be a twice continuously differentiable function, except for perhaps one internal point $\xi \in D$ possessing the following properties: (1) $F(x) < 0$, $x \in D$; $F(x) = 0$, $x \in S$; (2) $|\nabla F(x)| \neq 0$, $x \in \bar{D}$, $x \neq \xi$; and (3) for any point $\bar{x} \in D$, $\bar{x} \neq \xi$ the surface $F(x) = F(\bar{x})$ limits the simply connected domain.

Conditions (2) and (3) and the theorems on implicit functions show that a set of surfaces of the level $F(x) = C$ for $C < 0$ is regular, i.e. any surface $F(x) = C_1$ is contained inside the surface $F(x) = C_2$, if $C_2 > C_1$.

For a domain which is a unit sphere centred at the onset of the coordinates, one can choose the function $F(x) = |x|^2 - 1(\xi = 0)$ as the function $F(x)$, while in case of a semisphere $x_3 \geqslant 0$ the function $F(x) = -x_3$ can be used for this purpose.

A sufficient condition for the absence of boundary rays and waveguides is that of convexity of each of the surfaces $F(x) = C$, $C < 0$, with respect to the set of rays. For this purpose, in its turn, it is sufficient that all the rays touching surface $F(x) = C$ belong to domain $F(x) > C$, $C < 0$, with all their rest points. Assume that $\phi = F(x)$ and consider an arbitrary point $x^0 \in D$, $x^0 \neq \xi$, and the ray outgoing from this point at an angle v which is such that $v^0 F(x) = 0$. A derivative of ϕ with respect to the argument t along the ray can be calculated using (3.7). In particular

$$\phi_t = \nabla F x_t = \nabla F v \frac{1}{n(x)}$$

$$\phi_{tt} = \sum_{i,j=1}^{3} \frac{\partial^2 F}{\partial x_i \partial x_j} (x_i)_t (x_j)_t + \sum_{i=1}^{3} \frac{\partial F}{\partial x_i} (x_i)_{tt}$$

$$= \frac{1}{n^2(x)} \sum_{i,j=1}^{3} \frac{\partial^2 F}{\partial x_i \partial x_j} v_i v_j + \frac{1}{n^2(x)} \sum_{i=1}^{3} \frac{\partial F}{\partial x_i} \left[\frac{\partial}{\partial x_i} \ln n - \frac{2}{n} p_i (\nabla n x_t) \right].$$

Therefore, $\phi_t|_{x=x^0} = 0$. A sufficient condition for the ray $\Gamma(x, x^0)$ to lie in domain $F(x) > F(x^0)$ for $x \neq x^0$ is $\phi_{tt} > 0$. This condition is met if

$$\sum_{i,j=1}^{3} \frac{\partial^2 F}{\partial x_i \partial x_j} v_i v_j + \nabla F \nabla \ln n(x) > 0, \quad \forall v : v \nabla F = 0. \tag{3.23}$$

Indeed, under condition (3.23), at least in the vicinity of x^0, the value $\phi = F(x)$ along the ray increases, and hence, the ray lies in the domain $F(x) > F(x^0)$. If we assume that at its further extension the ray $\Gamma(x, x^0)$ can return to the surface $F(x) = F(x^0)$, then it means that the function $\phi(x)$ reaches a local maximum at a certain point $\bar{x}$ on the ray, and that at this point $\phi_t = 0$, and $\phi_{tt} \leqslant 0$. But it contradicts inequality (3.23). This contradiction demonstrates that any ray outgoing from the point x^0 in the direction of a tangential to the surface $F(x) = F(x^0)$ never returns to it again.

For a case of a limited domain it means that condition (3.23) is also sufficient for the ray $\Gamma(x, x^0)$ at its extension to go out onto the boundary of the domain D. For a case of an infinite domain a more rigid condition should be met

$$\sum_{i,j=1}^{3} \frac{\partial^2 F}{\partial x_i \partial x_j} v_i v_j + \nabla F \nabla \ln n(x) \geqslant \beta > 0, \quad \forall v : v\nabla F = 0. \tag{3.23'}$$

at a certain constant β.

Uniform fulfilment of condition (3.23) in this case is necessary to guarantee the absence of the rays asymptotically approaching any surface $F(x) = \text{const}$.

Condition (3.23) written for a unit sphere with use made of the function $F(x) = |x|^2 - 1$, results in the condition

$$x\nabla \ln n(x) + 1 > 0 \tag{3.24}$$

or, in the spherical system of coordinates r, θ, and ϕ

$$\frac{\partial}{\partial r}[rn(x)] > 0. \tag{3.24'}$$

Condition (3.23′) written for a semi-sphere $x_3 \geqslant 0$ is as follows:

$$-\frac{\partial}{\partial x_3} \ln n(x) \geqslant \beta > 0 \tag{3.25}$$

Note that conditions (3.23) and (3.23′) ensure the absence of boundary rays and waveguides, but even their fulfilment does not necessarily mean that there exists a one-to-one correspondence between a pair of points x, x^0 in the domain D and the rays which are the solutions to the system of Euler equations. When there is a one-to-one correspondence between the rays and the pairs of points from the domain D, consider a set of rays to be regular in D.

3.3. One-dimensional inverse kinematic problems

As has already been stated in Section 1.1, the inverse kinematic problem of seismology was considered by Herglotz and Wiehert under a supposition of the Earth's spherical symmetry as far back as 1905–7. In a mathematical sense it means that the velocity $v = v(r)$, $r = |x|$ (the centre of the coordinate system is considered to coincide with the centre of the Earth). In this case the iconical equation written in the spherical system of coordinates r, θ, and ϕ $(0 \leqslant \theta \leqslant \pi$, $0 \leqslant \phi \leqslant 2\pi)$, is as follows:

$$\tau_r^2 + \frac{1}{r^2}\left(\tau_\theta^2 + \frac{1}{\sin^2\theta}\tau_\phi^2\right) = n^2(r). \tag{3.26}$$

Indeed, if $x = r\alpha$, $\alpha = (\sin\theta\cos\phi, \sin\theta\sin\phi, \cos\theta)$, then

$$\tau_r = (\nabla\tau, \alpha) \qquad \tau_\theta = (r\nabla\tau, \alpha_\theta), \qquad \tau_\phi = (r\nabla\tau, \alpha_\phi).$$

The vectors α, α_θ, and $\alpha_\phi/\sin\theta$ form an orthogonal basis, hence

$$|\nabla\tau|^2 = (\nabla\tau, \alpha)^2 + (\nabla\tau, \alpha_\theta)^2 + (\nabla\tau, \alpha_\phi)^2/\sin^2\theta$$

and the iconical equation written in a general way gives (3.26). A system of the Euler equations for (3.26), written by the scheme suggested in Section 3.1, has the form

$$\frac{dr}{dt}=\frac{r^2 q}{m^2(r)},\qquad \frac{d\theta}{dt}=\frac{s}{m^2(r)},\qquad \frac{d\phi}{dt}=\frac{p}{m^2(r)\sin^2\theta}$$

$$\frac{dq}{dt}=\frac{m(r)m'(r)-rq^2}{m^2(r)},\qquad \frac{ds}{dt}=\frac{p^2\cos\theta}{m^2(r)\sin^3\theta},\qquad \frac{dp}{dt}=0. \tag{3.27}$$

Here the following notations are introduced

$$m(r)=rn(r),\quad q=\tau_r, s=\tau_\theta, p=\tau_\phi.$$

The parameter t is borrowed from condition $d\tau/dt=1$.

I shall now demonstrate that the rays determined by system (3.27) are plane curves. A necessary and sufficient condition for a curve to be plane is, for instance, the condition of t-independence of the unit vector $\beta=[x_t,x]/|[x_t,x]|$ ($[x_t,x]$ is a vector product of the vectors x_t and x). It must be made sure that $\beta_t=0$. From the equality $x=r\alpha$ and relation (3.27)

$$x_t=r_t\alpha+r(\alpha_\theta\theta_t+\alpha_\phi\phi_t)=\frac{r^2q}{m^2(r)}\alpha+\frac{s}{m^2(r)}\alpha_\theta+\frac{p}{m^2(r)\sin\theta}\alpha_\phi.$$

Hence

$$\beta=\frac{p}{\sqrt{(p^2+s^2\sin^2\theta)}}\alpha_\theta-\frac{s}{\sqrt{(p^2+s^2\sin^2\theta)}}\alpha_\phi.$$

Differentiating β with respect to t and using the easily verified equalities

$$\alpha_{\theta\theta}=-\alpha,\qquad \alpha_{\theta\phi}=\operatorname{cotan}\theta\alpha_\phi,\alpha_{\phi\phi}=-\sin^2\theta\alpha-\sin\theta\cos\theta\alpha_\theta$$

and relations (3.27)

$$\beta_t=\left[-\frac{p}{\sqrt{(p^2+s^2\sin^2\theta)}}\theta_t+\frac{s\sin^2\theta}{\sqrt{(p^2+s^2\sin^2\theta)}}\phi_t\right]\alpha+\left[\frac{s\sin\theta\cos\theta}{\sqrt{(p^2+s^2\sin^2\theta)}}\right.$$
$$\left.\times\phi_t+\frac{p_t(p^2+s^2\sin^2\theta)-p(pp_t+ss_t\sin^2\theta+s^2\sin\theta\cos\theta\cdot\theta_t)}{\sqrt{((p^2+s^2\sin^2\theta)^3)}}\right]\alpha_\theta$$
$$+\left[\frac{(p\phi_t-s\theta_t)\operatorname{cotan}\theta}{\sqrt{(p^2+s^2\sin^2\theta)}}\right.$$
$$\left.+\frac{(pp_t+ss_t\sin^2\theta+s^2\sin\theta\cos\theta\cdot\theta_t)s-(p^2+s^2\sin^2\theta)s_t}{\sqrt{((p^2+s^2\sin^2\theta)^3)}}\right]\alpha_\phi\equiv 0.$$

Thus, in case when $v=v(r)$ a seismic ray $\Gamma(x,x^0)$ is plane and lies in the plane passing through the points x, x^0 and the Earth's centre. The problem is therefore, plane. It is sufficient to consider an arbitrary cross-section of the Earth by the plane of a large circle and to locate the points x and x^0, for which we know the times $\tau(x,x^0)$, only on the boundary of this circle. Assume later that it is an equatorial cross-section and that the angle ϕ is read off the point x^0. In this case $\theta = \pi/2$, $s = 0$ and the system of Euler equations assumes a simpler form

$$\frac{dr}{dt} = \frac{r^2 q}{m^2(r)}, \qquad \frac{d\phi}{dt} = \frac{p}{m^2(r)}$$

$$\frac{dq}{dt} = \frac{m(r)m'(r) - rq^2}{m^2(r)}, \qquad \frac{dp}{dt} = 0. \tag{3.27'}$$

Herglotz and Wiehert have demonstrated that in the case of a one-dimensional model the point x^0 can be fixed, and the point x can be made run through a set of points of the boundary circumference. If the function $m(r)$ grows monotonically with growing r and is continuously differentiable, it is uniquely determined by setting the function $\tau = t(\Delta)$, called a hodograph of the wave. Here Δ is an angular distance between the points x, x^0; $t(\Delta)$ is the time of the signal run between these points. It appears that a local reconstruction of the function $v(r)$ also takes place, i.e. if the function $t(\Delta)$ is known for $\Delta \in [0, \Delta_0]$, $\Delta_0 > 0$, then the function $v(r)$ is located in a certain layer $r_0 \leqslant r \leqslant R$, where R is the Earth's radius, while r_0 is obtained through Δ_0.

To solve this problem, begin with the formulation of some more exact assumptions on the velocity $v(r)$ in terms of the function $m(r)$. Let $m(r) \in C^1[r_0, R]$, $0 < r_0 < R, m(r) > 0, m'(r) > 0, r \in [r_0, R]$. Note, the fulfilment of the last condition is, as shown in Section 3.25 a sufficient condition for the absence of boundary rays and waveguides in the layer $r_0 \leqslant r \leqslant R$. Besides, certain assumptions of the less constructive character of the function $m(r)$ have to be made. The point is that under the assumptions used for the function $m(r)$, the hodograph $t(\Delta)$ is a continuously differentiated single-valued function of the argument Δ for sufficiently small Δ. With growing Δ, however, it might occur that one and the same Δ is corresponded to by several rays connecting the points x, x^0 (or even by an uncountable set of rays which is the case of ray focusing) and hence, the

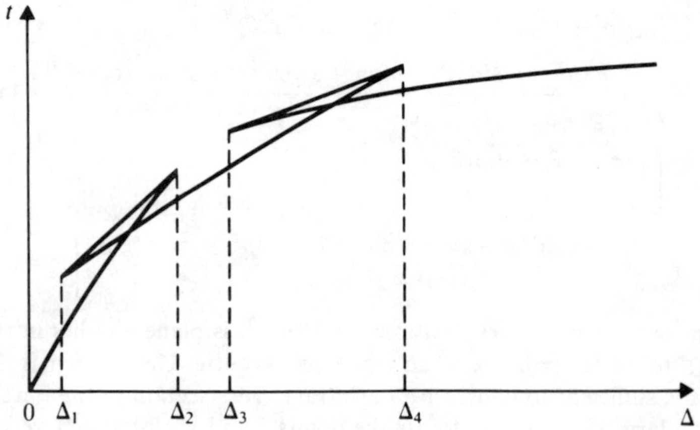

Figure 3.4.

uniqueness of the function $t(\Delta)$ is lost. Now make an additional supposition that the function $t(\Delta)$ consists of a finite number of continuously differentiable functions. A graph of the function $t(\Delta)$ in this case can have *a* finite number of return points $\Delta_1, \Delta_2, \ldots$ (Fig. 3.4). According to the graph, each value of Δ included between Δ_1 and Δ_2 or Δ_3 and Δ_4 is corresponded to by three different rays.

The equations determining the ray equation and the time it takes the signal to run along it will now be written out.

Equations (3.27′) demonstrate that along the ray $p = \text{const.}$ the relation

$$p^2 + r^2 q^2 = m^2(r) \tag{3.28}$$

resulting from (3.1) due to the notations assumed, is the first integral of system (3.27′). Hence

$$q = \pm \frac{1}{r} \sqrt{(m^2(r) - p^2)}$$

in which case the plus sign is to be taken at the points where $r_t > 0$, and the minus sign at the points where $r_t < 0$. The first two of relations (3.27′) result in the equation

$$\frac{\mathrm{d}\phi}{\mathrm{d}r} = \frac{p}{r^2 q} = \pm \frac{p}{r\sqrt{(m^2(r) - p^2)}}. \tag{3.29}$$

Let (ρ, ϕ_0) denote a point on the plane $\theta = \pi/2$ at which $r_\phi = 0$ and, hence,

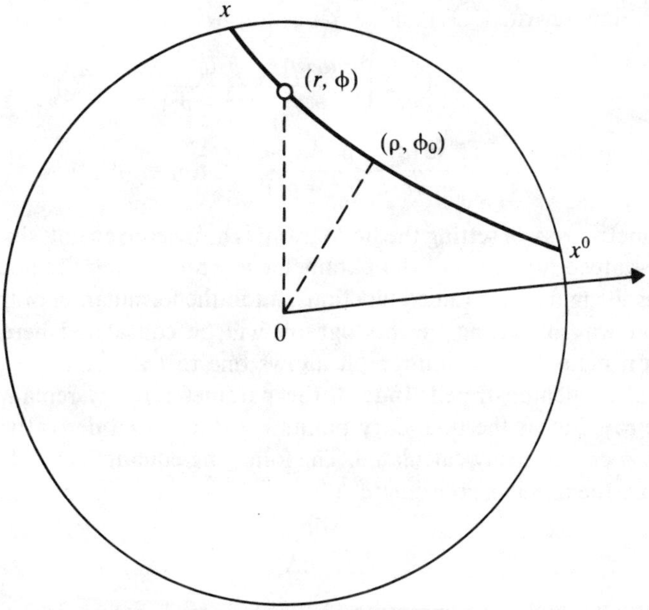

Figure 3.5.

$p = m(\rho)$. By integrating (3.29) we get the ray equation

$$\phi = \phi_0 \pm \int_\rho^r \frac{p\,\mathrm{d}r}{r\sqrt{(m^2(r) - p^2)}}, \quad p = m(\rho). \tag{3.30}$$

Here the minus sign corresponds to the branch from point x^0 up to the point (ρ, ϕ_0) (henceforth termed the top of the ray), and the plus sign to the branch of the ray from the top to the point x (Fig. 3.5). The ray is symmetrical with respect to the point (ρ, ϕ_0) which is the turning point of the ray. Formula (3.30) gives

$$\phi_0 = \frac{\Delta}{2}, \qquad \Delta = 2\int_\rho^R \frac{p\,\mathrm{d}r}{r\sqrt{(m^2(r) - p^2)}} \tag{3.31}$$

where Δ is an angular coordinate of the point x. Singular integrals in (3.30) and (3.31) converge due to the conditions imposed on the function. Indeed, in the vicinity of the point $r = \rho$

$$\sqrt{(m^2(r) - p^2)} = \sqrt{(r-\rho)}\sqrt{(2m(\rho)m'(\rho) + o(r-\rho)/(r-\rho))}.$$

The time it takes the signal to run from point x to x^0 is calculated, in line with the first of equalities (3.27′), by

$$t = 2\int_\rho^R \frac{m^2(r)\,\mathrm{d}r}{r\sqrt{(m^2(r) - m^2(\rho))}}. \tag{3.32}$$

Equations (3.31) and (3.32) are essential for solving a one-dimensional inverse kinematic problem. They afford a parametric setting of the hodograph $t = t(\Delta)$, the ray parameter p being convenient in this case. Here $\rho = \omega(p)$, where $\omega(p)$ is a function inverse to $m(\rho)$. Having introduced the additional notations $q = m(r)$, $q_0 = m(R)$, now rewrite (3.31) and (3.32) as follows:

$$\Delta(p) = 2\int_p^{q_0} \frac{\omega'(q)}{\omega(q)} \frac{p\,\mathrm{d}q}{\sqrt{(q^2 - p^2)}} \tag{3.33}$$

$$\tilde{t}(p) = 2\int_p^{q_0} \frac{\omega'(q)}{\omega(q)} \frac{q^2\,\mathrm{d}q}{\sqrt{(q^2 - p^2)}}, \qquad \tilde{t}(p) \equiv t(\Delta(p)). \tag{3.34}$$

A parametric way of setting the hodograph is more convenient, since $\Delta(p)$, $\tilde{t}(p)$ are single-valued functions of the parameter p, while setting the hodograph as $t = t(\Delta)$ results in a many-valued function. But in the formulation of the problem the second way of setting the hodograph will be considered here. That the supposition made on the hodograph allows one to transfer to its parametric setting will be demonstrated. Indeed, the parameter $p = \tau_\phi$ remains constant along the ray, but at the boundary points $r = R$ a partial derivative of τ with respect to ϕ can be easily calculated. The following equality is valid for a point $x \in S$ having the angular coordinate Δ

$$p = \frac{\mathrm{d}t(\Delta)}{\mathrm{d}\Delta}. \tag{3.35}$$

In accordance with the assumption on the absence of ray focusing at each of the separately taken branches of the hodograph formula (3.35) is valid.

Thus, by way of considering a point (Δ, t) on a multi-valued hodograph $t = t(\Delta)$ and calculating a derivative with respect to Δ at this point, a ray parameter p is obtained, and hence a pair of functions $\Delta(p)$, $\tilde{t}(p)$. Note that since $m(\rho) = p$, the calculation of the ray parameter p makes it possible to obtain the value of the wave velocity at $r = \rho$. However, the correlation between ρ and p is unknown, but can be found if the function $\omega(p)$ is determined. Therefore, the problem is reduced to obtaining $\omega(p)$ which can be found through either of two functions $\Delta(p)$, $\tilde{t}(p)$. Consider equality (3.23) which is an integral equation of the Abelian kind; having solved it, one obtains

$$\omega(p) = R \exp\left[\frac{1}{\pi}\int_p^{q_0} \frac{\Delta(z)\,\mathrm{d}z}{\sqrt{(z^2 - p^2)}}\right], \quad p \leqslant q_0. \tag{3.36}$$

This formula is valid until $\omega(p) \geqslant r_0$.

As is seen from the equations obtained, the solution of an inverse kinematic problem is reduced to two operations: calculation through the hodograph $t(\Delta)$ of the derivative by which $m(r)$ at the point $r = \rho$ can be found; calculation of integral (3.36) establishing a correlation between p and ρ. Thus, the solution to the problem proves continuous from C^1 to C. In practice, the function is set discretely at a number of points, and hence, the inverse kinematic problem is incorrect, the character of its incorrectness being similar to that in the problem of differentiating a tabularly set function [60].

Progress in investigating the inverse kinematic problem was associated with two points: refusal to use the condition of the function $m(r)$ monotonicity in the one-dimensional model and investigation of multidimensional models.

It has been demonstrated by geophysical studies that the supposition on the function $m(r)$ monotonicity is not generally fulfilled for the Earth. At the same time, without this assumption the inverse kinematic problem has no unique solution. The character of the solution ambiguity has been studied by Gerver and Markushevich [100, 101] (see also [102–105]). The ambiguity of the solution is

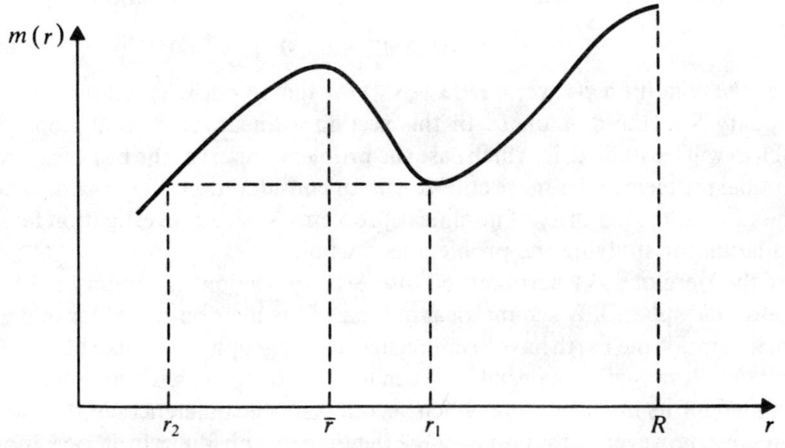

Figure 3.6.

accounted for in the following way. If it is assumed that the plots of the function $m(r)$ is of the type shown in Fig. 3.6, so that for $r \in (r_1, r_2)$ $m(\bar{r}) \geqslant m(r) > m(r_1) = m(r_2)$, $m(r_1) < m(R)$, then the rays corresponding to the values of the parameter $p \in (m(r_1), m(R))$ have their tops in the domain $r_1 < r < R$, in which case p changes from $m(R)$ to $m(r_1)$ correspond to ρ changes from R to r_1. The values $p < m(r_1)$ conform to the rays with the tops at which $\rho < r_2$. Thus, no rays exist that are outgoing from the point x^0 and turning in the domain $r_2 \leqslant r \leqslant r_1$. Therefore, having uniquely solved the problem on the section $[r_1, R]$, one can set 'almost arbitrarily' the function $m(r)$ on the section $[r_2, r_1]$ and then go on solving the problem on the following section of monotonicity at $r < r_2$. The domain $r_2 \leqslant r \leqslant r_1$ is a waveguide. If for any $r \in (r_2, r_1)$ a ray is let out from the point (r, ϕ) and this ray touches the circumference $r = \text{const.}$, then it will be 'winding' around the circumference $r = \bar{r}$, nowhere leaving the limits of the layer $r_2 \leqslant r \leqslant r_1$.

Of great interest from both a theoretical and practical standpoint are the necessary and sufficient conditions for the function $\tau(x, x^0)$, under which fulfilment a solution to the inverse kinematic problem exists and belongs to the given set. For a one-dimensional model such conditions have been obtained by Gerver and Markushevich [106, 107], and by Bukhgeim [108] for a multidimensional model in analytical media.

3.4. Linearized three-dimensional inverse problems

In the previous section the formulation of the inverse kinematic problem for the case when the velocity of the signal transfer in a medium depends only on a single coordinate was considered. More importantly from the point of view of practice is the case when the velocity of a signal transfer is essentially dependent on all the coordinates. One of the basic problems of modern geophysics is a quantitative estimation of horizontal nonhomogeneities in the velocities of seismic waves.

The inverse kinematic problem is nonlinear; its multidimensional formulation is as follows: find the function $n(x)$ in the domain D of the space R^3 if one knows the function $\tau(x, x^0)$ which is the solution to the iconical equation

$$|\nabla_x \tau(x, x^0)|^2 = n^2(x) \tag{3.37}$$

under the condition $\tau(x, x^0) = O(|x - x^0|)$ for the points x, x^0 belonging to the boundary S of the domain D. In this section a linearized formulation of the problem will be studied, in which case the problem preserves the basic features of the initial problem and is more convenient for consideration. At the same time, as is the case for the majority of nonlinear problems, such an investigation lays the foundation for studying the problem as a whole.

By the Herglotz–Wiehert method discussed in Section 3.3, a number of one-dimensional spherically symmetrical models of the distribution of seismic wave velocities inside the Earth have been constructed by geophysicists (see [109–118]). Different seismological material has been used for their basis, different ways have been used for its interpretation which accounts for the differences in the models. Of interest, however, is that on average the models only differ from one another by about 10%—they are fairly close.

To date geophysics has accumulated a number of facts demonstrating the existence of nonhomogeneities inside the Earth along the geographical coordinates or, in other words, of horizontal nonhomogenities. Such facts as systematic deviations in the hodographs of the waves, constructed for different mainlands, from the averaged hodograph, asymmetry of the gravitational and electromagnetic fields. In this case deviations from the hodographs corresponding to the spherically symmetrical distribution of the elastic wave velocities are rather small. For instance, at the times of the seismic wave run from the source to the receiver of about 10–20 min, deviations in the hodographs do not exceed 5–10 s. This fact allows one to conclude that deviations of the velocity structure of the Earth from the spherically symmetrical model are also small. Nonetheless, these small fluctuations in the distribution of the wave velocities are of great interest for geophysists, since they can help to explain the mechanism of the development of the Earth's crust, the problem of the mainland drift and so on. The problem of determining the three-dimensional velocity structure of the Earth is at present one of the most urgent in geophysics.

With the view of linearizing the problem assume that the function $n(x)$ can be written as

$$n(x) = n_0(x) + n_1(x)$$

where function $n_0(x)$ is given and $n_0(x) > 0$, and $n_1(x)$ is small as compared with $n_0(x)$ in the norm $C^2(D)$. In this case finding the function $n(x)$ is reduced to finding the function $n_1(x)$. Its smallness can be used to linearize the problem. For this purpose introduce the parameter of smallness λ

$$n(x) = n_0(x) + \lambda n_1(x) \tag{3.38}$$

and write the function $\tau(x, x^0)$ as a series with respect to the small parameter

$$\tau(x, x^0) = \sum_{n=0}^{\infty} \lambda^n \tau_n(x, x^0). \tag{3.39}$$

Substituting formally expressions (3.38) and (3.39) into the iconical equation and equating the terms at the same degrees of λ gives

$$|\nabla_x \tau_0(x, x^0)|^2 = n_0^2(x) \tag{3.40}$$

$$\nabla_x \tau_1(x, x^0) \nabla_x \tau_0(x, x^0) = n_0(x) n_1(x). \tag{3.41}$$

Thus an equation for any $\tau_k(x, x^0)$ can easily be written. It is clear, however, that all these equations, beginning with $k = 2$, depend on n_1 in a nonlinear way and have the order of smallness n_1^k. To the accuracy of the small quantities of the order of n_1^2 one has the approximate equality

$$\tau(x, x^0) = \tau_0(x, x^0) + \tau_1(x, x^0). \tag{3.42}$$

As is seen from (3.40), the function $\tau_0(x, x^0)$ depends only on $n_0(x)$ and shows the time of the signal run between the points x, x^0 in the medium characterized by the velocity of the signal transfer $v_0(x) = 1/n_0(x)$. Since the function $n_0(x)$ is given, then, having solved the direct kinematic problem (i.e. having constructed the solutions of the corresponding system of the Euler equations), one finds $\tau_0(x, x^0)$. In this case from equality (3.41) one gets $\tau_1(x, x^0)$ for x, x^0 belonging to S to the accuracy of the small quantities of the order of n_1^2.

Now consider (3.41) for the function $\tau_1(x, x^0)$, beginning with dividing both its parts by $n_0(x)$. The vector $\nabla_x \tau_0(x, x^0)/n_0(x)$ is a unit vector of the tangential at the point x to the ray $\Gamma_0(x, x^0)$ [the ray $\Gamma_0(x, x^0)$ is constructed for the function $n_0(x)$]. Therefore the expression in the left-hand part of the transformed equality (3.41) is a derivative along the ray $\Gamma_0(x, x^0)$. By way of integrating both parts of the resulting equality one gets

$$\tau_1(x, x^0) = \int_{\Gamma_0(x, x^0)} n_1(\xi)\, ds. \tag{3.43}$$

Here ds is an euclidean element of a length of the curve $\Gamma_0(x, x^0)$.

Formula (3.43) is a basic one for further studies. As has already been mentioned, the function $\tau_1(x, x^0)$ can be considered known for x, x^0 belonging to S. Hence, the following problem arises: the integrals of the function $n_1(x)$ along the given set of curves $\Gamma_0(x, x^0)$ are known for the points x, x^0 belonging to the boundary S of the domain D; the task is to find the function $n_1(x)$. This problem refers to the type of problems called the problems of integral geometry. In a general case the problems of integral geometry are those where the task is to find a function by its known integrals along the given set of curves or surfaces; it is also quite possible that the integrals are taken with a certain given weight multiplier (see [41, 42, 119]).

Equation (3.43) can be sufficiently easily investigated if the function $n_0(x)$, determining the set of curves $\Gamma_0(x, x^0)$, depends only on one coordinate or on the distance to the fixed point [41–43, 120]. Consider the case when the domain D is a sphere of a radius R and $n_0 = n_0(r)$, $r = |x|$. In this case the equations for the rays $\Gamma_0(x, x^0)$ from Section 3.3 can be used. Since the rays $\Gamma_0(x, x^0)$ are plane, (3.43) can be considered separately in each cross-section of the sphere passing through its centre. Let, as in Section 3.3, it be an equatorial cross-section. In this case the problem of finding the function $n_1(x)$ becomes plane and it can be assumed that $x = (x_1, x_2)$. The function $\tau(x, x^0)$ can be set only for the points x, x^0 lying on the boundary of the equatorial cross-section.

Theorem 3.1. Let function $m(r) = rn_0(r)$ satisfy the conditions

$$m(r) > 0, \quad m'(r) > 0, \quad m(r) \in C^2[r_0, R]. \tag{3.44}$$

in the domain $D = \{x : r_0 \leqslant r \leqslant R, r_0 > 0\}$. In this case the continuous function $n_1(x)$, $x = (x_1, x_2)$ is uniquely defined in the domain D by the function $\tau_1(x, x^0)$ known for those points x, x^0 of the boundary $r = R$ for which the rays $\Gamma_0(x, x^0)$ are contained in D.

To prove the theorem, use the polar system of coordinates r, ϕ for the equatorial cross-section. To each ray $\Gamma_0(x, x^0)$ contained in D one can put the coordinates (ρ, ϕ_0) of its top in a one-to-one correspondence. In these terms (3.43) can be written as follows:

$$\sum_{j=1}^{2} \int_{\rho}^{R} n_1(r, \phi_0 + (-1)^j \phi(r, \rho)) \frac{m(r)\, dr}{\sqrt{(m^2(r) - m^2(\rho))}} = \tau_1(\rho, \phi_0),$$

$$r_0 \leqslant \rho \leqslant R, \quad 0 \leqslant \phi_0 \leqslant 2\pi. \tag{3.45}$$

Here, according to (3.30) one has

$$\phi(r,\rho)=\int_{\rho}^{r}\frac{m(\rho)\,\mathrm{d}r}{r\sqrt{(m^2(r)-m^2(\rho))}}. \tag{3.46}$$

Denote for an arbitrary 2π-periodic with respect to variable ϕ the function $v(r,\phi)$ its Fourier coefficients with respect to the variable ϕ through $v_k(r)$ at $k=0$, $\pm 1, \pm 2, \ldots$

$$v_k(r)=\frac{1}{2\pi}\int_0^{2\pi} v(r,\phi)\mathrm{e}^{-ik\phi}\,\mathrm{d}\phi.$$

Multiplying both parts of equality (3.45) by $(1/2\pi)\mathrm{e}^{-ik\phi_0}$ and integrating them over ϕ_0 from 0 to 2π, one gets

$$\int_{\rho}^{R} n_{1k}(r)\cos k\phi(r,\rho)\frac{2m(r)\,\mathrm{d}r}{\sqrt{(m^2(r)-m^2(\rho))}}=\tau_{1k}(\rho),$$

$$k=0,\pm 1,\pm 2,\ldots; r_0\leqslant\rho\leqslant R. \tag{3.47}$$

At every fixed k (3.47) are the Volterra equations of the first kind with the peculiarity on the diagonal $r=\rho$. They can easily be reduced to the equation of the second kind by the operation of half differentiating. In this case it is convenient to transfer to new variables

$$p=m(\rho), \qquad q=m(r).$$

Denoting the inverse to $m(r)$ function through $\omega(q)$ and assuming that

$$m_k(q)=qn_{1k}(\omega(q))\omega'(q), \qquad q_0=m(R), q_1=m(r_0)$$

$$\tilde{\phi}(q,p)=\int_p^q\frac{\omega'(q)}{\omega(q)}\frac{p\,\mathrm{d}q}{\sqrt{(q^2-p^2)}}$$

write equality (3.47) in the following way

$$\int_p^{q_0} m_k(q)\cos k\tilde{\phi}(q,p)\frac{2\,\mathrm{d}q}{\sqrt{(q^2-p^2)}}=\tau_{1k}(\omega(p)),$$

$$k=0,\pm 1,\pm 2,\ldots; q_1\leqslant p\leqslant q_0. \tag{3.48}$$

Now multiply both parts of this equality by $p/\pi\sqrt{(p^2-s^2)}$ $(q_1\leqslant s\leqslant q_0)$ and integrate them over p from s to q_0. By way of changing the order of integration in the left-hand part of the resulting equality

$$\int_s^{q_0} m_k(q)R_k(q,s)\,\mathrm{d}q=\frac{1}{\pi}\int_s^{q_0}\frac{p\tau_{1k}(\omega(p))\,\mathrm{d}p}{\sqrt{(p^2-s^2)}} \tag{3.49}$$

$$R_k(q,s)=\frac{1}{\pi}\int_s^q\frac{2p\cos k\tilde{\phi}(q,p)\,\mathrm{d}p}{\sqrt{((p^2-s^2)(q^2-p^2))}}. \tag{3.50}$$

Now demonstrate that the kernels $R_k(q,s)$ possess the following properties: (1) $R_k(s,s)=1$; (2) the derivative $(\partial/\partial s)R_k(q,s)$ exists and is continuous in the domain $q_1\leqslant s\leqslant q\leqslant q_0$. For this purpose substitute a variable θ for the integration variable p in integral (3.50)

$$p = (q^2 \cos^2 \theta + s^2 \sin^2 \theta)^{1/2}.$$

In this case the formula for calculating $R_k(q, s)$ assumes the form

$$R_k(q, s) = \frac{2}{\pi} \int_0^{\pi/2} \cos k\tilde{\phi}(q, \sqrt{(q^2 \cos^2 \theta + s^2 \sin^2 \theta)})\, d\theta, \quad k = 0, \pm 1, \pm 2, \ldots. \tag{3.51}$$

At q tending to s the integrand function tends to unity, which results in property 1. The formula to calculate the derivative $(\partial/\partial s)R_k(q, s)$, as is seen from (3.51) is as follows:

$$\frac{\partial}{\partial s} R_k(q, s) = -\frac{2k}{\pi} \int_0^{\pi/2} \left[\sin k\tilde{\phi}(q, p) \frac{\partial \tilde{\phi}}{\partial p} \right]_{p = \sqrt{(q^2 \cos^2 \theta + s^2 \sin^2 \theta)}} \times \frac{s \sin^2 \theta \, d\theta}{\sqrt{(q^2 \cos^2 \theta + s^2 \sin^2 \theta)}} \tag{3.51'}$$

To calculate the derivative $\partial\tilde{\phi}/\partial p$ one can use the formula for $\tilde{\phi}(q, p)$, having carried out integration by parts in it

$$\tilde{\phi}(q, p) = \frac{\omega'(q)}{q\omega(q)} p\sqrt{(q^2 - p^2)} - \int_p^q p\sqrt{(q^2 - p^2)} \frac{d}{dq} \left[\frac{\omega'(q)}{q\omega(q)} \right] dq.$$

The formula shows that the derivative $\partial\tilde{\phi}/\partial p$ is continuous everywhere except for the diagonal $q = p$, in the vicinity of which it has a peculiarity of type $(q^2 - p^2)^{-1/2}$. At the same time, the function $\tilde{\phi}(q, p)$ itself, standing under the sign of sine in formula (3.51′), is a small quantity of the order of $(q^2 - p^2)^{1/2}$ at $p \to q$, therefore, generally the function $(\partial/\partial s)R_k(q, s)$ is continuous within the triangle $q_1 \leqslant s \leqslant q \leqslant q_0$ up to its diagonal. Thus, property 2 of the kernels $R_k(q, s)$ is stated.

Using the properties of the kernels $R_k(q, s)$ one can easily reduce (3.49) to those of the second kind by way of differentiating the equalities over s. As a result of this operation (3.49) assumes the form

$$-m_k(s) + \int_s^{q_0} m_k(q) \frac{\partial}{\partial s} R_k(q, s)\, dq = \frac{1}{\pi} \frac{\partial}{\partial s} \int_s^{q_0} \frac{\tau_{1k}(\omega(p))\, dp}{\sqrt{(p^2 - s^2)}},$$

$$k = 0, \pm 1, \pm 2, \ldots; \quad q_1 \leqslant s \leqslant q_0. \tag{3.52}$$

At any fixed k (3.52) is the Volterra equation of the second kind with continuous kernels. Such equations are known (see [83, 121, 122]) to have unique solutions that can be found, for instance, by the method of successive approximations. Therefore, each function $m_k(r)$ and, hence, each function $n_{1k}(r)$, can be uniquely found. Since Fourier coefficients are known to define a continuous function uniquely, $n_1(r, \phi)$ can be also found uniquely and thus the theorem is proved.

One more important thing concerning (3.52) should be remembered. These equations can be used to estimate the function $n_1(x)$ through the function $\tau_1(x, x^0)$. It can be achieved by way of preliminary estimation of each of the functions $m_k(s)$. As is known, the estimation of $m_k(s)$ contains an exponential multiplier, the exponent of which includes the norm of the integral equation kernel. But it follows from (3.51) that the norm of the kernels at $|k| \to \infty$ infinitely increases which results in the fact that stability estimation turns out to be logarithmic. The latter fact is associated with the incorrectness of the problem in question.

The discussed above algorithm of finding a solution can be also employed in numerical calculations [42, 123–125]. In this case one should bear in mind the problem incorrectness and approach its numerical realization taking into account the general methods developed within the theory of incorrect problems; the linearization method having been also used in [126–128].

Now consider a general case when $n_0(x)$ is a function of three variables. The method to be used was first employed by Mukhometov [129–133] to investigate the inverse kinematic problem and its linearized model—the problem of integral geometry. The present approach is based on the paper [134] in which the Mukhometov's method was used to study the problems of integral geometry in the case of an n-dimensional space. Analogous results have been obtained by Bernshtejn and Gerver [135, 136].

For the sake of convenience, investigate the problem of integral geometry which in essence coincides with problem (3.43) but differs from it as far as notation is concerned.

Consider a set of rays $\Gamma(x, \xi)$ resulting from a certain function $n(x) \geqslant \lambda_0 > 0$ of class $C^2(\bar{D})$ in an open limited simply connected domain D of a three-dimensional space. Assume that the boundaries S of the domain D are given as $F(x) = 0$, where the function $F(x) \in C^3(\bar{D})$ and has the property $F(x) < 0$, $x \in D$; $|\nabla F|_S = 1$. Also set that the surface S is concave with respect to the rays $\Gamma(x, \xi)$ and the set of rays inside D is regular. As has been shown in Section 3.1, the equation of the geodesic line passing through any point $\xi \in \bar{D}$ in the direction of the unit vector v^0, can be written as $x = f(\zeta, \xi)$, $\zeta = v^0 \tau / n(x^0)$, where $f(\zeta, \xi)$ is a continuously differentiable function of the arguments ζ, ξ and $f(0, \xi) = \xi$, $(\partial f / \partial \zeta)|_{\zeta = 0} = E$, where E is a unit matrix. The condition of regularity means that there exists a unique inverse function $\zeta = g(x, \xi)$ continuously differentiable for $x \in \bar{D}$, $\xi \in \bar{D}$. This results in the fact that the determinant $|\partial f / \partial \zeta|$ does not turn to zero and is therefore positive.

For the function $u(x) \in C^1(D)$ consider the function $w = w(x, \xi)$ that is defined by

$$w(x, \xi) = \int_{\Gamma(x, \xi)} u(y) |\mathrm{d}y|, \qquad |\mathrm{d}y| = \left(\sum_{i=1}^{3} \mathrm{d}y_i^2 \right)^{1/2}. \tag{3.53}$$

The essence of the problem to be investigated is to find the function $u(x)$ in the domain D by the known $x \in S$, $\xi \in S$ function $w(x, \xi)$.

Theorem 3.2. For the problem stated under the supposition that $n(x) \in C^2(\bar{D})$, $n(x) \geqslant \lambda_0 > 0$, $u(x) \in C^1(\bar{D})$ *the following estimate of stability is valid*

$$\int_D u^2(x)\,\mathrm{d}x \leqslant \frac{1}{8\pi\lambda_0} \int_S \mathrm{d}S_\xi \int_S \Phi(w, \tau)\,\mathrm{d}S_x \tag{3.54}$$

where

$$\Phi(w,\tau)=\begin{vmatrix} 0 & 0 & w_{x_1} & w_{x_2} & w_{x_3} \\ 0 & 0 & F_{x_1}(x) & F_{x_2}(x) & F_{x_2}(x) \\ w_{\xi_1} & F_{\xi_1}(\xi) & \tau_{x_1\xi_1} & \tau_{x_2\xi_1} & \tau_{x_3\xi_1} \\ w_{\xi_2} & F_{\xi_2}(\xi) & \tau_{x_1\xi_2} & \tau_{x_2\xi_2} & \tau_{x_3\xi_2} \\ w_{\xi_3} & F_{\xi_3}(\xi) & \tau_{x_1\xi_3} & \tau_{x_2\xi_3} & \tau_{x_3\xi_3} \end{vmatrix}. \tag{3.55}$$

It should be recalled that the function $\Phi(w, \tau)$ only contains the derivatives of the function w with respect to x and ξ in the directions tangential to the surface S. It becomes evident when the determinant (3.55) is expanded with respect to the minors included in the first two rows and columns.

Proof of the theorem is based on consideration of an auxiliary differential equation for the function $w(x, \xi)$. For any fixed point $\xi \in \bar{D}$ a derivative of the function $w(x, \xi)$, calculated at the point x in the direction tangential to $\Gamma(x, \xi)$, obviously coincides with $u(x)$. Thus, if a unit vector of the tangential to $\Gamma(x, \xi)$ at the point x is denoted through $v(x, \xi)$, $v = (v_1, v_2, v_3)$, then

$$(\nabla_x w, v(x,\xi)) = u(x). \tag{3.56}$$

This equality can be written in a modified form using formula $\nabla_x \tau(x, \xi) = n(x)v(x, \xi)$

$$(\nabla_x w, \nabla_x \tau) = n(x)u(x). \tag{3.56'}$$

Equation (3.56′) is used to obtain a certain energetic estimate of its solutions. Consider the matrix

$$T=\begin{Vmatrix} 0 & \tau_{x_1} & \tau_{x_2} & \tau_{x_3} \\ F_{\xi_1}(\xi) & \tau_{x_1\xi_1} & \tau_{x_2\xi_1} & \tau_{x_3\xi_1} \\ F_{\xi_2}(\xi) & \tau_{x_1\xi_2} & \tau_{x_2\xi_2} & \tau_{x_3\xi_2} \\ F_{\xi_3}(\xi) & \tau_{x_1\xi_3} & \tau_{x_2\xi_3} & \tau_{x_3\xi_3} \end{Vmatrix}.$$

Let T_{kj} denote the co-factors of the element t_{kj}, $k, j = 0, 1, 2, 3$, of the matrix T.

Lemma 3.5. Let $n(x) \in C^3(D)$, $w(x, \xi)$ is a continuous for $(x, \xi) \in D \times D$ function having continuous second-order derivatives at $x \neq \xi$, in which case

$$|w_{x_i}| \leqslant C, \qquad |w_{\xi_i}| \leqslant C, \qquad \left|\frac{\partial^2}{\partial x_i \partial \xi_j} w\right| \leqslant C/|x-\xi|$$

$$x \neq \xi, \qquad (x,\xi) \in D \times D.$$

Then the following integral equality is valid

$$\begin{aligned} &2\int_D \frac{dx}{n^2(x)} \int_S \sum_{k,j=1}^{3} T_{kj} w_{xj} \frac{\partial}{\partial \xi_k} (\nabla_x w, \nabla_x \tau)\, dS_\xi \\ &= \int_D n^2(x) \left\{ \int_{|v|=1} [|\nabla_x w(x,\xi)|^2 + (\nabla_x w, v)^2]_{\xi=\xi(x,v)}\, d\omega_v \right\} dx \\ &- \int_S dS_\xi \int_S \Phi(w,\tau)\, dS_x \end{aligned} \tag{3.57}$$

where $d\omega_\nu$ is an element of the area of the unit sphere $|\nu| = 1$, and $\xi(x, \nu)$ is a point at which the surface S is intersected with the ray drawn from the point x in the direction $-\nu$.

Let q_{skj0} denote a minor corresponding to the matrix obtained from matrix T provided the sth and kth lines and the zero and jth columns are withdrawn. In this case the following identity is valid for $x \neq \xi$

$$\frac{1}{n^2(x)} \sum_{k,j=1}^{3} T_{kj} w_{x_j} \frac{\partial}{\partial \xi_k} (\nabla_x w, \nabla_x \tau) \equiv \frac{1}{n(x)} \sum_{s=1}^{3} (-1)^s F_{\xi_s} \left\{ \sum_{k=1}^{s-1} \sum_{j=1}^{3} (-1)^{k+j} w_{x_j} \right.$$

$$\left. \times\, q_{skj0} \frac{\partial}{\partial \xi_k} (\nabla_x w, \nu) - \sum_{k=s+1}^{3} \sum_{j=1}^{3} (-1)^{k+j} w_{x_j} q_{skj0} \frac{\partial}{\partial \xi_k} (\nabla_x w, \nu) \right\}$$

$$\equiv \frac{1}{n(x)} \sum_{s=1}^{3} (-1)^s F_{\xi_s} \left\{ \sum_{k=1}^{s-1} \frac{\partial}{\partial \xi_k} \left[(\nabla_x w, \nu) \sum_{j=1}^{3} (-1)^{k+j} w_{x_j} q_{skj0} \right] \right.$$

$$\left. - \sum_{k=s+1}^{3} \frac{\partial}{\partial \xi_k} \left[(\nabla_x w, \nu) \sum_{j=1}^{3} (-1)^{k+j} w_{x_j} q_{skj0} \right] \right\} - \frac{1}{n(x)} (\nabla_x w, \nu)$$

$$\times \sum_{k,j=1}^{3} T_{kj} w_{x_j \xi_k} - \frac{1}{n(x)} (\nabla_x w, \nu) \sum_{j=1}^{3} w_{x_j} \left\{ \sum_{s=1}^{3} (-1)^s F_{\xi_s} \right.$$

$$\left. \times \left[\sum_{k=1}^{s-1} (-1)^{k+j} \frac{\partial}{\partial \xi_k} q_{skj0} - \sum_{k=s+1}^{3} (-1)^{k+j} \frac{\partial}{\partial \xi_k} q_{skj0} \right] \right\}$$

$$\equiv (\nabla_\xi F(\xi), \mathrm{rot}_\xi q) - \frac{1}{n^2(x)} (\nabla_x w, \nabla_x \tau) \sum_{k,j=1}^{3} T_{kj} w_{x_j \xi_k} + 2(\nabla_x w, \nu)^2 |T| \frac{1}{n^2(x)}.$$

Here $|T|$ is a determinant of the matrix T; q is the vector with the components q_1, q_2, q_3

$$q = -\frac{1}{n(x)} (\nabla_x w, \nu) \left(\sum_{j=1}^{3} (-1)^j w_{x_j} q_{23j0}, \sum_{j=1}^{3} (-1)^j w_{x_j} q_{13j0}, \sum_{j=1}^{3} (-1)^j w_{x_j} q_{12j0} \right).$$

When deducing the identity use has been made of the obvious equalities $q_{skj0} = q_{ksj0}$ and the easily verified relations

$$\sum_{s=1}^{3} (-1)^s F_{\xi_s} \left[\sum_{k=1}^{s-1} (-1)^{k+j} \frac{\partial}{\partial \xi_k} q_{skj0} - \sum_{k=s+1}^{3} (-1)^{k+j} \frac{\partial}{\partial \xi_k} q_{skj0} \right]$$

$$= -2T_{0j} = -2 \frac{\nu_j}{n(x)} |T|.$$

On the other hand, at $x \neq \xi$ the following identity holds

$$\frac{1}{n^2(x)} \sum_{k,j=1}^{3} T_{kj} w_{x_j} \frac{\partial}{\partial \xi_k} (\nabla_x w, \nabla_x \tau) \equiv \frac{1}{n^2(x)} \sum_{i,k,j=1}^{3} T_{kj} w_{x_j} \tau_{x_i} w_{x_i \xi_k}$$

$$+ \frac{1}{n^2(x)} \sum_{i,j=1}^{3} w_{x_j} w_{x_i} \left(\sum_{k=1}^{3} T_{kj} \tau_{x_i \xi_k} \right) \equiv \frac{1}{n^2(x)} \sum_{i,k,j=1}^{3} T_{kj} w_{x_j} \tau_{x_i} w_{x_i \xi_k}$$

$$+ \frac{1}{n^2(x)} |T| [|\nabla_x w|^2 - (\nabla_x w, \nu)^2].$$

By way of summing up the two resulting identities the new identity is obtained

$$\frac{2}{n(x)}\sum_{k,j=1}^{3} T_{kj}w_{x_j}\frac{\partial}{\partial\xi_k}(\nabla_x w, v) \equiv (\nabla_\xi F, \mathrm{rot}_\xi q)$$

$$+\frac{1}{n^2(x)}|T|[|\nabla_x w|^2 + (\nabla_x w, v)^2] + \frac{1}{n^2(x)}\sum_{i,k,j=1}^{3} w_{x_j}w_{x_i\xi k}$$

$$\times (T_{kj}\tau_{x_i} - T_{ki}\tau_{x_j}). \tag{3.58}$$

Now demonstrate that the last part in the above identity can be written in a divergent form with respect to the variable x. Taking into account the fact that the expression $(T_{kj}\tau_{x_i} - T_{ki}\tau_{x_j})$ changes its sign when changing the places of indices i, j, this part can be written as

$$\frac{1}{n^2(x)}\sum_{i,j=1}^{3}\left[\frac{\partial}{\partial x_i}(w_{x_j}w_{\xi k})\right](T_{kj}\tau_{x_i} - T_{ki}\tau_{x_j})$$

$$= \sum_{i=1}^{3}\frac{\partial}{\partial x_i}\left[\sum_{k,j=1}^{3} w_{x_i}w_{\xi k}\frac{1}{n^2(x)}(T_{kj}\tau_{x_i} - T_{ki}\tau_{x_j})\right]$$

$$-\sum_{k,j=1}^{3} w_{x_j}w_{\xi k}\left\{\sum_{i=1}^{3}\frac{\partial}{\partial x_i}\left[\frac{1}{n^2(x)}(T_{kj}\tau_{x_i} - T_{ki}\tau_{x_j})\right]\right\}. \tag{3.59}$$

Let q_{0kij} denote a minor of the matrix obtained from the matrix T by way of withdrawing the zero and kth rows and the ith and jth columns. Then

$$\frac{1}{n^2(x)}(T_{kj}\tau_{x_i} - T_{ki}\tau_{x_j}) = (-1)^{k+i+j}q_{0kij}\begin{cases} 1, & j > i, \\ -1, & j < i, \end{cases} \quad i, j = 1, 2, 3. \tag{3.60}$$

One can easily verify these equalities by carrying out a number of successive transformations which will be described below. In the expression for T_{kj} multiply the ith column by τ_{x_i} and in the expression for T_{ki}—the jth column by τ_{x_j}. Then replace the columns with numbers i and j (note that the column numbers correspond to the matrix T) with one another and add the two resulting determinants according to the rules of their addition. In this case the expression $T_{kj}\tau_{x_i} - T_{ki}\tau_{x_j}$ will coincide to the accuracy of the sign with the determinant of the matrix resulting from the matrix T by way of replacing the ith column with the column containing the elements $\tau_{x_i}^2 + \tau_{x_j}^2, \tau_{x_i}\tau_{x_i\xi_1} + \tau_{x_j}\tau_{x_j\xi_1}, \ldots, \tau_{x_i}\tau_{x_i\xi_3} + \tau_{x_j}\tau_{x_j\xi_3}$ and by withdrawing the kth row and the jth column. Now transform the obtained determinant by adding to the ith column all the sth columns, but the zero one, previously multiplied by τ_{x_s}. As a result, the elements of the ith column will have the form

$$\sum_{i=1}^{3}\tau_{x_i}^2, \sum_{i=1}^{3}\tau_{x_i}\tau_{x_i\xi_1}, \ldots, \sum_{i=1}^{3}\tau_{x_i}\tau_{x_i\xi_3}.$$

But

$$\sum_{i=1}^{3}\tau_{x_i}^2 = n^2(x)$$

and hence

$$\sum_{i=1}^{3} \tau_{x_i}\tau_{x_i\xi k}=0, \quad k=1,2,3.$$

Thus, after transformations the ith column assumes the form of the column, the first element of which equals $n^2(x)$, the rest columns are noughts. Expanding this determinant in the elements of the ith column, one comes to equalities (3.60).

Using equalities (3.60) one can easily verify that

$$\sum_{i=1}^{3} \frac{\partial}{\partial x_i}\left[\frac{1}{n^2(x)}(T_{kj}\tau_{x_i} - T_{ki}\tau_{x_j})\right]=0, \quad k,j=1,2,3$$

and that the last part in expression (3.59) turns to zero. In this case identity (3.58) has the form

$$\frac{2}{n^2(x)}\sum_{k,j=1}^{3} T_{kj}w_{x_j}\frac{\partial}{\partial \xi_k}(\nabla_x w, \nabla_x \tau) \equiv \nabla_\xi F \operatorname{rot}_\xi q$$
$$+\frac{1}{n^2(x)}|T|[|\nabla_x w|^2 + (\nabla_x w, v)^2] - \sum_{i=1}^{3}\frac{\partial}{\partial x_i}\Phi_i. \tag{3.61}$$

Here Φ_i denotes the expressions

$$\Phi_i = \sum_{k=1}^{3}(-1)^{k+i}w_{\xi k}\left[\sum_{j=1}^{i-1}(-1)^j w_{x_j}q_{0kij} - \sum_{j=i+1}^{3}(-1)^j w_{x_j}q_{0kij}\right], \quad i=1,2,3.$$

it can be readily checked that Φ_i, $i=1,2,3$, coincide with the co-factors of the elements of the second line in the matrix standing under the sign of the determinant in (3.55).

Use equality (3.61) to obtain (3.58). For this purpose multiply both parts of this equality by $\delta(F(\xi))\,d\xi\,dx$ and integrate the product over the domain resulting from the direct product of the domains $\bar{D}\times\bar{D}$ by throwing away the points x, ξ, satisfying the inequality $|x-\xi|\leqslant\varepsilon$, where ε is a small positive number. In this case we get

$$2\int_D \frac{dx}{n(x)}\int_{D_\varepsilon(x)}\sum_{k,j=1}^{3} T_{kj}w_{x_j}\frac{\partial}{\partial\xi_k}(\nabla_x w, v)\delta(F(\xi))\,d\xi$$
$$=\int_D dx\int_{D_\varepsilon(x)}(\nabla_\xi F, \operatorname{rot}_\xi q)\delta(F(\xi))\,d\xi + \int_D dx\int_{D_\varepsilon(x)}[|\nabla_x w|^2 + (\nabla_x w, v)^2]$$
$$\times\frac{1}{n^2(x)}|T|\delta(F(\xi))\,d\xi - \int_D \delta(F(\xi))\,d\xi\int_{D_\varepsilon(\xi)}\sum_{i=1}^{3}\frac{\partial}{\partial x_i}\Phi_i\,dx. \tag{3.62}$$

In this formula

$$D_\varepsilon(x)=\{\xi:\xi\in D, |\xi - x|>\varepsilon\},$$
$$D_\varepsilon(\xi)=\{x:x\in D, |\xi - x|>\varepsilon\}.$$

Using the property of the delta function and the Stocks and Gauss–Ostrogradskij formulas, equality (3.62) can be reduced to

$$2\int_D \frac{dx}{n(x)}\int_{S_\varepsilon(x)} \sum_{k,j=1}^{3} T_{kj} w_{x_j} \frac{\partial}{\partial \xi_k}(\nabla_k w, v)\, dS_\xi = \int_D dx \int_{Q_\varepsilon(x)} \sum_{i=1}^{3} q_i\, d\xi_i$$

$$+\int_D dx \int_{D_\varepsilon(x)} [|\nabla_x w|^2 + (\nabla_x w, v)^2] \frac{1}{n^2(x)} |T| \delta(F(\xi))\, d\xi$$

$$-\int_S dS_\xi \int_{S_\varepsilon(\xi)} \sum_{i=1}^{3} \Phi_i F_{x_i}(x)\, dS_x$$

$$Q_\varepsilon(x) = \{\xi : \xi \in S, |\xi - x| = \varepsilon\}, S_\varepsilon(\xi) = \{x : x \in S, |x - \xi| \geqslant \varepsilon\}.$$

Note that for all the x located from the boundary S of the domain D at distances greater than ε, the set $Q_\varepsilon(x)$ is empty. In the obtained equality let us go over to the limit by tending ε to zero. For the derivatives of the function $\tau(\xi, x)$ the following inequalities are valid [see (3.18)]

$$|\tau_{x_i}| \leqslant C, |\tau_{\xi_i}| \leqslant C, |\tau_{x_i \xi_k}| \leqslant C/|x - \xi|.$$

Therefore, the following estimates hold

$$|q_i| \leqslant C/|x - \xi|, |\Phi_i| \leqslant C/|x - \xi|, |\det T| \leqslant C/|x - \xi|.$$

Due to these estimates, at $\varepsilon \to 0$ the first of the integrals tends to zero, while the second and the third ones have finite limits and define improper integrals. As a result of passage to the limit one gets

$$2\int_D \frac{dx}{n(x)}\int_S \sum_{k,j=1}^{3} T_{kj} w_{x_j} \frac{\partial}{\partial \xi_k}(\nabla_x w, v)\, dS_\xi = \int_D dx \int_D [|\nabla_x w|^2 + (\nabla_x w, v)^2]$$

$$\frac{1}{n^2(x)} |T| \delta(F(\xi))\, d\xi - \int_S dS_\xi \int_S \Phi(w, \tau)\, dS_x. \tag{3.63}$$

The task now is to demonstrate that the first of the integrals standing in the right-hand parts of (3.57) and (3.63) coincide. For this purpose introduce, at a fixed x, the curvilinear coordinates of the point ξ. For such coordinates use the spherical coordinates θ, ϕ of the unit vector $v(x, \xi)$ and $\psi = F(\xi)$. The coordinate ψ has a sense at least for the points ξ located sufficiently close to surface S, which is quite enough since practically all the integration in the inner integral is carried out for $\xi \in S$. Calculating the derivatives with respect to ξ_k included in the determinant of the matrix T, through θ, ϕ, ψ and using the equalities $\nabla_x \tau = n(x) v$, $F_\theta = F_\phi = 0$, $F_\psi = 1$, $v_\psi = 0$ one obtains

$$T = \begin{vmatrix} 1 & 0 & 0 & 0 \\ 0 & \dfrac{\partial \theta}{\partial \xi_1} & \dfrac{\partial \phi}{\partial \xi_1} & \dfrac{\partial \psi}{\partial \xi_1} \\ 0 & \dfrac{\partial \theta}{\partial \xi_2} & \dfrac{\partial \phi}{\partial \xi_2} & \dfrac{\partial \psi}{\partial \xi_2} \\ 0 & \dfrac{\partial \theta}{\partial \xi_3} & \dfrac{\partial \phi}{\partial \xi_3} & \dfrac{\partial \psi}{\partial \xi_3} \end{vmatrix} \begin{vmatrix} 0 & n(x) v \\ F_\theta & n(x) v_\theta \\ F_\phi & n(x) v_\phi \\ F_\psi & n(x) v_\psi \end{vmatrix} =$$

$$= -n^3(x)|v_\phi|\frac{\partial(\theta,\phi,\psi)}{\partial(\xi_1,\xi_2,\xi_3)} = -n^3(x)\sin\theta\frac{\partial(\theta,\phi,\psi)}{\partial(\xi_1,\xi_2,\xi_3)}.$$

But due to the equalities $\xi = f(\zeta, x)$, $\zeta = -v\tau/n(x)$ and the condition of $\Gamma(x,\xi)$ convexity with respect to S

$$-n^3(x)\frac{\partial(\theta,\phi,\psi)}{\partial(\xi_1,\xi_2,\xi_3)} = \left[\left|\frac{\partial f}{\partial \zeta}\right|\tau^2\frac{\partial \tau}{\partial \psi}\right]^{-1} > 0, \quad \xi \in S.$$

Therefore

$$|T|\,\mathrm{d}\xi = n^3(x)\sin\theta\,\mathrm{d}\theta\,\mathrm{d}\phi\,\mathrm{d}\psi = n^3(x)\,\mathrm{d}\omega_v\,\mathrm{d}\psi$$

and the integrals in (3.63) and (3.57) coincide.

Thus, the lemma is proved. Estimate (3.54) can be easily derived from equality (3.57). Indeed, in the solutions of (3.56) the left-hand part of equality (3.57) becomes zero and besides, for the solutions of (3.56)

$$|\nabla_x w| \geqslant |(\nabla_x w, v)| = u(x).$$

Replacing in equality (3.57) the sum $|\nabla_x w|^2 + (\nabla_x w, v)^2$ with $2u^2(x)$ and taking into account the fact that the inner integral remaining after the multiplier $2u^2(x)$ has been put outside the brackets, is numerically equal to the area of the unit sphere

$$8\pi\int_D n(x)u^2(x)\,\mathrm{d}x \leqslant \int_S \mathrm{d}S_\xi \int_S \Phi(w,\tau)\mathrm{d}S_x$$

and estimate (3.54) becomes obvious.

This estimate, however, is proved when the lemma conditions that the functions $n(x)$, $u(x)$ have smoothness a unity greater than that given in Theorem 3.2, are fulfilled. Validity of estimate (3.54) in the case when the theorem conditions are met results from the fact that the right-hand part of inequality (3.54) is defined in these conditions too. Indeed, for any pair of the functions $u(x)\in C^1(D)$, $n(x)\in C^2(D)$ one can construct such a pair of sequences $u_k(x)\in C^2(D)$, $n_k(x)\in C^3(D)$ that $u_k(x)\to u(x)$, $n_k(x)\to n(x)$ in the norm of the spaces $C^1(D)$ and $C^2(D)$, respectively. Going over to the limit at $k\to\infty$, estimate (3.54) is obtained for $u(x)\in C^1(D)$, $n(x)\in C^2(D)$.

By way of conclusion note that the estimate of stability (3.54) is much more valid than the estimate resulting from (3.52) obtained under the supposition that $n_0(x) = n_0(|x|)$. As has been already mentioned, equalities (3.52) can give only a logarithmic estimate of stability. This essential difference in estimates, as has been shown by Bukhgeim [137], can be accounted for by the following fact. In the case when $n_0 = n_0(|x|)$ the problem was considered in the domain $D_{r0} = \{x: r_0 \leqslant |x| \leqslant R\}$, in which case for any fixed $x\in D_{r0}$ the set of the ends of the unit vectors $v(x,\xi)$ corresponding to the rays $\Gamma(x,\xi)$, $\xi\in S$, which are entirely contained within D_{r_0}, fills in only a part of the area of the unit sphere. It appears that in this case the problem of integral geometry can have no Hölder estimate of stability. The situation is quite contrary when the rays connecting any pair of points of a closed domain D are considered. In this case the condition of regularity of the set of rays in D guarantees that the set of the ends of the unit vectors $v(x, \xi)$ fills in the whole of the unit sphere.

Thus, there is an essential difference in the stability of a linearized problem when it is considered for all possible rays inside D or only for a part of the rays filling in only a part of the domain D.

3.5. Nonlinear three-dimensional inverse problems

The multi-dimensional inverse kinematic problem became the subject of investigation only comparatively recently. In the course of the last 15 years the results have been obtained in two directions: (1) the inverse kinematic problem in special classes of functions has been studied as far as the uniqueness of its solution; (2) an estimate of the problem stability has been obtained for the case of a regular ray behaviour if the times of the run for the rays connecting all possible pairs of points of the boundary of a limited simply connected domain are known.

It was Anikonov [35, 138–143] who obtained the first result associated with the investigation of the multi-dimensional inverse kinematic problem. He demonstrated that the function $n(x, y)$, $x = (x_1, \ldots, x_n)$ which is positive and analytical in the semi-sphere $y \geqslant 0$ and such that $n_y(x, y) < 0$, is uniquely determined in this domain provided the times of the run between the points of the domain boundary are given. Later the requirement of analyticity over all the variables was substituted for by that of uniform analyticity only with respect to the variable y

$$n(x, y) = \sum_{k=0}^{\infty} n_k(x) y^k.$$

Moreover, the theorem of uniqueness was expanded to the class of piecewise analytical with respect to y functions. Consider the proof of only the first of the results obtained.

Theorem 3.3. Let $n(x)$ be a positive analytical function in the three-dimensional domain D with the analytical boundary $S = \{x : F(x) = 0\}$, and for $x \in S$ the inequality

$$\sum_{i,j=1}^{3} \frac{\partial^2 F}{\partial x_i \partial x_j} v_i v_j + \nabla F \nabla \ln n(x) > 0, \quad \forall v : v \nabla F = 0. \tag{3.64}$$

holds. In this case $n(x)$ is uniquely defined inside D with the function $\tau(x, \xi)$ for $x \in S$, $\xi \in S$.

To prove the theorem it is sufficient to demonstrate that different functions $\tau(x, \xi)$ correspond to different functions $n(x)$. Consider two functions $n_1(x)$, $n_2(x)$ satisfying the conditions of the theorem. It follows from their analyticity that there exists such a point $\xi \in S$ and such is its open neighbourhood $Q_\varepsilon(\xi) = \{x : |x - \xi| < \varepsilon, x \in D\}$, that in this neighbourhood one of the functions n_1, n_2 is strictly greater than the other. Let, for the sake of definiteness, $n_1(x) > n_2(x)$. By way of decreasing ε one can always reach the point when inequality (3.64) holds for all $x \in Q_\varepsilon(\xi)$. In this case (see Section 3.2) the part of the boundary S adjoining the set $Q_\varepsilon(\xi)$ will be convex with respect to the set of rays $\Gamma(\xi, x)$ entirely contained in the closure of the set $Q_\varepsilon(\xi)$. Henceforth only such rays will be considered. If ε is sufficiently small this set of rays is regular. Hence, denoting through $\tau_k(x, \xi)$ the time of the signal run corresponding to the function $n_k(x)$ and through $\Gamma_k(x, \xi)$ the ray connecting x, ξ one finds from the inequality $n_1 > n_2$

$$\tau_1(x, \xi) = \int_{\Gamma_1(x,\xi)} n_1(x)|\mathrm{d}x| > \int_{\Gamma_1(x,\xi)} n_2(x)|\mathrm{d}x|.$$

Since out of all the curves connecting the points x, ξ in the medium corresponding to the function $n_2(x)$ the minimum time of the signal run along the curve is reached on the rays $\Gamma_2(x, \xi)$, then

$$\int_{\Gamma_1(x,\xi)} n_2(x)|\mathrm{d}x| \geqslant \int_{\Gamma_2(x,\xi)} n_2(x)|\mathrm{d}x| = \tau_2(x, \xi).$$

Thus, $\tau_1(x, \xi) > \tau_2(x, \xi)$ and functions $\tau(x, \xi)$ corresponding to different $n(x)$ do not coincide. The theorem is proved.

It follows from the proof of the theorem that for the function $n(x)$ to be uniquely determined in D it is sufficient to know the function $\tau(x, \xi)$ for all the points $x \in S$, $\xi \in S$, located from one another at a distance less than ε, where ε is an arbitrary but fixed positive number.

Theorems of uniqueness of the solution to the inverse kinematic problem have been also proved in the class of functions $n(x) \in C^3(\bar{D})$ which can be written for the domain $D = \{x : x_3 \geqslant 0\}$ in the following way

$$n(x) = f\left(\sum_{k=1}^{N} a_k(x_1, x_2) b_k(x_3) \right)$$

where f is a given monotonic function (see [144]). Uniqueness of the 'small-scale' problem solution has also been established [98, 99].

Markushevich and Reznikov [145] considered a special case when the velocity of the signal transfer in a three-dimensional space is independent of one of the variables and the system of observations is chosen in a special way. The authors have found the sufficient conditions for uniqueness of the solution to the problem in this case.

As has been mentioned earlier, an estimate of stability of the inverse kinematic problem has been found for a simply connected limited domain D with the boundary S in the condition of regular behaviour of the rays, if the function $\tau(x, \xi)$ is known for all $x \in S$, $\xi \in S$. It has been obtained by Mukhometov [132, 133] for a plane case and for a case of a space of higher dimensionality by Mukhometov and Romanov [146]. Bernshtejn and Gerver [135, 136] and Beil'kin [147]. In a number of works [129–131, 136, 148] generalizations of the developed method for the solution of the problem on restoration of the structure of a more complex anisotropic medium are reported.

Now set forth, in line with [146], the method of obtaining an estimate of stability for a three-dimensional space. It is based on (3.57), obtained in the preceding section, for the problem of integral geometry. Consider a class of functions $\Lambda(n_0, n_{00})$ consisting of the functions $n(x) \in C^2(D)$ which at fixed n_0, n_{00}

satisfy the conditions: (1) $0 < n_0 \leqslant n(x)$; $\|n\|_{C^2(D)} \leqslant n_{00} < \infty$; (2) a set of the rays $\Gamma(x, \xi)$ corresponding to the function $n(x)$ is regular in D.

Note that with the function $\tau(x, \xi)$ given on $S \times S$ and using the iconical equation, one can easily calculate the derivatives $\tau_{x_i}, \tau_{\xi_i}, \tau_{x_i\xi_j}$ for any i, j on the set $S \times S$. Indeed, the derivatives in the directions tangential to this set are found with the function $\tau(x, \xi)$, and for finding the derivatives in the directions normal to $S \times S$ one has the equations

$$|\nabla_x \tau(x, \xi)| = n(x), \qquad |\nabla_x \tau(\xi, x)| = n(\xi).$$

Denote

$$\|\tau\|^2 = \int_S dS_\xi \int_S \left[|\nabla_x \tau|^2 + |\nabla_\xi \tau|^2 + |x - \xi|^2 \sum_{i,j=1}^{3} \tau^2_{x_i\xi_j} \right] \frac{1}{|x - \xi|} dS_x. \tag{3.65}$$

Theorem 3.4. For any two functions $n, \tilde{n}$ of the class $\Lambda(n_0, n_{00})$ and the corresponding functions $\tau, \tilde{\tau}$ the estimate

$$\|n - \tilde{n}\|^2_{L_2(D)} \leqslant C \|\tau - \tilde{\tau}\|^2 \tag{3.66}$$

is valid, where the constant C depends only on the constants n_0, n_{00}.

Assume that $w_1 = \tau - \tilde{\tau}$, $w_2 = \tilde{\tau}$, $w_3 = \tau$. In the solutions of the iconical equation $|\nabla_x \tau|^2 = n^2(x)$, $|\nabla_x \tilde{\tau}|^2 = \tilde{n}^2(x)$ one can easily verify the identity

$$\sum_{k,j=1}^{3} T_{kj}(w_1)_{x_j} \frac{\partial}{\partial \xi_k} (\nabla_x w_1, \nabla_x \tau) - \sum_{k,j=1}^{3} T_{kj}(w_2)_{x_j} \frac{\partial}{\partial \xi_k} (\nabla_x w_2, \nabla_x \tau)$$
$$+ \sum_{k,j=1}^{3} T_{kj}(w_3)_{x_j} \frac{\partial}{\partial \xi_k} (\nabla_x w_3, \nabla \tau) \equiv 0. \tag{3.67}$$

Indeed, the third addent in the above identity turns to zero since the expression $(\nabla_x w_3, \nabla_x \tau) = n^2(x)$ is ξ-independent. To check whether the sum of the first two addents equals zero one should use the obvious identity

$$\sum_{k,j=1}^{3} T_{kj} \tau_{x_j} \frac{\partial}{\partial \xi_k} (\nabla_x w_2, \nabla_x \tau) = 0.$$

Using this identity one can reduce the first two addents to the form

$$\sum_{k,j=1}^{3} T_{kj}(w_1)_{x_j} \frac{\partial}{\partial \xi_k} [(\nabla_x w_1, \nabla_x \tau) + (\nabla_x w_2, \nabla \tau)].$$

The expression in the square brackets coincides with $n^2(x)$ and hence its derivatives calculated with respect to the variables ξ_k become zero.

Multiply both parts of identity (3.67) by $2\delta(F(\xi))\, d\xi\, dx / n^2(x)$, integrate the product over $D \times D$ and apply to each of the addents of identity (3.67) equality (3.57). As a result

$$4 \int_D n^2(x)\, dx \int_{|\nu|=1} \{n(x) - \tilde{n}(x)[\nu(x, \xi), \tilde{\nu}(x, \xi)]\}_{\xi = \xi(x,\nu)}\, dw_\nu$$
$$= \int_S dS_\xi \int_S [\Phi(\tau - \tilde{\tau}, \tau) - \Phi(\tilde{\tau}, \tau) + \Phi(\tau, \tau)]\, dS_x$$

where ν is a unit vector $\nabla_x \tilde{\tau}(x, \xi)/n(x)$. Therefore, estimating the difference $n(x) - \tilde{n}(x)\nu\tilde{\nu}$ from below

$$\int_D n^2(x)[n(x) - \tilde{n}(x)]\,\mathrm{d}x \leqslant \frac{1}{16\pi}\int_S \mathrm{d}S_\xi \int_S [\Phi(\tau - \tilde{\tau}, \tau) - \Phi(\tilde{\tau}, \tau) + \Phi(\tau, \tau)]\,\mathrm{d}S_x. \tag{3.68}$$

Since the functions $n(x)$, $\tilde{n}(x)$ enjoy equal rights, an inequality analogous to (3.68) can be written by way of substituting $\tilde{n}$ for n, $\tilde{\tau}$ for τ, and vice versa. By adding this new inequality to inequality (3.68) one gets

$$\int_D [n^2(x) - \tilde{n}^2(x)][n(x) - \tilde{n}(x)]\,\mathrm{d}x \leqslant \frac{1}{16\pi}\int_S \mathrm{d}S_\xi \int_S \Psi(\tau, \tilde{\tau})\,\mathrm{d}S_x$$

$$\Psi(\tau, \tilde{\tau}) = \Phi(\tau - \tilde{\tau}, \tau) - \Phi(\tilde{\tau}, \tau) + \Phi(\tau, \tau) + \Phi(\tilde{\tau} - \tau, \tilde{\tau}) - \Phi(\tau, \tilde{\tau}) + \Phi(\tilde{\tau}, \tilde{\tau}).$$

But, obviously

$$\int_D [n^2(x) - \tilde{n}^2(x)][n(x) - \tilde{n}(x)]\,\mathrm{d}x \geqslant 2n_0 \int_D (n - \tilde{n})^2\,\mathrm{d}x = 2n_0 \|n - \tilde{n}\|^2_{L_2(D)}.$$

Hence

$$\|n - \tilde{n}\|_{L_2(D)} \leqslant \frac{1}{32\pi n_0}\int_S \mathrm{d}S_\xi \int_S \Psi(\tau, \tilde{\tau})\,\mathrm{d}S_x. \tag{3.69}$$

Now demonstrate that the right-hand part in inequality (3.69) can be estimated using the right-hand part of inequality (3.66) at a certain value of the constant C. By way of expanding the determinants contained in the expression for $\Psi(\tau, \tilde{\tau})$ with respect to the elements of the first column and the first line, one gets

$$\Psi(\tau, \tilde{\tau}) = \sum_{k,j=1}^{3} (-1)^{k+j+1}\{(\tau_{xk} - \tilde{\tau}_{xk})[\tau_{\xi_j}\Phi_{kj}(\tau) - \tilde{\tau}_{\xi_j}\Phi_{kj}(\tilde{\tau})] + (\tau_{\xi_j} - \tilde{\tau}_{\xi_j})[\tau_{xk}\Phi_{kj}(\tau) - \tilde{\tau}_{xk}\Phi_{kj}(\tilde{\tau})]\}.$$

Here $\Phi_{kj}(\tau)$ is a minor obtained from the matrix for $\Phi(w, \tau)$ [see (3.55)] by crossing out the first and the $j+$ second lines and the first and the $k+$ second columns.

The difference $\tau_{\xi_j}\Phi_{kj}(\tau) - \tilde{\tau}_{\xi_j}\Phi_{kj}(\tilde{\tau})$ can be written as follows:

$$\tau_{\xi_j}\Phi_{kj}(\tau) - \tilde{\tau}_{\xi_j}\Phi_{kj}(\tilde{\tau}) = \sum_{l,s=1}^{3} c_{ls}^{kj}(x, \xi)[\tau_{\xi_j}\tau_{x_l\xi_s} - \tilde{\tau}_{\xi_j}\tilde{\tau}_{x_l\xi_s}]$$

$$= \sum_{l,s=1}^{3} c_{ls}^{kj}(x, \xi)[(\tau_{\xi_j} - \tilde{\tau}_{\xi_j})\tau_{x_l\xi_s} + (\tau_{x_l\xi_s} - \tilde{\tau}_{x_l\xi_s})\tilde{\tau}_{\xi_j}], \quad c_{ks}^{kj} = c_{lj}^{kj} = 0$$

in which case c_{ls}^{kj} coincide to the accuracy of the sign with certain products of the first derivatives of the functions $F(\xi)$, $F(x)$, an analogous equality being valid for the difference $\tau_{xk}\Phi_{kj}(\tau) - \tilde{\tau}_{xk}\Phi_{kj}(\tilde{\tau})$. Since $|\nabla_x \tau| = n(x) \leqslant n_{00}$, $|\nabla_x \tilde{\tau}| = \tilde{n}(x) \leqslant n_{00}$, $|\tau_{x_l\xi_s}| \leqslant C/|x - \xi|$ and $|\nabla F|_s = 1$, then at a certain value of the numerical constant

C the following estimate holds

$$\Psi(\tau,\tilde{\tau}) \leqslant C\left\{|\nabla_x(\tau-\tilde{\tau})|^2 + |\nabla_\xi(\tau-\tilde{\tau})|^2 + |x-\xi|^2 \times \sum_{i,j=1}^{3} [(\tau-\tilde{\tau})_{x_i\xi_j}]^2\right\} \Big/ |x-\xi|$$

and thus the validity of inequality (3.66) is stated.

3.6. Inverse problems using inner sources

There exist some other variations of formulating the inverse kinematic problem and it is first of all associated with the fact that under real conditions a source of disturbances x^0 is not necessarily located on the boundary of the domain D. For instance, earthquakes have been recorded occurring as deep as 600 km from the Earth's surface. Moreover, such areas exist on the globe (so-called focal zones) where the centres of earthquakes fill in the whole domains reaching deep down into the Earth. It was natural to try to use this information to approach the determination of the velocity of seismic waves from a new angle. In particular, the problem arises if it is possible to get rid of ambiguity of velocity determination in the presence of waveguides in the medium by way of using the data from subsurface sources. The problem of existing subsurface waveguides is of principal importance for characterizing the Earth's matter, since there are reasons to believe, judging by the model experiments, that the matter in the waveguide zones possesses some special properties. It accounts for geophysicists' interest to the problem of the velocity structure of the Earth that would take into account the possibility of waveguide presence.

A peculiarity of the formulation of the problem with inner sources is associated with the fact that knowing nothing about the velocity structure we have no right to consider the coordinates of sources x^0 given. Moreover, for every fixed x^0 the times of the run $\tau(x,x^0)$ are known only to the accuracy of the moment of reference (the beginning of an earthquake), i.e. to the accuracy of an additive constant.

For a one-dimensional model the problem using an inner source has been considered by Gerver and Markushevich [107] with respect to the domain D which is a semi-space. For a sphere with a radius R the result obtained is reduced to the following: if $|x^0| = r_0 < R$ and the function $m(r) = r/v(r)$ is strictly monotonic on the intercept $[r_0, R]$, then by the hodograph (known to the accuracy of an additive constant) one can find r_0 uniquely and velocity $v(r)$ for $r \in [r_0, R]$, while in a case when the function $m(r)$ is not monotonic, there is no uniqueness in determining r_0 and $v(r)$. Note that in a one-dimensional model an epicentre coordinates can be easily determined since the hodograph $\tau(x, x^0)$, $x \in S$ is symmetrical with respect to the straight line passing through the sphere centre and the point x^0. From the above said one can, in particular, deduce that if the function $m(r)$ has a finite number of extremums and there are hodographs of sources x^0 corresponding to the extremal points of the function $m(r)$, then the function $m(r)$ is uniquely determined by this set of hodographs. Later Markushevich [149] described the necessary and sufficient properties of the hodographs corresponding to deep-focal earthquakes.

Within a multi-dimensional approach the formulation of the inverse kinematic problem using inner sources has been considered by Anikonov [35, 150, 151], and it is given in the following. It is assumed that a certain domain D_0 contained inside a sphere D is filled in with sources x^0 from each of which the hodographs $\tau(x, x^0)$, $x \in S$, $x^0 \in D_0$ are given on the surface S of the sphere D (in which case x^0 are unknown, while the hodographs are set to the accuracy of an additive constant); the task is to find $v(x)$ inside D. Under the supposition that the set of hodographs $\tau(x, x^0)$ allows parametrization (so that there exists a one-to-one and differentiated correspondence between x^0 and a certain parameter u) Anikonov has shown that the velocity $v(x)$ is determined in the domain D_0 to the accuracy of the conformal transformation of this domain. It means that in a case when the domain D adjoins the surface S the velocity inside D_0 is uniquely determined. A numerical algorithm for the solution of the problem is given in [152].

Another possible variation of formulating the inverse kinematic problem is associated with considering it within the class of piecewise-continuous velocities. If on a certain surface $S_0 \subset D$ the velocity changes in a jump fashion when going over from one side of the surface to another, then on S_0 alongside with the refracted waves having passed through S_0 there arise the waves reflected from S_0. The times of the wave run along the rays reflected from S_0 provide an additional information on the velocity structure of the medium and on the geometry of the surface S_0 itself. Hence, here arises the problem of determining the velocity and the surface S_0 (see [153, 154]). Sometimes a simplified variation of the problem formulation is considered, when the velocity is assumed given and it is only the surface S_0 that is to be determined. Using the hodographs of reflected waves is especially characteristic of seismic surveying which primary concern is studying the Earth's core structure.

3.7. Inverse kinematic problems for an anisotropic medium

Differential ray equations

In the case when the process of oscillation propagation is described by

$$\begin{gathered} u_{tt} = \sum_{i,j=1}^{3} a_{ij}(x) u_{x_i x_j} \\ a_{ij} = a_{ji}, \quad \mu_0 \sum_{i=1}^{3} \alpha_i^2 \leqslant \sum_{i,j=1}^{3} a_{ij}(x) \alpha_i \alpha_j \leqslant \mu_{00} \sum_{i=1}^{3} \alpha_i^2 \\ 0 < \mu_0 \leqslant \mu_{00} < \infty \end{gathered} \tag{3.70}$$

the iconical equation is substituted for by the characteristic equation

$$p = \nabla_x \tau(x, x^0), \qquad p = (p_1, p_2, p_3). \tag{3.71}$$

In this case the velocity of signal transfer appears to be dependent not only on the point x but also on the direction in which the disturbance propagates. Thus, (3.70) describes a wave process in an anisotropic medium.

Obtain a system of differential equations to construct the characteristic conoids $\tau(x, x^0) = t$. As is the case in isotropic media, the surface is formed by way of constructing the separate lines which are the bicharacteristics of (3.71) projections of which onto the space define the trajectories of the signal propagation, i.e. the rays. For this purpose denote

$$\sum_{i,j=1}^{3} a_{ij}(x)\tau_{x_i}\tau_{x_j} = 1. \tag{3.72}$$

Differentiating the equation

$$\sum_{i,j=1}^{3} a_{ij}(x)p_i p_j = 1 \tag{3.73}$$

over the variable x_k and using the equality $(p_i)_{x_k} = (p_k)_{x_i}$, one gets

$$2\sum_{i,j=1}^{3} a_{ij}p_j\frac{\partial p_k}{\partial x_i} + \sum_{i,j=1}^{3} p_i p_j \frac{\partial}{\partial x_k} a_{ij}(x) = 0, \quad k = 1, 2, 3. \tag{3.74}$$

From this equation one can derive a system of equations for finding p_k and the characteristics of (3.74)

$$\frac{dx_k}{dt} = \sum_{j=1}^{3} a_{kj}p_j, \qquad \frac{dp_k}{dt} = -\frac{1}{2}\sum_{i,j=1}^{3} p_i p_j \frac{\partial}{\partial x_k} a_{ij}(x), \quad k = 1, 2, 3. \tag{3.75}$$

Note, here that the parameter t coincides with τ to the accuracy of an additive constant along each of the solutions to system (3.75). Indeed

$$\frac{d\tau}{dt} = (\nabla_x \tau, x_t) = \sum_{k=1}^{3} p_k \sum_{j=1}^{3} a_{kj}p_j = 1.$$

Setting $\tau = 0$ at $t = 0$, one has $\tau = t$ and the value of the parameter t denotes the time of the signal run from the point x^0 to the point x.

In order to construct all the bicharacteristics of equation (3.71) it is sufficient to solve for system (3.75) the Cauchy problem with the following data

$$x|_{t=0} = x^0, \qquad p|_{t=0} = p^0, \tag{3.76}$$

corresponding to all possible values $p^0 = (p_1^0, p_2^0, p_3^0)$

$$\sum_{i,j=1}^{3} a_{ij}(x^0)p_i^0 p_j^0 = 1. \tag{3.77}$$

A solution to (3.75) and (3.76) defines the ray outgoing from the point x^0 in the direction α^0

$$\alpha^0 = x_t|_{t=0} = p^0 A(x^0).$$

In this formula $A(x)$ denotes a symmetrical matrix composed of the elements $a_{ij}(x)$, $i, j = 1, 2, 3$.

Expressing the vector p through equality (3.75), one gets

$$p = x_t B(x), \qquad B = A^{-1}, \qquad B = (b_{ij}).$$

Substituting the expression for p in (3.73), the condition for the components of the vector x_t is obtained

$$\sum_{i,j=1}^{3} b_{ij}(x)(x_i)_t(x_j)_t = 1.$$

Hence

$$dt = d\tau = \left(\sum_{i,j=1}^{3} b_{ij}(x)\, dx_i\, dx_j \right)^{1/2}. \tag{3.78}$$

The expression $ds/dt = |x_t|$ (ds is an element of the arc length) is known to define the velocity of a signal transfer. Thus, at the point x in the direction determined by the unit vector $v = dx/ds$, the velocity $v(x, v)$ can be found by the formula

$$v(x, v) = \left(\sum_{i,j=1}^{3} b_{ij}(x) v_i v_j \right)^{-1/2}. \tag{3.79}$$

Due to positiveness of the matrix B (3.78) defines a Riemann metric with a length element $d\tau$, while the system of equalities (3.75) defines the geodesic lines of this metric. On the geodesic lines of metric (3.78) the functional

$$\tau(L) = \int_{L(x,x^0)} \left[\sum_{i,j=1}^{3} b_{ij}(x)\, dx_i\, dx_j \right]^{1/2}$$

reaches its minimum (which is generally local).

Differential properties of the solutions to (3.75) and (3.76) are quite identical to those of the solutions to the system discussed in Section 3.1. Therefore, instead of repeating the corresponding considerations, note that a solution to (3.75) and (3.76) can be written as

$$x = f(\zeta, x^0), \quad \zeta = \alpha^0 t \tag{3.80}$$

in which case the function $f(\zeta, x^0)$ is an s times continuously differentiable function of its arguments in a certain neighbourhood of the point $(0, x^0)$ provided that the coefficients $a_{ij}(x)$ are $(s + 1)$ times continuously differentiable. As regards $f(\zeta, x^0)$, there exists an inverse function $\zeta = g(x, x^0)$ having the same smoothness as the function $f(\zeta, x^0)$. Using the function $g(x, x^0)$ the time of the run $\tau(x, x^0)$ between the points x, x^0 is found by the formula [resulting from (3.80) and (3.77)]

$$\tau(x, x^0) = \left[\sum_{i,j=1}^{3} b_{ij}(x^0) g_i(x, x^0) g_j(x, x^0) \right]^{1/2} \tag{3.81}$$

Here g_i are the components of the function $g(x, x^0)$.

Now derive a sufficient condition for the absence of waveguides and boundary rays in a medium. For this purpose, as was the case in Section 3.2, consider a domain D whose boundary is described by the equation $F(x) = 0$; in this case $F(x) < 0$ inside D and the set of rays of the level $F(x) = C (C < 0)$ is regular in the same sense as in Section 3.2. In this case a sufficient condition for the waveguide absence for the function $\phi = F(x(t))$ is the condition that the inequality $\phi_{tt} \geqslant \delta$

holds for the rays with $\phi_t|_{t=0}=0$ at a certain positive δ, i.e. for

$$\forall p: \quad \sum_{i,j=1} F_{x_i} a_{ij} p_j = 0$$

the inequality

$$\sum_{i,j,k,s=1}^{3} \left[F_{x_i x_k} a_{ij} a_{ks} - \tfrac{1}{2} F_{x_i} a_{ik} \frac{\partial}{\partial x_k} a_{js} + F_{x_i} a_{ks} \frac{\partial}{\partial x_k} a_{ij} \right] p_j p_s \geqslant \delta > 0.$$

must be fulfilled.

Now consider the problem of determining the coefficients $a_{ij}(x)$ in the domain $D=\{x{:}x_3 \geqslant 0\}$, first under the supposition that the coefficients depend only on the variable x_3(a one-dimensional inverse problem) and then under the supposition that the coefficients a_{ij} can be presented as known coefficients $a_{ij}^0(x_3)$ plus such small unknown functions $a_{ij}^1(x)$ that the problem allows linearization.

A one-dimensional case

Assume that the coefficients $a_{ij} \equiv a_{ij}(x_3)$, $i, j=1,2,3$, the domain $D=\{x{:}0 \leqslant x_3 \leqslant H\}$ the point $x^0 \in S=\{x{:}x_3=0\}$ is fixed and coincides with the origin of the coordinates, $a_{ij} \in C^1[0,H]$. The problem to be discussed is to find the function a_{ij} by the function $\tau(x,x^0)$ known for $x \in S$. The discussion will be carried out under the supposition that no waveguides and boundary rays are present in the domain D, in which case a sufficient condition has the form $[F(x)=-x_3]$

$$\tfrac{1}{2} a_{33} \sum_{k,s=1}^{3} \left(\frac{\partial}{\partial x_3} a_{ks} \right) p_k p_s \geqslant \delta > 0, \qquad \forall p: \sum_{k=1}^{3} a_{3k} p_k = 0. \tag{3.82}$$

Later this condition will be considered fulfilled.

Condition (3.82) has a certain physical sense. First demonstrate that the following formulas are valid

$$\frac{\partial}{\partial x_k} \ln v(x,\nu) = \frac{1}{2} \sum_{s,j=1}^{3} (a_{sj})_{x_k} p_j p_s, \quad k=1,2,3. \tag{3.83}$$

Note that $x_t = \nu v(x,\nu)$ and hence

$$p = v(x,\nu)\nu B. \tag{3.84}$$

Differentiating the equality $AB=E$, one gets

$$B_{x_k} = -BA_{x_k}B, \quad k=1,2,3. \tag{3.85}$$

Equations (3.79), (3.84), and (3.85) give a chain of equalities

$$-2v^{-3}v_{x_k} = \sum_{i,l=1}^{3} \nu_i \nu_l (b_{il})_{x_k} = - \sum_{i,j,l,s=1}^{3} \nu_i \nu_l b_{is} b_{jl} (a_{sj})_{x_k}$$
$$= -v^2 \sum_{s,j=1}^{3} p_j p_s (a_{sj})_{x_k}$$

resulting in (3.83). Taking into account (3.83), condition (3.82) can be rewritten as

$$\frac{a_{33}}{v(x_3,\nu)}\frac{\partial}{\partial x_3}v(x_3,\nu)\geqslant\delta>0,\quad \forall \nu:\nu_3=0. \tag{3.86}$$

The last expression is obviously equivalent to the condition

$$\frac{\partial}{\partial x_3}v(x_3,\nu)>0,\quad \forall \nu:\nu_3=0, 0\leqslant x_3\leqslant H. \tag{3.87}$$

Thus, a monotonic increase in the velocity of the signal transfer with growing coordinate x_3 for all the directions ν, which are orthogonal to the axis x_3, guarantees the absence of waveguides in the domain D. In this case each of the rays $\Gamma(x,x^0)$, $x\in S$, $x^0\in S$ is an arc lying in the domain $x_3\geqslant 0$.

Now write an equation of rays $\Gamma(x,x^0)$. From (3.75) and the condition of independence of the coefficients a_{ij} on x_1 and x_2 the equalities

$$\mathrm{d}p_1/\mathrm{d}t=0,\qquad \mathrm{d}p_2/\mathrm{d}t=0$$

follow and from them, in their turn, results that p_1 and p_2 remain constant along the ray. They can be easily found by the inverse problem data. Indeed

$$p_1=\frac{\partial}{\partial x_1}\tau(x,x^0)\Big|_{x_3=0},\qquad p_2=\frac{\partial}{\partial x_2}\tau(x,x^0)\Big|_{x_3=0}. \tag{3.88}$$

Equalities (3.88) have a sense under the condition of regularity of the ray field, which impose additional restrictions on $a_{ij}(x_3)$. These restrictions, however, can be formulated in terms of the inverse problem data, i.e. assume that τ_{x_1}, τ_{x_2} are continuous functions of the point $x\in S$ at $x_3=0$. It can also be assumed that for $x\in S$ the function $\tau(x,x^0)$ consists of a finite number of unique branches.

Thus, p_1 and p_2 can be considered known for every ray. The component p_3 is found from (3.73). Solving it with respect to p_3, one gets

$$p_3=-\frac{1}{a_{33}}\sum_{j=1}^{2}a_{3j}p_j\pm\frac{1}{\sqrt{(a_{33})}}\left(1-\sum_{i,j=1}^{2}c_{ij}p_ip_j\right)^{1/2} \tag{3.89}$$

where

$$c_{ij}=a_{ij}-\frac{1}{a_{33}}a_{3i}a_{3j},\quad i,j=1,2. \tag{3.90}$$

In (3.89) the plus sign should be chosen at the portions of the ray where $(x_3)_t>0$, and the minus sign where $(x_3)_t<0$.

In the case in question it is more convenient to pass from the parametric presentation of the ray $[x=x(t)]$ over to its presentation as $x_1=x_1(x_3)$, $x_2=x_2(x_3)$. In this case one should remember that any ray $\Gamma(x,x^0)$, belonging for $x\in S, x^0\in S$ to the domain D, is crossed, provided conditions (3.82) are satisfied, by a plane parallel to the plane $x_3=0$ at two points, and hence, the functions $x_1(x_3)$ and $x_2(x_3)$ are two-valued functions. The differential equations determining them are obtained from (3.75)

$$\mathrm{d}x_k/\mathrm{d}x_3=\sum_{j=1}^{3}a_{kj}p_j\Big/\sum_{j=1}^{3}a_{3j}p_j,\quad k=1,2. \tag{3.91}$$

Using (3.89) reduce (3.91) to the form

$$\frac{dx_k}{dx_3} = \frac{a_{k3}}{a_{33}} \pm \frac{\sum_{j=1}^{2} c_{kj} p_j}{(a_{33})^{1/2}\left(1 - \sum_{i,j=1}^{2} c_{ij} p_i p_j\right)^{1/2}}, \quad k = 1, 2. \tag{3.92}$$

As earlier, refer to the most distant from the boundary S of the domain D point on the ray as the ray top. Such a point exists for any ray $\Gamma(x, x^0)$, $x^0 \in S$, $x \in S$, when condition (3.82) is fulfilled. Denote its coordinates through $\xi = (\xi_1, \xi_2, \xi_3)$. It is obvious that at the point ξ dx_k/dx_3, $k = 1, 2$, turn to infinity. Therefore

$$\sum_{i,j=1}^{2} c_{ij}(\xi_3) p_i p_j = 1. \tag{3.93}$$

On the portion of the ray $\Gamma(x, x^0)$ from the point x^0 to the point ξ one should choose the plus sign in (3.92) since in this case $dx_3/dt > 0$, while on the portion from the point ξ to the point x the minus sign should be chosen. By way of integrating (3.92) the following equation is obtained for the ray

$$x_k = \int_0^{x_3} \left[\frac{a_{k_3}(z)}{a_{33}(z)} + \frac{\sum_{j=1}^{2} c_{kj}(z) p_j}{(a_{33}(z))^{1/2}\left(1 - \sum_{i,j=1}^{2} c_{ij}(z) p_i p_j\right)^{1/2}} \right] dz, \quad k = 1, 2.$$

This equation, in particular, results in

$$\xi_k = \int_0^{\xi_3} \left[\frac{a_{k3}(z)}{a_{33}(z)} + \frac{\sum_{j=1}^{2} c_{kj}(z) p_j}{(a_{33}(z))^{1/2}\left(1 - \sum_{i,j=1}^{2} c_{ij}(z) p_i p_j\right)^{1/2}} \right] dz, \quad k = 1, 2.$$

The equation of the ray $\Gamma(x, \xi)$ [a portion of the ray $\Gamma(x, x^0)$ from the point ξ to the point x] is as follows:

$$x_k = \xi_k - \int_{x_3}^{\xi_3} \left[\frac{a_{k3}(z)}{a_{33}(z)} - \frac{\sum_{j=1}^{2} c_{kj}(z) p_j}{(a_{33}(z))^{1/2}\left(1 - \sum_{i,j=1}^{2} c_{ij}(z) p_i p_j\right)^{1/2}} \right] dz, \quad k = 1, 2.$$

From this equation for the points $x \in S$ follows:

$$x_k = 2 \int_0^{\xi_3} \left[\frac{\sum_{j=1}^{2} c_{kj}(z) p_j}{(a_{33}(z))^{1/2}\left(1 - \sum_{i,j=1}^{2} c_{ij}(z) p_i p_j\right)^{1/2}} dz, \right. \left. k = 1, 2 \right] \tag{3.94}$$

which will be used to investigate the inverse problem.

Note, that there is no necessity to write an equation to find the times $\tau(x, x^0)$, as this information has already been used when calculating the ray parameters p_1, p_2

by (3.88). Therefore, all the information that can be used when solving the inverse problem is concentrated in (3.94) [one should not also forget relation (3.93)].

At a fixed vector $\alpha=(\alpha_1,\alpha_2)$, $\alpha\neq 0$, introduce into consideration the function

$$\phi_\alpha(x_3)=\sum_{i,j=1}^{2} c_{ij}(x_3)\alpha_i\alpha_j. \tag{3.95}$$

Now demonstrate that the function $\phi_\alpha(x_3)$ is a monotonic increasing function of x_3. For this purpose make sure that the matrix $C=(c_{ij}, i,j=1,2)$ coincides with the matrix $\bar{B}^{-1}$ which is inverse to the matrix $\bar{B}=(b_{ij}, i,j=1,2)$. The simplest way to do it is to write the matrices A, B as follows:

$$A=\begin{Vmatrix} \bar{A} & & a_{13} \\ & & a_{23} \\ a_{31} & a_{32} & a_{33} \end{Vmatrix}, \qquad B=\begin{Vmatrix} \bar{B} & & b_{13} \\ & & b_{23} \\ b_{31} & b_{32} & b_{33} \end{Vmatrix}.$$

Using the condition $AB=E$ two matrix correlations are obtained

$$\overline{AB}+\begin{Vmatrix} a_{13} \\ a_{23} \end{Vmatrix}\|b_{31} \quad b_{32}\|=E$$

$$\|a_{31} \quad a_{32}\|\bar{B}+a_{33}\|b_{31} \quad b_{32}\|=\|0,0\|.$$

Hence

$$\bar{B}^{-1}=\bar{A}-\begin{Vmatrix} a_{13} \\ a_{23} \end{Vmatrix}\|a_{31} \quad a_{32}\|.$$

The matrix in the right-hand part of this equality coincides with C, and hence, $C=\bar{B}^{-1}$, which means that the matrix C is positively determined. Now see if $\phi'_\alpha(x_3)>0$. Calculating a derivative of the symmetrical matrix C by the equality which is analogous to equalities (3.85)

$$C_{x_3}=-C\bar{B}_{x_3}C$$

one has

$$\phi'_\alpha(x_3)=\alpha C_{x_3}\alpha^*=-\alpha C\bar{B}_{x_3}C\alpha^*=\left.\frac{2(\partial/\partial x_3)v(x_3,v)}{v^3(x_3,v)}\right|_{v=(\alpha C,0)/|\alpha C|}>0.$$

Thus, the sought monotonicity of the function $\phi_\alpha(x_3)$ is stated.

On the plane of p_1, p_2 now consider a set of the points lying on a fixed half-line passing through the origin of the coordinates in the direction of the unit vector $\alpha=(\alpha_1,\alpha_2)$

$$p_1=s\alpha_1, p_2=s\alpha_2, \qquad s=\sqrt{(p_1^2+p_2^2)}.$$

From equality (3.93) it follows that for any unit vector α the s values from the intercept $[(\phi_\alpha(H))^{-1/2},(\phi_\alpha(0))^{-1/2}]$ are corresponded to by the rays $\Gamma(x,x^0)$, $x\in S$ with the parameters p_1, p_2, these rays completely included within D. Indeed, if

$$\phi_\alpha(\xi_3)=1/s^2 \tag{3.96}$$

then the ray with the parameters $p_1=s\alpha_1$, $p_2=s\alpha_2$, $s=[\phi_\alpha(\xi_3)]^{-1/2}$, is in the domain $0 \leqslant x_3 \leqslant \xi_3$ and has its top at $x_3 = \xi_3$. But, as was stated earlier, the function $\phi_\alpha(x_3)$ monotonically increases with growing x_3, and hence, for any $\xi_3 \in [0, H]$

$$\phi_\alpha(0) \leqslant \phi_\alpha(\xi_3) \leqslant \phi_\alpha(H)$$

any point s from the intercept $[(\phi_\alpha(H))^{-1/2}, (\phi_\alpha(0))^{-1/2}]$ is corresponded to by a certain ray $\Gamma(x, x^0)$ entirely contained in D.

Thus, a surface hodograph corresponding to the rays $\Gamma(x, x^0)$ contained in D must possess the following important property: a set of (p_1, p_2) values resulting under condition (3.88) must contain at any α, $|\alpha| = 1$ a certain intercept of the half-line $(p_1, p_2) = \alpha s$, $s > 0$. Having constructed a map of the points $x \in S$ using (3.88) onto the plane of p_1, p_2 one can easily find such curves L_α on the plane for which the vector $\alpha = (p_1^2 + p_2^2)^{-1/2} \cdot (p_1, p_2)$ is constant. The parametric equation of such lines, as is seen from (3.94), is as follows:

$$L_\alpha : x_k(s, \alpha) = 2 \int_0^{\xi_3} \frac{s \sum_{j=1}^{2} c_{kj}(z) \alpha_j}{\sqrt{(a_{33}(z))} \sqrt{(1 - s^2 \phi_\alpha(z))}} \, \mathrm{d}z, \quad k = 1, 2. \tag{3.97}$$

In this formula ξ_3 and s are related to one another with the monotonic correlation (3.96).

Note, that one can readily find $\phi_\alpha(0)$ since at $s \to [\phi_\alpha(0)]^{-1/2}$ the ray $\Gamma(x, x^0)$ contracts to the point x^0. Therefore, using the inverse problem data to calculate τ_{x_1}, τ_{x_2} on the plane S and going over to the limit at $x \to x^0$ so that vector α remains constant, one gets the limit the vector (p_1^0, p_2^0) with the length $s_0 = [\phi_\alpha(0)]^{-1/2}$.

Replace the integration variable z in (3.97) with

$$\zeta = \int_0^z \frac{\mathrm{d}z}{\sqrt{(a_{33}(z))}}. \tag{3.98}$$

Then equalities (3.97) take the form

$$x_k(s, \alpha) = 2 \int_0^{\mu_\alpha(s)} \frac{s \sum_{j=1}^{2} \tilde{c}_{kj}(\zeta) \alpha_j}{\sqrt{(1 - s^2 \tilde{\phi}_\alpha(\zeta))}} \, \mathrm{d}\zeta, \quad k = 1, 2 \tag{3.99}$$

where $\tilde{\phi}_\alpha(\zeta) = \phi_\alpha(z(\zeta))$; $\tilde{c}_{kj}(\zeta) = c_{kj}(z(\zeta))$; $z(\zeta)$ is the function defined relation (3.89); $\mu_\alpha(s)$ is the root of the equation

$$1 - s^2 \tilde{\phi}_\alpha(\zeta) = 0.$$

Equalities (3.99) demonstrate that $x_k(s, \alpha)$ depend only on the coefficients $\tilde{c}_{kj}(\zeta)$, and hence, only they can be found by the inverse problem data. For this purpose calculate

$$\rho_\alpha(s) = \sum_{k=1}^{2} x_k(s, \alpha) \alpha_k,$$

which is the vector (x_1, x_2) projection on the vector α, by

$$\rho_\alpha(s) = \int_0^{\mu_\alpha(s)} \frac{2s\tilde{\phi}_\alpha(\zeta)\,\mathrm{d}\zeta}{\sqrt{(1 - s^2\tilde{\phi}_\alpha(\zeta))}} \tag{3.100}$$

$$s_1 \leqslant s \leqslant s_0, \qquad s_1 = [\phi_\alpha(H)]^{-1/2}, \qquad s_0 = [\phi_\alpha(0)]^{-1/2}.$$

It is obvious that $\mu_\alpha(s_0) = 0$. Besides, by differentiating the identity

$$\tilde{\phi}_\alpha(\mu_\alpha(s)) \equiv 1/s^2 \tag{3.101}$$

one gets

$$\mu'_\alpha(s) = -\frac{1}{s^3}[\phi'_\alpha(\mu_\alpha(s))]^{-1} < 0. \tag{3.102}$$

At any fixed α (3.100) can be easily solved with respect to the function $\mu_\alpha(s)$. Indeed

$$\int_t^{s_0} \frac{\rho_\alpha(s)\,\mathrm{d}s}{\sqrt{(s^2 - t^2)}} = \int_t^{s_0} \frac{\mathrm{d}s}{\sqrt{(s^2 - t^2)}} \int_0^{\mu_\alpha(s)} \frac{2s\tilde{\phi}_\alpha(\zeta)\,\mathrm{d}\zeta}{\sqrt{(1 - s^2\tilde{\phi}_\alpha(\zeta))}}$$
$$= \int_t^{s_0} \frac{\mathrm{d}s}{\sqrt{(s^2 - t^2)}} \int_{s_0}^{s} \frac{2su^{-2}\mu'_\alpha(u)\,\mathrm{d}u}{\sqrt{(1 - s^2/u^2)}} = \int_t^{s_0} \frac{\mu'_\alpha(u)}{u}\mathrm{d}u$$
$$\times \int_t^{u} \frac{2s\,\mathrm{d}s}{\sqrt{((s^2 - t^2)(u^2 - s^2))}} = -\pi \int_t^{s_0} \frac{\mu'_\alpha(u)\,\mathrm{d}u}{u}.$$

Therefore

$$\mu_\alpha(s) = -\frac{1}{\pi}\int_s^{s_0} t\left[\frac{\partial}{\partial t}\int_t^{s_0} \frac{\rho_\alpha(u)\,\mathrm{d}u}{\sqrt{(u^2 - t^2)}}\right]\mathrm{d}t = \frac{1}{\pi}\left[s\int_s^{s_0} \frac{\rho_\alpha(u)\,\mathrm{d}u}{\sqrt{(u^2 - s^2)}}\right.$$
$$\left. + \int_s^{s_0} \rho_\alpha(u) \arccos\frac{s}{u}\,\mathrm{d}u\right], \quad s \in [s_1, s_0].$$

Since the function $\zeta = \mu_\alpha(s)$ is monotonic, one can easily find its inverse function $s = \gamma_\alpha(\zeta)$ and calculate $\tilde{\phi}_\alpha(\zeta)$ by (3.101)

$$\tilde{\phi}_\alpha(\zeta) = 1/\gamma_\alpha^2(\zeta), \quad \zeta \in [0, \zeta_0].$$

Here ζ_0 corresponds to the value $z = H$ under condition (3.98).

Thus, one can obtain $\tilde{\phi}_\alpha(\zeta)$ from (3.100). It is obvious that the coefficients $\tilde{c}_{ij}(\zeta) = c_{ij}(z(\zeta))$, $i, j = 1, 2$, can be found by the function $\tilde{\phi}_\alpha(\zeta)$ obtained at various α. Theoretically, for this purpose it is sufficient to know $\tilde{\phi}_\alpha(\zeta)$ for three different vectors α. If the coefficient $a_{33}(z)$ is known for $z \in [0, H]$, a correspondence between ζ and z can also be considered known and in this case all c_{ij} are found as functions of z. When the coefficient $a_{33}(z)$ is unknown, the coefficients c_{ij} are found only as functions of ζ.

Note, that the result obtained has a direct physical meaning. First of all, the coordinate ζ has a sense of the time of the signal run from the plane $x_3 = 0$ to the plane $x_3 = z$. Indeed, under the supposition that $a_{kj} = a_{kj}(x_3)$ it follows from the system of equalities (3.75) that along the ray drawn from the point $x^0 \in S$ in the direction defined by the value $p^0 = [0, 0, 1/\sqrt{(a_{33}(0))}]$, the relations $p_1 = p_2 = 0$, and hence, $p_3 = 1/\sqrt{(a_{33}(x_3))}$, are valid. Therefore, a time of the signal run along this ray $\tau = t$ can be found by the relation

$$dx_3/dt = a_{33}p_3 = \sqrt{(a_{33}(x_3))}.$$

Integrating this relation one finds that t is expressed through the right-hand part of (3.98), i.e. $t = \zeta$. At the same time, along the ray $\nabla\tau = (0, 0, p_3)$, and hence, a tangential plane to the surface $\tau(x, x^0) = t$, constructed at the point of its crossing with the ray considered, coincides with the plane $x_3 = \text{const}$. Therefore, the time of the run along the ray is equal to that of the signal run between the planes $x_3 = 0$ and $x_3 = \text{const}$. Then, the matrix $C = (c_{ij},\ i, j = 1, 2)$, as has been earlier stated, coincides with the matrix $\bar{B}^{-1}$, and thus, $\bar{B} = C^{-1}$. Hence, if the matrix $C(\zeta)$ is known one can find the matrix $\bar{B}$ as a function of the parameter ζ. But, according to (3.79), the matrix $\bar{B}$ defines the value of the velocity of signal propagation $v(x_3(\zeta), v)$ in the direction v which is orthogonal to the axis x_3.

Therefore, a physical sense of the result obtained is in the fact that through the times of the signal run from a fixed point $x^0 \in S$ to an arbitrary point $x \in S$, we can find the velocity of signal propagation in any direction that is orthogonal to the plane $x_3 = z$.

By way of summarizing, the obtained result can be presented as a theorem.

Theorem 3.5. Let the coefficients a_{ij}, $i, j = 1, 2, 3$, *depend only on the variable* x_3 *and such that (1) condition (3.82) is met; (2) the vector* (τ_{x_1}, τ_{x_2}) *is a continuous function of a point of the plane* $x_3 = 0$.

In this case in the domain $D = \{x : 0 \leqslant x_3 \leqslant H\}$ *through the function* $\tau(x, x^0)$ *where the point* $x^0 \in S$ *is fixed and the point* $x \in S$ *is arbitrary, one can uniquely define the elements* c_{ij}, $i, j = 1, 2$, *determined by (3.90) as functions of the variable* ζ *related to the variable* $z = x_3$ *through correlation (3.98).*

By way of conclusion note that in the case when waveguides are present in the medium, uniqueness in determining c_{ij} is violated. The nature of ambiguity can be studied by the scheme discussed in [101], while in [42] the case when the point x^0 is an inner point of the domain D is considered.

A linearized three-dimensional problem

Now consider for (3.71) a linearized formulation of the inverse kinematic problem analogous to that discussed in Section 3.4. In this case assume that the coefficients $a_{ij}(i, j = 1, 2, 3)$ can be written as

$$a_{ij}(x) = a^0_{ij}(x) + a^1_{ij}(x), \qquad a^0_{ij} = a^0_{ji}, a^1_{ij} = a^1_{ji} \tag{3.103}$$

where $a^0_{ij} \in C^2(\bar{D})$ are known, and $a^1_{ij}(x)$ are small in the norm of the space $C^2(\bar{D})$. Also assume that the matrix $A^0 = (a^0_{ij}, i, j = 1, 2, 3)$, is positively determined. If the parameter of λ smallness is introduced

$$a_{ij} = a^0_{ij}(x) + \lambda a^1_{ij}(x) \tag{3.104}$$

the function $\tau(x, x^0)$ can be written as a series with respect to the small parameter

$$\tau(x, x^0) = \sum_{k=0}^{\infty} \lambda^k \tau_k(x, x^0). \tag{3.105}$$

Substituting the expressions for a_{ij}, $\tau(x, x^0)$ from (3.104) and (3.105) into (3.71) and equating the terms at the same degrees of λ the equations for the functions $\tau(x, x^0)$ are obtained. In particular, the equation for τ_0, τ_1 are as follows:

$$\sum_{i,j=1}^{3} a_{ij}^0(x)(\tau_0)_{x_i}(\tau_0)_{x_j} = 1 \tag{3.106}$$

$$2\sum_{i,j=1}^{3} a_{ij}^0(\tau_0)_{x_i}(\tau_1)_{x_j} + \sum_{i,j=1}^{3} a_{ij}^1(\tau_0)_{x_i}(\tau_0)_{x_j} = 0. \tag{3.107}$$

Equation (3.106) demonstrates that the value of the function $\tau_0(x, x^0)$ has a physical sense of the time of the signal run in the medium with the coefficients a_{ij}^0 and therefore can be considered known, while in a geometrical sense it denotes the distance between the points x, x^0 in the metric where the length element $d\tau_0$ is obtained by the formula

$$d\tau_0 = \left[\sum_{i,j=1}^{3} b_{ij}^0(x)\, dx_i\, dx_j\right]^{1/2} \tag{3.108}$$

where $b_{ij}^0(x)$ are the elements of the symmetrical matrix $B_0 = A_0^{-1}$.

Along the geodesic lines $\Gamma_0(x, x^0)$ of this metric the following equalities [see (3.75)] are valid

$$\frac{dx_i}{d\tau_0} = \sum_{j=1}^{3} a_{ij}^0 p_j^0, \quad i = 1, 2, 3; \quad p^0 = \nabla_x \tau_0.$$

Taking the above equalities into account, (3.107) can be written in the following way

$$2\frac{d\tau_1}{d\tau_0} + \sum_{i,j=1}^{3} a_{ij}^1 p_i^0 p_j^0 = 0.$$

Integrating this equality along the geodesic line $\Gamma_0(x, x^0)$

$$\tau_1(x, x^0) = -\frac{1}{2}\int_{\Gamma_0(x,x^0)} \sum_{i,j=1}^{3} a_{ij}^1 p_i^0 p_j^0\, d\tau_0. \tag{3.109}$$

It is a basic equation for investigating the linearized problem. Assume that the function $\tau_1(x, x^0)$ is known for various points $x^0 \in S$, $x \in S$, $S = \{x : x_3 = 0\}$ and introduce a limitation: when $a_{ij}^0 \equiv a_{ij}^0(x_3)$, $i, j = 1, 2, 3$. Besides, as has been shown earlier, all the coefficients a_{ij} cannot be uniquely found, and hence, assume that the coefficients $a_{3i}^1 = 0$, $i = 1, 2, 3$. Thus, only $a_{ij}^1(x)$ $i, j = 1, 2$, are to be determined. Also suppose that the functions $a_{ij}^1(x)$ are other than zero only in a certain finite domain contained in $D = \{x : 0 \leqslant x_3 \leqslant H\}$, and the coefficients a_{ij}^0 meet condition (3.82) in D. Also demonstrate that under these suppositions $a_{ij}^1(x)$ $(i, j = 1, 2)$ can be uniquely found by the function $\tau_1(x, x^0)$ which is known only for such rays $\Gamma_0(x, x^0)$, $x^0 \in S$, $x \in S$, that are entirely contained in D.

Consider for every fixed point $x^0 \in S$ a line $L_\alpha(x^0)$ lying on the plane S, passing through the point x^0 and corresponding to the fixed value

$$\alpha = \frac{1}{s}(p_1^0, p_2^0), \qquad s = \sqrt{((p_1^0)^2 + (p_2^0)^2)}$$

and the value of the parameter $s \in [s_1, s_0]$

$$s_1 = [\phi_\alpha^0(H)]^{-1/2}, \qquad s_0 = [\phi_\alpha^0(0)]^{-1/2}, \qquad \phi_\alpha^0(x_3) = \sum_{i,j=1}^{2} c_{ij}^0(x_3)\alpha_i\alpha_j.$$

The parametric equation of this line is as follows:

$$x_k = x_k^0 + \int_0^{\xi_3} \frac{2s\sum_{j=1}^{2} c_{kj}^0(z)\alpha_j}{(a_{33}(z))^{1/2}(1 - s^2\phi_\alpha^0(z))^{1/2}}\,dz, \quad k = 1, 2 \tag{3.110}$$

where $\xi_3 = \xi_3(s)$ is the root of the equation $1 - s^2\phi_\alpha^0(\xi_3) = 0$.

For the points $x \in L_\alpha(x^0)$ equation (3.109) can be written in the following form

$$\tau_1(x, x^0) = -\tfrac{1}{2}s^2 \int_{\Gamma_0(x,x^0)} \phi_\alpha^1(x')\,d\tau_0. \tag{3.111}$$

Now introduce the coordinates of the top $\xi = (\xi_1, \xi_2, \xi_3)$ of the curve $\Gamma_0(x, x^0)$. Earlier the formulas determining the coordinates ξ_1, ξ_2 have already been considered for the case when $x^0 = 0$. In a general case the formulas for $\xi_k (k = 1, 2)$ are as follows:

$$\xi_k = x_k^0 + \int_0^{\xi_3} \left[\frac{a_{k3}^0(z)}{a_{33}^0(z)} + \frac{s\sum_{j=1}^{2} c_{kj}^0(z)\alpha_j}{(a_{33}^0(z))^{1/2}(1 - s^2\phi_\alpha^0(z))^{1/2}}\right] dz, \quad k = 1, 2.$$

In a parametric form the equation for the ray $\Gamma_0(x, x^0)$ can be written using coordinate x_3

$$x_k = \xi_k - \int_{x_3}^{\xi_3} \left[\frac{a_{k3}^0(z)}{a_{33}^0(z)} \pm \frac{s\sum_{j=1}^{2} c_{kj}^0(z)\alpha_j}{(a_{33}^0(z))^{1/2}(1 - s^2\phi_\alpha^0(z))^{1/2}}\right] dz, \quad k = 1, 2. \tag{3.112}$$

In this formula the plus sign corresponds to the portion of the ray from the point x^0 to the point ξ, and the minus sign to the portion of the ray from the point ξ to the point x. Judging by (3.112), a ray $\Gamma_0(x, x^0)$ can be completely characterized by setting the parameters ξ_1, ξ_2, s, α. Indeed, one can easily find ξ_3 through s and α. By setting in (3.112) $x_3 = 0$ one finds the coordinates of a point $x^0 \in S$ or $x \in S$ depending on the chosen sign. Thus, there is a one-to-one correspondence between a set of points $(x, x^0) \in S \times S$ and a set of parameters $\xi_1, \xi_2, \alpha, s(|\alpha| = 1, s \in [s_1, s_0])$. Subjecting both parts of (3.111) to Fourier transform with respect to the variables ξ_1, ξ_2 and taking into account (3.112)

$$\tilde{\tau}_1(\lambda, \alpha, s) = \int_0^{\xi_3} \tilde{\Phi}_\alpha^1(\lambda, x_3)K(\lambda, \alpha, x_3, \xi_3)\,dx_3, \quad \xi_3 \in [0, H] \tag{3.113}$$

where

$$K(\lambda,\alpha,x_3,\xi_3) = -\frac{s^2}{(a_{33}^0(x_3))^{1/2}(1-s^2\phi_\alpha^0(x_3))^{1/2}}$$

$$\times \exp\left[-i\int_{x_3}^{\xi_3}\frac{\sum_{k=1}^{2}\lambda_k a_{k3}^0(z)}{a_{33}^0(z)}\,\mathrm{d}z\right]\cdot\cos\left[s\int_{x_3}^{\xi_2}\frac{\sum_{k,j=1}^{2}c_{kj}^0(z)\alpha_j\lambda_k}{(a_{33}^0(z))^{1/2}(1-s^2\phi_\alpha^0(z))^{1/2}}\,\mathrm{d}z\right]$$

and $\tilde{\tau}_1(\lambda,\alpha,s)$, $\tilde{\Phi}_\alpha^1(\lambda,x_3)$ are the images of the Fourier functions $\tau_1(x,x^0)$, $\phi_\alpha^1(x)$

$$\tilde{\tau}_1(\lambda,\alpha,s) = \iint_S \tau_1(x,x^0)\exp[i(\lambda_1\xi_1+\lambda_2\xi_2)]\,\mathrm{d}\xi_1\,\mathrm{d}\xi_2$$

$$\tilde{\Phi}_\alpha^1(\lambda,x_3) = \iint_S \phi_\alpha^1(x)\exp[i(\lambda_1x_1+\lambda_2x_2)]\,\mathrm{d}x_1\,\mathrm{d}x_2,\quad \lambda=(\lambda_1,\lambda_2).$$

When deducing (3.113) the fact that along the ray $\Gamma_0(x,x^0)$ the element of time $\mathrm{d}\tau_0$ is calculated by the following formula is taken into account

$$\mathrm{d}\tau_0 = \frac{\mathrm{d}x_3}{\sum_{j=1}^{3}a_{3j}^0(x_3)p_j^0} = \frac{|\mathrm{d}x_3|}{(a_{33}^0(x_3))^{1/2}(1-s^2\phi_\alpha^0(x_3))^{1/2}}.$$

Equation (3.113) is an integral Volterra equation of the first kind at every fixed α,λ. On the diagonal the kernel of this equation has a peculiarity like that of the Abelian equation. Indeed

$$\sqrt{(1-s^2\phi_\alpha^0(x_3))} = s\sqrt{(\phi_\alpha^0(\xi_3)-\phi_\alpha^0(x_3))} = s\sqrt{(\xi_3-x_3)}$$
$$\times\left[\int_0^1\frac{\mathrm{d}}{\mathrm{d}z}[\phi_\alpha^0(z)]|_{z=x_3+t(\xi_3-x_3)}\,\mathrm{d}t\right]^{1/2}$$

The above relation shows that the product $\sqrt{(\xi_3-x_3)}K(\lambda,\alpha,x_3,\xi_3)$ is a function continuous up to the diagonal $x_3=\xi_3$ and has a continuous in the same domain derivative with respect to the variable ξ_3. At the same time

$$\sqrt{(\xi_3-x_3)}K(\lambda,\alpha,x_3,\xi_3)\xrightarrow[x_3\to\xi_3]{}\frac{-s}{\sqrt{(a_{33}^0(\xi_3))}\sqrt{((\phi_\alpha^0(\xi_3))')}}\neq 0,\quad \xi_3\in[0,H].$$

These properties of the kernel ensure uniqueness of the solution to (3.113). Therefore, at every fixed α,λ one can uniquely find $\tilde{\Phi}_\alpha^1(\lambda,x_3)$ and then $\phi_\alpha^1(x)$, $x\in D$ and $a_{ij}^1(x)$. Thus, the theorem is proved.

Theorem 3.6. Let the coefficients a_{ij} allow in the domain $D=\{x:0\leqslant x_3\leqslant H\}$ the notation

$$a_{ij}(x) = a_{ij}^0(x_3)+a_{ij}^1(x),\quad i,j=1,2,3,$$

and let the functions $a_{ij}^0\in C^2[0,H]$ be known, forming a positively determined matrix and obeying condition (3.82), and the coefficients $a_{ij}^1(x)$ be small in the norm of the space $C^2(D)$. In this case if $a_{3j}^1=0, j=1,2,3$, then the finite in the domain D

functions $a^1_{ij}(x), i, j = 1, 2$, *are uniquely defined through the function* $\tau_1(x, x^0)$, $x \in S$, $x^0 \in S$, *which is known for all the rays* $\Gamma_0(x, x^0)$ *entirely contained in* D.

It would be of interest to study the linearized problem for a general case, when a^0_{ij} depend on all the variables x, but such investigations are still unavailable at present.

Note a result associated with a particular formulation of the problem on determining the structure of an anisotropic medium. Let there be two Riemann metrics of type (3.78) arising from two sets of the coefficients $a^1_{ij}(x)$, $a^2_{ij}(x)$ and let it be known that $a^2_{ij}(x) = \lambda(x) a^1_{ij}(x)$, $i, j = 1, 2, 3$. It has been demonstrated in [131, 148] that when the functions $\tau_1(x, x^0)$, $\tau_2(x, x^0)$ calculated by the coefficients a^1_{ij}, a^2_{ij}, respectively, coincide for the points $x \in S$, $x^0 \in S$ of the boundary of a compact domain D, then $\lambda = 1$ in D, i.e. $a^1_{ij} = a^2_{ij}$. An analogous result has been formulated [135, 136] for the case of a more general, the so-called Finsler, metric.

Chapter 4

Second-order equations of a hyperbolic type and related inverse problems

The subject of this chapter is the equation

$$u_{tt} - Lu = F(x, t) \tag{4.1}$$

where L is a uniformly elliptical operator with its coefficients depending only on the space variable $x = (x_1, x_2, x_3)$

$$Lu = \sum_{i,j=1}^{3} a_{ij}(x)u_{x_i x_j} + \sum_{i=1}^{3} b_i(x)u_{x_i} + c(x)u \tag{4.2}$$

$$\mu \sum_{i=1}^{3} \alpha_i^2 \leqslant \sum_{i,j=1}^{3} a_{ij}(x)\alpha_i\alpha_j \leqslant \frac{1}{\mu} \sum_{i=1}^{3} \alpha_i^2, \quad 0 < \mu < \infty.$$

In the inverse problems which will be considered here for (4.1) the task is to define either all or a part of the operator L coefficients, or the right-hand part of (4.1). In the latter case the right-hand part will have a peculiar structure

$$F(x, t) = f(x)\phi(x, t) \tag{4.3}$$

where $f(x)$ is to be determined, while $\phi(x, t)$ is known. As will be shown later, investigation of the problems of uniqueness and stability when determining the operator L coefficients is reduced to studying the problem of defining the right-hand part of type (4.3). In this case, under rather general suppositions on the functions $f(x)$, $\phi(x, t)$, the problem of determining the right-hand part of (4.1) is of principal significance.

When investigating inverse problems it is convenient to make use of the notion of a fundamental solution to (4.1) (see [77, 79, 155, 156]). Therefore, the chapter begins with considerations on the structure of the fundamental solution.

4.1. Fundamental solution and its differential properties

In its basic features the structure of the fundamental solution to (4.1) is well known [94, 95, 155, 157], it can be easily traced in the works of Hadamard [48] and Sobolev [25]. The fundamental solution consists of a singular part with the carrier on a characteristic conoid of (4.1) and a regular part with the carrier concentrated in the closure of the internal part of this conoid. The regular part of the fundamental solution is a smooth function inside the conoid, provided the operator L coefficients are sufficiently smooth. A qualitative estimate of the smoothness is given by the theorem formulated below [158].

Let D be a certain open domain of the space R^3, where the Riemann metric is considered

$$d\tau = \left(\sum_{i,j=1}^{3} b_{ij}(x)\, dx_i\, dx_j \right)^{1/2} \tag{4.4}$$

where b_{ij} are the elements of the matrix B which is inverse to the matric $A = (a_{ij}, i, j = 1, 2, 3)$ composed of the coefficients at higher derivatives of the operator L. Also assume that $a_{ij} \in C^{l+4}(D)$, $b_i \in C^{l+2}(D)$, $c \in C^l(D)$ at sufficiently great l and that any pair of points x, x^0 in the domain D can be connected with a single in R^3 geodesic line $\Gamma(x, x^0)$ of metric (4.4) entirely belonging to D. Let the distance between x and x^0 be denoted through $\tau(x, x^0)$.

As has been shown in Section 3.1, under such suppositions and at the point x^0 fixed, the point x can be given with the Riemann coordinates $\zeta = (\zeta_1, \zeta_2, \zeta_3)$, i.e. $x = f(\zeta, x)$. It should be remembered that the Riemann coordinates have the following sense: $\zeta = \tau(x, x^0)\alpha$, where $\alpha = (\alpha_1, \alpha_2, \alpha_3)$ is a vector directed at the point x^0 along the tangential to the geodesic line $\Gamma(x, x^0)$ towards the point x and such that

$$\sum_{i,j=1}^{3} b_{ij}(x^0)\alpha_i \alpha_j = 1.$$

With respect to the variable ζ the function $f(\zeta, x^0)$ has the inverse $\zeta = g(x, x^0)$, in which case $g(x, x^0) \in C^{l+3}(D \times D)$. A Jacobian of the transition from the Riemann to the Cartesian coordinates is $|dg/dx| > 0$, and at $x = x^0$ it equals 1. In this case the square distance between the points x, x^0 can be calculated using the function $g(x, x^0)$ by

$$\tau^2(x, x^0) = \sum_{i,j=1}^{3} b_{ij}(x^0) g_i(x, x^0) g_j(x, x^0) \tag{4.5}$$

and hence, $\tau^2(x, x^0) \in C^{l+3}(D \times D)$.

As is known, a fundamental solution to (4.1) is the function $H(x, t, x^0, t^0)$ which obeys, being a function of the variables x, t, the equation $u_{tt} - Lu = \delta(x - x^0, t - t^0)$ and the condition $u|_{t < t^0} \equiv 0$. This function is introduced because when the function $H(x, t, x^0, t^0)$ is known, the solution to the nonhomogeneous equation (4.1) with the zero Cauchy data is written as the following

$$u(x, t) = \int_{R^4} F(x^0, t^0) H(x, t, x^0, t^0)\, dx^0\, dt^0 \tag{4.6}$$

where the integration is carried out throughout the whole space of the variables x^0, t^0. Indeed, since the domain of influence of the point (x^0, t^0) is the conoid $t - t_0 \geqslant \tau(x, x^0)$ (see [155, p. 642]), then for $\tau(x, x_0) < t - t^0$ $H(x, t, x^0, t^0) = 0$, and hence, for a fixed point (x, t) the integration in (4.6) is realized only over the finite domain of the space.

Since the operator L coefficients are only x-dependent, then $H(x, t, x^0, t^0) = H(x, t - t^0, x^0, 0)$ and it can be also considered that $t^0 = 0$. In this case denote the fundamental solution through $H(x, t, x^0)$ and the Riemann ellipsoid $S(x, x^0, t) = \{\xi : \tau(\xi, x^0) + \tau(\xi, x) = t\}$ through $S(x, x^0, t)$. This ellipsoid is determined for

$t \geqslant \tau(x, x^0)$ and at $t \to \tau(x, x^0)$ it contracts to the geodesic line $\Gamma(x, x^0)$. Let G be a set of points (x, t, x^0), such that $x \in D$, $x^0 \in D$, $t > \tau(x, x^0)$ and $S(x, x^0, t) \subset D$; $G_0 = G \cup G'$, $G' = \{(x, t, x^0)\colon x \in D,\ x^0 \in D,\ 0 \leqslant t \leqslant \tau(x, x^0)\}$.

Theorem 4.1. If $l \geqslant 2s + 7$, $s \geqslant -1$, the fundamental solution to (4.1) is of the following structure

$$H(x, t, x^0) = \frac{1}{2\pi}\theta(t) \sum_{k=-1}^{s} \sigma_k(x, x^0)\theta_k(t^2 - \tau^2(x, x^0)) + v_s(x, t, x^0) \tag{4.7}$$

where $\theta(t)$ is the Heavyside function

$$\theta_{-1}(t) = \delta(t), \qquad \theta_s(t) = \frac{t^s}{s!}\theta(t), \quad s \geqslant 0,$$

the functions σ_k are calculated with the recurrence relations

$$\sigma_{-1}(x, x^0) = \left(\frac{1}{|A(x_0)|}\left|\frac{\partial}{\partial x} g(x, x^0)\right|\right)^{1/2}$$
$$\times \exp\left\{-\frac{1}{2}\int_0^1 \sum_{i,k,s=1}^{3} \left[b_i(\xi) - \sum_{j=1}^{3} \frac{\partial}{\partial \xi_j} a_{ij}(\xi)\right] b_{ks}(x^0) g_k(x, x^0)\right.$$
$$\left.\times \frac{\partial}{\partial \xi_i} g(\xi, x^0)|_{\xi = f(tg(x,x^0),x^0)}\,\mathrm{d}t\right\}, \quad \sigma_k(x, x^0) = \tfrac{1}{4}\sigma_{-1}(x, x^0)$$
$$\times \int_0^1 [\sigma_{-1}(\xi, x^0)]^{-1} t^k L_\xi \sigma_{k-1}(\xi, x^0)|_{\xi = f(tg(x,x^0),x^0)}\,\mathrm{d}t, \quad k \geqslant 0.$$

In this case $\sigma_k \in C^{l-2k}(D \times D)$, at $s = -1$ the function $v_s(x, t, x^0)$ is continuous in G and piecewise-continuous in G_0; at $s = 0$ $v_s(x, t, x^0)$ is continuous in G_0, and at $s \geqslant 1$ it is continuous together with the derivatives

$$D_x^\alpha D_t^\beta D_{x^0}^\lambda v_s, \qquad \alpha = (\alpha_1, \alpha_2, \alpha_3), \qquad \gamma = (\gamma_1, \gamma_2, \gamma_3)$$
$$|\alpha| + |\gamma| \leqslant l - 2s - 3, \qquad |\alpha| + \beta + |\gamma| \leqslant s - 1, \qquad |\alpha| = \alpha_1 + \alpha_2 + \alpha_3.$$

Besides, each of the functions $v_s (s \geqslant -1)$ can be written as

$$v_s(x, t, x^0) = \theta_{s+1}(t^2 - \tau^2(x, x^0)) v_s^0(x, t, x^0) \tag{4.8}$$

where $v_s^0(x, t, x^0)$ is a function limited in any closed domain contained in G_0.

Divide the proof of the theorem into two parts: firstly demonstrate that

$$u_s(x, t, x^0) \equiv \frac{1}{2\pi}\theta(t) \sum_{k=-1}^{s} \sigma_k(x, x^0)\theta_k(t^2 - \tau^2(x, x^0)) \tag{4.9}$$

satisfies the equation

$$\left(\frac{\partial^2}{\partial t^2} - L\right) u_s = -\frac{1}{2\pi}\theta(t)\theta_s(t^2 - \tau^2(x, x^0)) L\sigma_s + \delta(x - x^0, t) \tag{4.10}$$

and hence, is the solution to the Cauchy problem

$$\left(\frac{\partial^2}{\partial t^2} - L\right)v_s = \frac{1}{2\pi}\theta(t)\theta_s(t^2 - \tau^2(x, x^0))L\sigma_s \tag{4.11}$$

$$v_s|_{t<0} \equiv 0. \tag{4.12}$$

Then show that v_s possesses the necessary smoothness and can be represented as (4.8).

Introduce the following notation

$$\Gamma = t^2 - \tau^2(x, x^0).$$

By way of calculating $(\partial^2/\partial t^2 - L)u_s$, one obtains

$$\begin{aligned}\left(\frac{\partial^2}{\partial t^2} - L\right)u_s = {}& \frac{1}{2\pi}\sum_{k=-1}^{s}\left[\delta'(t)\theta_k(\Gamma) + 2\delta(t)\frac{\partial}{\partial t}\theta_k(\Gamma)\right]\sigma_k(x, x^0)\\ &+ \frac{1}{2\pi}\theta(t)\sum_{k=-1}^{s}\left\{\sigma_k\left[\theta_k''(\Gamma)\left(\Gamma_t^2 - \sum_{i,j=1}^{3} a_{ij}\Gamma_{x_i}\Gamma_{x_j}\right) + \theta_k'(\Gamma)\Gamma_{tt}\right]\right.\\ &- \theta_k(\Gamma)L\sigma_k - \theta_k'(\Gamma)\left[2\sum_{i,j=1}^{3} a_{ij}\Gamma_{x_j}\frac{\partial\sigma_k}{\partial x_i}\right.\\ &\left.\left.+ \sigma_k\sum_{i=1}^{3} b_i\Gamma_{x_i} + \sigma_k\sum_{i,j=1}^{3} a_{ij}\Gamma_{x_i x_j}\right]\right\}.\end{aligned}$$

Use the following easily verified equalities here

$$\Gamma_t^2 - \sum_{i,j=1}^{3} a_{ij}\Gamma_{x_i}\Gamma_{x_j} = 4\Gamma, \qquad \Gamma_{tt} = 2$$

$$\Gamma\theta_k''(\Gamma) = (k-1)\theta_{k-1}(\Gamma), \qquad \theta_k'(\Gamma) = \theta_{k-1}(\Gamma),$$

$$\delta'(t)\theta_k(\Gamma) + 2\delta(t)\frac{\partial}{\partial t}\theta_k(\Gamma) = 2\frac{\partial}{\partial t}[\delta(t)\theta_k(\Gamma)]$$

$$-\delta'(t)\theta_k(\Gamma) = -\delta'(t)\theta_k(\Gamma) = \frac{1}{t}\delta(t)\theta_k(\Gamma)$$

$$\frac{1}{t}\delta(t)\theta_k(\Gamma) = 2\pi|A(x^0)|^{1/2}\delta(t)\delta(x - x^0)\begin{cases}1, & k = -1,\\ 0, & k \geqslant 0.\end{cases}$$

Clarify the last of the above equalities. For any infinitely differentiated and finite function $\psi(x, t)$ one has

$$\int_{R^4}\frac{1}{t}\delta(t)\theta_k(\Gamma)\psi(x, t)\,\mathrm{d}x\,\mathrm{d}t = \lim_{t\to+0}\frac{1}{t}\int_{R^3}\theta_k(t^2 - \tau^2(x, x^0))\psi(x, t)\,\mathrm{d}x.$$

When calculating the obtained integral introduce the curvilinear coordinates τ, θ, ϕ of the point x using the equalities

$$x = f(\zeta, x^0), \qquad \zeta = \zeta' B^{-1/2}(x^0), \qquad \zeta' = \tau(\sin\theta\cos\phi, \sin\theta\sin\phi, \cos\theta).$$

In this case the curvilinear coordinate τ coincides with $\tau(x, x^0)$, and then

$$\lim_{t\to+0}\frac{1}{t}\int_{R^3}\theta_k(t^2-\tau^2(x,x^0))\psi(x,t)\,\mathrm{d}x = \lim_{t\to+0}\frac{1}{t}$$

$$\times\int_0^{2\pi}\int_0^{\pi}\int_0^{\infty}(t+\tau)^k\theta_k(t-\tau)\psi(x,t)\left|\frac{\partial f}{\partial\zeta}\right||B(x^0)|^{-1/2}\tau^2\sin\theta\,\mathrm{d}\tau\,\mathrm{d}\theta\,\mathrm{d}\phi$$

$$= 2\pi\psi(x^0,0)|A(x^0)|^{1/2}\begin{cases}1, & k=-1\\ 0, & k\geqslant 0.\end{cases}$$

This equality confirms the validity of the formula given above.

Making use of the above equalities

$$\left(\frac{\partial^2}{\partial t^2}-L\right)u_s = |A(x^0)|^{1/2}\sigma_{-1}(x^0,x^0)\delta(x-x^0,t)$$

$$+\frac{1}{2\pi}\theta(t)\sum_{k=-1}^{s}\left\{\theta_{k-1}(\Gamma)\left[-2\sum_{i,j=1}^{3}a_{ij}\Gamma_{x_j}\frac{\partial\sigma_k}{\mathrm{d}x_i}\right.\right.$$

$$\left.\left.-\sigma_k\left(\sum_{i,j=1}^{3}a_{ij}\Gamma_{x_i,x_j}+\sum_{i=1}^{3}b_i\Gamma_{x_i}\right)+(4k-2)\sigma_k\right]-\theta_k(\Gamma)L\sigma_k\right\}.$$

Now choose σ_k from the conditions of the coefficients at $\theta_k(\Gamma)$ becoming zero for $-2\leqslant k\leqslant s-1$ and the condition

$$\sigma_{-1}(x^0,x^0)|A(x^0)|^{1/2}=1 \tag{4.13}$$

which results in (4.10). The conditions of choosing σ_k result in the equations

$$2\sum_{i,j=1}^{3}a_{ij}\Gamma_{x_j}\frac{\partial\sigma_k}{\partial x_i}+\sigma_k(L'\Gamma-4k+2)=\begin{cases}0, & k=-1\\ -L\sigma_{k-1}, & k>-1.\end{cases} \tag{4.14}$$

Here

$$L'\Gamma=\sum_{i,j=1}^{2}a_{ij}\Gamma_{x_ix_j}+\sum_{i=1}^{3}b_i\Gamma_{x_i}. \tag{4.15}$$

Now it must be made sure that (4.13) and (4.14) result in the equations for σ_k written above.

Along the geodesic line $\Gamma(x,x^0)$ the following equality is valid [see (3.72) and (3.75)]

$$\frac{\mathrm{d}x_i}{\mathrm{d}\tau}=\sum_{j=1}^{3}a_{ij}\tau_{x_j},\quad i=1,2,3. \tag{4.16}$$

In this case the parameter τ coincides with the Riemann length of $\Gamma(x,x^0)$, i.e. with $\tau(x,x^0)$. Therefore, along $\Gamma(x,x^0)$ one has

$$\sum_{i,j=1}^{3}a_{ij}\Gamma_{x_j}\frac{\partial\sigma_k}{\partial x_i}=-2\tau\frac{\partial\sigma_k}{\mathrm{d}\tau}. \tag{4.17}$$

On the other hand the lemma resulting from Lemmas 1 and 2 given in [83, pp. 447–449] is valid.

Lemma 4.1. Along every geodesic line the following equality holds

$$\sum_{i,j=1}^{3}\frac{\partial}{\partial x_i}(a_{ij}\Gamma_{x_j}) = -6 + 2\tau\frac{\partial}{\partial \tau}\ln\left|\frac{\partial}{\partial x}g(x,x^0)\right|. \tag{4.18}$$

Now prove the lemma, denoting

$$\sum_{j=1}^{3} a_{ij}\Gamma_{x_j} = F_i(x,x^0), \quad i=1,2,3.$$

At $\zeta = \tau\alpha^0$ the function $x = f(\zeta, x^0)$ is an integral of system (4.16). Therefore, for the components f_i of the function f the identity

$$-2\frac{\partial f_i}{\partial \zeta}\zeta \equiv F_i(f(\zeta,x^0),x^0), \quad i=1,2,3.$$

is valid. Hence

$$\tau\frac{\partial}{\partial \tau}\left(\frac{\partial f_i}{\partial \zeta_k}\right) = \frac{\partial}{\partial \zeta}\left(\frac{\partial f_i}{\partial \zeta_k}\right)\zeta$$

$$= -\frac{1}{2}\sum_{j=1}^{3}\frac{\partial F_i}{\partial x_j}\frac{\partial f_j}{\partial \zeta_k} - \frac{\partial f_i}{\partial \zeta_k}, \quad i,k=1,2,3. \tag{4.19}$$

Calculating

$$\tau\frac{\partial}{\partial \tau}\left|\frac{\partial f}{\partial \zeta}\right|$$

by way of differentiating the determinant with respect to the columns and using (4.19) gives

$$\tau\frac{\partial}{\partial \tau}\left|\frac{\partial f}{\partial \zeta}\right| = \begin{vmatrix} \tau\frac{\partial}{\partial \tau}\left(\frac{\partial f_1}{\partial \zeta_1}\right) & \frac{\partial f_2}{\partial \zeta_1} & \frac{\partial f_3}{\partial \zeta_1} \\ \tau\frac{\partial}{\partial \tau}\left(\frac{\partial f_1}{\partial \zeta_2}\right) & \frac{\partial f_2}{\partial \zeta_2} & \frac{\partial f_3}{\partial \zeta_2} \\ \tau\frac{\partial}{\partial \tau}\left(\frac{\partial f_1}{\partial \zeta_3}\right) & \frac{\partial f_2}{\partial \zeta_3} & \frac{\partial f_3}{\partial \zeta_3} \end{vmatrix} + \begin{vmatrix} \frac{\partial f_1}{\partial \zeta_1} & \tau\frac{\partial}{\partial \tau}\left(\frac{\partial f_2}{\partial \zeta_1}\right) & \frac{\partial f_3}{\partial \zeta_1} \\ \frac{\partial f_1}{\partial \zeta_2} & \tau\frac{\partial}{\partial \tau}\left(\frac{\partial f_2}{\partial \zeta_2}\right) & \frac{\partial f_3}{\partial \zeta_2} \\ \frac{\partial f_1}{\partial \zeta_3} & \tau\frac{\partial}{\partial \tau}\left(\frac{\partial f_2}{\partial \zeta_3}\right) & \frac{\partial f_3}{\partial \zeta_3} \end{vmatrix}$$

$$+ \begin{vmatrix} \frac{\partial f_1}{\partial \zeta_1} & \frac{\partial f_2}{\partial \zeta_1} & \tau\frac{\partial}{\partial \tau}\left(\frac{\partial f_3}{\partial \zeta_1}\right) \\ \frac{\partial f_1}{\partial \zeta_2} & \frac{\partial f_2}{\partial \zeta_2} & \tau\frac{\partial}{\partial \tau}\left(\frac{\partial f_3}{\partial \zeta_2}\right) \\ \frac{\partial f_1}{\partial \zeta_3} & \frac{\partial f_2}{\partial \zeta_3} & \tau\frac{\partial}{\partial \tau}\left(\frac{\partial f_3}{\partial \zeta_3}\right) \end{vmatrix} = -\left|\frac{\partial f}{\partial \zeta}\right|\left(\frac{1}{2}\sum_{i=1}^{3}\frac{\partial F_i}{\partial x_i} + 3\right).$$

The resulting expression results in (4.18).

Write $L'\Gamma$ as

$$L'\Gamma = \sum_{i,j=1}^{3}\frac{\partial}{\partial x_i}(a_{ij}\Gamma_{x_j}) + \sum_{i=1}^{3}\left(b_i - \sum_{j=1}^{3}\frac{\partial a_{ij}}{\partial x_j}\right)\Gamma_{x_i}.$$

Taking into account (4.17) and (4.18), equalities (4.14) along $\Gamma(x, x^0)$ can be written as follows:

$$\frac{\partial \sigma_k}{\partial \tau} + \sigma_k \left[\frac{k+1}{\tau} - \frac{1}{2}\frac{\partial}{\partial \tau} \ln \left| \frac{\partial}{\partial x} g(x, x^0) \right| \right.$$

$$\left. - \frac{1}{4\tau} \sum_{i=1}^{3} \left(b_i - \sum_{j=1}^{3} \frac{\partial a_{ij}}{\partial x_j} \right) \Gamma_{x_i} \right] = \frac{1}{4\tau} L\sigma_{k-1} \begin{cases} 0, & k = -1 \\ 1, & k > -1 \end{cases}$$

or

$$\frac{\partial}{\partial \tau}(\tau^{k+1}\psi\sigma_k) = \tfrac{1}{4}\tau^k \psi L\sigma_{k-1} \begin{cases} 0, & k = -1 \\ 1, & k > -1. \end{cases} \tag{4.20}$$

In the last formula

$$\psi = \psi(x, x^0) = \left| \frac{\partial}{\partial x} g(x, x^0) \right|^{-1/2} \exp\left[\frac{1}{2} \int_{\Gamma(x,x^0)} \sum_{i=1}^{3} \left(b_i - \sum_{j=1}^{3} \frac{\partial a_{ij}}{\partial x_j} \right) \tau_{x_i} \, d\tau \right].$$

Choosing an integration constant at $k = -1$ from condition (4.13), and at $k > -1$ from the condition of σ_k limitedness, results in

$$\sigma_{-1}(x, x^0) = 1/\psi(x, x^0) |A(x^0)|^{1/2}$$

$$\sigma_k(x, x^0) = [(\tau(x, x^0))^{-(1+k)}/4\psi(x, x^0)]$$

$$\times \int_{\Gamma(x,x^0)} \tau^k \psi(\xi, x^0) L_\xi \sigma_{k-1}(\xi, x^0) \, d\tau, \quad k > -1.$$

The formulas obtained coincide with those already given above. To ensure this it is sufficient to replace the integration variable τ with $t = \tau/\tau(x, x^0)$ in the integrals for σ_k, ψ and to use (4.5) for calculating τ_{x_i}.

Now demonstrate that the function v_s possesses the properties outlined in the theorem. For this purpose introduce the following notation

$$\frac{1}{2\pi} \theta(t) \theta_s(t^2 - \tau^2(x, x^0)) L\sigma_s = f_s(x, t, x^0). \tag{4.21}$$

The function v_s is the solution of the non-homogeneous Cauchy problem

$$\left(\frac{\partial^2}{\partial s^2} - L \right) v_s = f_s(x, t, x^0) \tag{4.22}$$

$$v_s|_{t<0} \equiv 0. \tag{4.23}$$

From (4.21) $f_s(x, t, x^0) \equiv 0$, $\tau(x, x^0) < t$, $x \in D$, $x^0 \in D$. Therefore (see [155]) an analogous property is inherent to the solution to problem (4.22) and (4.23)

$$v_s(x, t, x^0) \equiv 0, \quad \tau(x, x^0) < t, \quad x \in D, \quad x^0 \in D. \tag{4.24}$$

The other differential properties of the function v_s result from the corresponding properties of the function f_s. Let $s \geqslant 1$. Since $\sigma_s \in C^{l-2s}(D \times D)$, $\tau^2 \in C^{l+3}(D \times D)$, then

$$f_s^{\alpha\beta\gamma} = D_x^\alpha D_t^\beta D_{x^0}^\gamma f_s \in C(G_0),$$
$$|\alpha| + |\gamma| \leqslant l - 2s - 2, \quad |\alpha| + \beta + |\gamma| \leqslant s - 1$$

and at $|\alpha| + \beta + |\gamma| = s$ the functions $f_s^{\alpha\beta\gamma}$ are piecewise-continuous in G_0.
Denote

$$f_s^{\beta\gamma} = D_t^\beta D_{x^0}^\gamma f_s, \qquad v_s^{\beta\gamma} = D_t^\beta D_{x^0}^\gamma v_s, \beta + |\gamma| \leqslant s, \qquad |\gamma| \leqslant l - 2s - 2.$$

As follows from (4.22) and (4.23), each of the functions $v_s^{\beta\gamma}$ is a solution to the problem

$$\left(\frac{\partial^2}{\partial t^2} - L\right) v_s^{\beta\gamma} = f_s^{\beta\gamma} \tag{4.25}$$

$$v_s^{\beta\gamma}|_{t<0} \equiv 0. \tag{4.26}$$

Demonstrate that the functions $v_s^{\beta\gamma}$, $\beta + |\gamma| \leqslant s - 1$, $|\gamma| \leqslant l - 2s - 3$ are continuous in G_0 together with the derivatives

$$D_x^\alpha v_s^{\beta\gamma}, \quad |\alpha| \leqslant \min(l - 2s - |\gamma| - 3, s - \beta - |\gamma| - 1).$$

For this purpose use the energetic inequalities and the theorem of embedding (see for instance [49, 155, 159, 160]). Let (ξ, τ, ξ^0) be an arbitrary fixed point of the set G; Q_δ is its δ-vicinity $Q_\delta = \{(x, t, x^0): |x - \xi| < \delta, |t - \tau| < \delta, |x^0 - \xi^0| < \delta\}$. Since G is an open set, we can consider that $Q_\delta \subset G$ provided δ is chosen in a proper way. Then, in the space of the variables x, t there exists such a point (x', t'), $x' \in D$, $t' > \tau + \delta$ that the frustrum of the conoid $K(x', t') = \{(x, t): 0 \leqslant t \leqslant t' - \tau(x, x'), x \in D\}$ contains inside the point (ξ, τ) together with its δ-vicinity, the point ξ^0 together with its δ-vicinity, in which case for any point x^0 from the δ-vicinity of

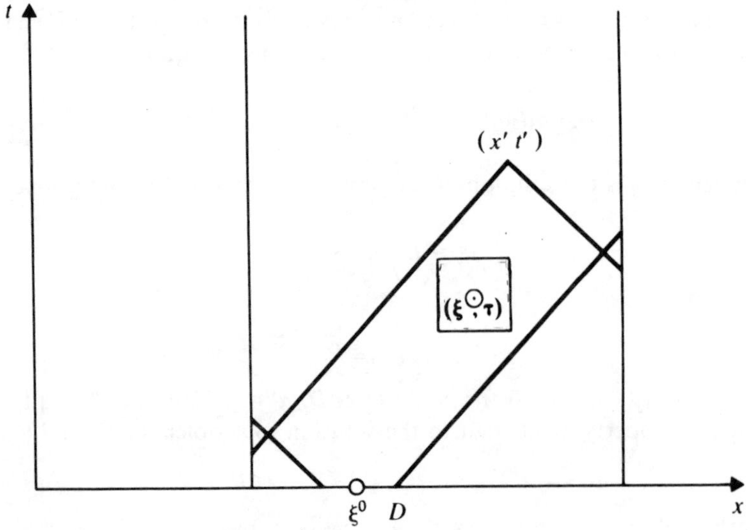

Figure 4.1.

the point ξ^0 the ellipsoid $S(x_0, x', t') \subset D$ (Fig. 4.1). Denote a cross-section of the domain $K(x', t')$ by the plane $t = \tau$, $0 \leqslant \tau \leqslant t'$ through $N(\tau)$. In this case the following energetic inequality is valid

$$\int_{N(t)} \left(\left| \frac{\partial}{\partial t} v_s^{\beta\gamma} \right|^2 + |\nabla_x v_s^{\beta\gamma}|^2 \right) \mathrm{d}x \leqslant C \int_0^t \mathrm{d}\tau \int_{N(\tau)} (f_s^{\beta\gamma}(x, \tau, x^0))^2 \, \mathrm{d}x,$$
$$0 \leqslant t \leqslant t', |x^0 - \xi^0| < \delta, \quad \beta + |\gamma| \leqslant s, |\gamma| \leqslant l - 2s - 2$$

where the constant C depends on the constant μ evaluating the uniform ellipticity of the operator L, and on the norm of the coefficients c, b_i, $\nabla_x a_{ij}$ included in the operator L.

By way of differentiating equalities (4.25) and (4.26) with respect to the variable x and using the inequalities analogous to that written above for the functions

$$v_s^{\alpha\beta\gamma} = D_x^\alpha D_t^\beta D_{x^0}^\gamma v_s$$

one can easily obtain a more general inequality

$$\sum_{|\alpha| \leqslant \rho} \int_{N(t)} \left(\left| \frac{\partial}{\partial t} v_s^{\alpha\beta\gamma} \right|^2 + |\nabla_x v_s^{\alpha\beta\gamma}|^2 \right) \mathrm{d}x$$
$$\leqslant C_1 \sum_{|\alpha| \leqslant \rho} \int_0^t \mathrm{d}\tau \int_{N(\tau)} (f_s^{\alpha\beta\gamma}(x, t, x^0))^2 \, \mathrm{d}x,$$
$$0 \leqslant t \leqslant t', |x^0 - \xi^0| < \delta$$
$$\rho = \min(l - 2s - |\gamma| - 2, s - \beta - |\gamma|), \beta + |\gamma| \leqslant s, |\gamma| \leqslant l - 2s. \tag{4.27}$$

In this case the constant C_1 depends on the norm of the coefficients a_{ij}, b_i, c in $C^\rho(D)$.

Inequality (4.27) demonstrates that at fixed x^0, t the function $v_s^{\beta\gamma} \in w_2^{\rho+1}$. It follows from the theorems of embedding [325] for a three-dimensional space that at $\rho \geqslant 1$

$$|v_s^{\alpha\beta\gamma}| \leqslant C_2 \left[\sum_{|\alpha| \leqslant \rho} \int_0^t \mathrm{d}\tau \int_{N(\tau)} (f_s^{\alpha\beta\gamma}(x, \tau, x^0))^2 \, \mathrm{d}x \right]^{1/2},$$
$$x \in N(t), \quad 0 \leqslant t \leqslant t', \quad |x^0 - \xi^0| < \delta, \quad |\alpha| \leqslant \rho - 1. \tag{4.28}$$

The condition $\rho \geqslant 1$ results in more rigid limitations for β, γ

$$\beta + |\gamma| \leqslant s - 1, \qquad |\gamma| \leqslant l - 2s - 3.$$

It is obvious from estimate (4.28) that for the written above values of α, β, γ the functions $v_s^{\alpha\beta\gamma}$ are continuous in the domain $Z_\delta(\xi, \tau, \xi^0) = \{(x, t, x^0)$: $(x, t) \in K(x', t')$, $|x^0 - \xi^0| \leqslant \delta\}$. Since the combination of all the sets $Z_\delta(\xi, \tau, \xi^0)$ coincides with G_0, they are continuous in the domain G_0.

Therefore for $s \geqslant 1$ the differential properties of the function v_s are proved. Continuity of the function v_0 in G_0 results from the equality

$$v_0(x, t, x^0) = \frac{1}{2\pi} \theta(t) \theta_1(t^2 - \tau^2(x, x^0)) \sigma_1(x, x^0) + v_1(x, t, x^0). \tag{4.29}$$

From an analogous equality for the function v_{-1}

$$v_{-1}(x,t,x^0)=\frac{1}{2\pi}\theta(t)\theta_0(t^2-\tau^2(x,x^0))\sigma_0(x,x^0)+v_0(x,t,x^0) \tag{4.30}$$

it follows that the function $v_{-1}(x,t,x^0)$ is continuous in G and piecewise-continuous in G_0.

To complete the proof of the theorem one must make sure that (4.8) is valid. Assuming again that $s\geqslant 1$, prove continuity of the derivatives $D_t^\beta v_s$ up to the values $\beta\leqslant s+1$. Note that the conditions of smoothness imposed on the coefficients allow two more steps in expanding the function $u(x,t,x')$ with respect to singularities. Therefore

$$v_s(x,t,x^0)=\frac{1}{2\pi}\theta(t)\sum_{k=1}^{2}\theta_{s+k}(t^2-\tau^2(x,x^0))\sigma_{s+k}(x,x^0)+v_{s+2}(x,t,x^0). \tag{4.31}$$

Since $D_t^{s+1}v_{s+2}$ is, in line with the above proved, a continuous function in G_0 and the same property is characteristic of the first two terms in (4.31), $D_t^{s+1}v_s$ is also continuous in G_0.

On the other hand, (4.24) has been proved earlier which, taken in conjunction with the Taylor formula, gives

$$\begin{aligned}v_s(x,t,x^0)=\theta_{s+1}(t-\tau(x,x^0))D_t^{s+1}v(x,\tau(x,x^0)\\+\theta(t-\tau(x,x^0)),x^0),\quad 0<\theta<1.\end{aligned}$$

This is qualitatively different from (4.8) through the multiplier $(t+\tau(x,x^0))^{s+1}$, which at $x\to x^0$ has the order t^{s+1}. Hence (4.8) will be proved if it can be established that at $t\to 0$, $x\to x^0$ the following inequality is valid for $D_t^{s+1}v_s$

$$|D_t^{s+1}v_s|\leqslant Ct^{s+1}. \tag{4.32}$$

This inequality is obvious to hold for the first two terms of (4.29). Now demonstrate that it holds for the third term too.

For this purpose make use of (4.28) at $\alpha=\gamma=0$, having substituted in it $s+2$ for s. In this case

$$|D_t^\beta v_{s+2}\leqslant C_2\left[\sum_{|\alpha|\leqslant 1}\int_0^t \mathrm{d}\tau\int_{N(\tau)}(f_{s+2}^{\alpha\beta 0}(x,\tau,x^0))^2\,\mathrm{d}x\right]^{1/2},\quad \beta=s+1.$$

But it is obvious that for small t

$$|D_x^\alpha D_t^{s+1}f_{s+2}(x,t,x^0)|\leqslant C_4 t^{s+1}\theta(t-\tau(x,x^0)),\quad |\alpha|\leqslant 1.$$

Therefore

$$D_t^{s+1}v_{s+2}|\leqslant C_5\left(\int_0^t \tau^{2(s+1)}\int_{\tau(x,x^0)=\tau}\mathrm{d}x\,\mathrm{d}\tau\right)^{1/2}\leqslant C_6 t^{s+3}.$$

In obtaining the last estimate, use is made of the fact that at $\tau\to 0$ the volume of a Riemann sphere with the radius τ is of the order τ^3.

Therefore, the third term in (4.29) allows an estimate that does not contradict (4.32), thus showing that (4.8) holds for $s \geqslant 1$. At $s = -1$ and 0 it in an obvious way results from (4.29) and (4.30).

Remark 1. The proof of the representation of the function v_s as (4.8) has demanded overstating the smoothness coefficients by four units. It could have been avoided provided use had been made of an integral equation for the function v_s. The latter can be derived by way of applying (4.6) to (4.22) and (4.23)

$$v_s(x, t, x^0) = \int_{R^4} f_s(\xi, t^0, x^0)[u_s(x, t - t^0, \xi) + v_s(x, t - t^0, \xi)]\, d\xi\, dt^0. \quad (4.33)$$

The domain of actual integration in (4.33) is the interior of the Riemann ellipsoid $S(x, x^0, t)$ and hence the investigation of (4.33) is associated with a detailed studying of the analytical structure of the set of surfaces of $S(x, x^0, t)$. This task is far from being easy, since in this case we should choose such a parametrization of the set of surfaces of $S(x, x^0, t)$ under which the analytical peculiarities of the representation of the set of surfaces at $t \to \tau(x, x^0)$, when the surfaces contract to the geodesic line $\Gamma(x, x^0)$ can be traced. Investigation of this kind is beyond the scope of this book, but it should be remembered that this is the way of obtaining the most accurate estimates of smoothness of the fundamental solution structure.

Remark 2. The function $\sigma_{-1}(x, x^0) \neq 0$. However, in the calculation of σ_k it might happen that for a $k = s - 1$ $L_\xi \sigma_k(\xi, x^0) = 0$, in which case the function $\sigma_s = 0$. Formulas (4.11), (4.12), and (4.7) show that in this case $v_s = 0$ and the function $H(x, t, x^0)$ coincides with u_s. This is, in particular, the case when $a_{ij} = \text{const.}$, $b_i = 0$, $c = 0$. It means that $\sigma_{-1} = \text{const.}$ and hence $\sigma_0 = 0$ and

$$H(x, t, x^0) = \frac{1}{2\pi}\theta(t)\delta(t^2 - \tau^2(x, x^0))\sigma_{-1}$$

$$\sigma_{-1} = |A|^{-1/2}$$

$$\tau^2(x, x^0) = \sum_{i,j=1}^{3} b_{ij}(x_i - x_i^0)(x_j - x_j^0).$$

4.2. Ray formulation of inverse problems for the coefficients at minor derivatives

The inverse kinematic problem considered in Chapter 3 can be treated as an inverse problem as regards the Cauchy problem

$$u_{tt} - Lu = \delta(x - x^0, t) \quad (4.34)$$

$$u|_{t<0} \equiv 0. \quad (4.35)$$

A solution to (4.34) and (4.35) describes the process of disturbance propagation from a point source concentrated at the point x^0. At every fixed moment of time t the disturbance induced by the source affects all the points of the space R^3 lying inside the Riemann sphere $\tau(x, x^0) = t$. The surface of the Riemann sphere $\tau(x, x^0) = t$ is a front of the wave from the source x^0. The function $\tau(x, x^0)$ and the coefficients a_{ij} at higher derivatives of the operator L are interrelated by the iconical equation

$$\sum_{i,j=1}^{3} a_{ij}(x)\tau_{x_i}\tau_{x_j} = 1. \tag{4.36}$$

Single out in the space R^3 a certain smooth surface S limiting the domain D. Let $x^0 \in S$, x being a fixed point of S where the solution to (4.34) and (4.35) is considered to be. At the moment of time $t < \tau(x, x^0)$ the solution equals zero. At the moment of time $t = \tau(x, x^0)$ (when the wave front reaches point x) the solution becomes other than zero and is, generally speaking, dependent on all the coefficients of the operator L. Thus, the time of the signal run $\tau(x, x^0)$ is a functional of the solution to problem (4.34) and (4.35). Therefore, in the case when the solution is known as a function of time t for all the points $x \in S$ at all possible $x^0 \in S$, this information contains the data on the times $\tau(x, x^0)$ which start for the solution of the inverse kinematic problem.

Thus, the inverse kinematic problem uses the information only on the carrier of the function $u(x, t, x^0)$.

It would be interesting, however, to use the solution itself at the points of the surface S as a function of time to determine the operator L coefficients. This leads one to an inverse dynamic problem for (4.34) and (4.35). Some ways of formulating this problem will be considered below while it will now be proved that even some elements of the dynamics make it possible to know a fair amount on the coefficients of the operator L.

A disturbance from the point x^0 gets to point x through the ray $\Gamma(x, x^0)$ connecting these two points [here I assume that the rays $\Gamma(x, x^0)$ are regular inside D and are the shortest in R^3 lines in the sense of metric (4.4)]. Obviously all the values of the operator L coefficients affect the solution $u(x, t, x^0)$. Therefore, to find the coefficients b_i, c one can use the information on the solution $u(x, t, x^0)$ in the vicinity of the moment of the wave arriving. As will be shown below, I shall only consider the functions $\sigma_{-1}(x, x^0)$, $\sigma_0(x, x^0)$ in expansion (4.7) of the solution $u(x, t, x^0)$ with respect to singularities. It would be natural to refer to such a formulation of the inverse problem in the same way as the ray one, since the functions $\sigma_{-1}(x, x^0)$, $\sigma_0(x, x^0)$ depend on the geometry of the rays. The function $\sigma_{-1}(x, x^0)$ is related to the ray amplitude $\sigma(x, x^0)$, which is the amplitude of the wave front through a simple relation $\sigma(x, x^0) = \sigma_{-1}(x, x^0)/2\tau(x, x^0)$. The functions σ_{-1}, σ_0 are the elements of the dynamic information on the solution $u(x, t, x^0)$.

Thus, let the following information on the solution of problem (4.34) and (4.35) be given

$$u(x, t, x^0) = f(x, t, x^0), \quad x \in S, x^0 \in S, -\infty < t \leqslant \tau(x, x^0) + \varepsilon, \varepsilon > 0. \tag{4.37}$$

As has been stated earlier, this information defines $\tau(x, x^0)$. Assume that the operator L coefficients are such that at $s = 0$ representation (4.7) holds. Then from

(4.7) one has

$$\sigma_{-1}(x, x^0) = \frac{1}{4\pi} \lim_{t \to \tau(x,x^0)+0} \int_{-\infty}^{t} \tau(x, x^0) f(x, \tau, x^0)\, \mathrm{d}\tau, \ x \in S, x^0 \in S \tag{4.38}$$

$$\sigma_0(x, x^0) = \lim_{t \to \tau(x,x^0)+0} \left[f(x, t, x^0) - \frac{1}{2\pi} \sigma_{-1}(x, x^0) \delta(t^2 - \tau^2(x, x^0)) \right], \quad x \in S, \quad x^0 \in S. \tag{4.39}$$

Note, that the expression written in the last equality in square brackets is a function continuous for the values $t \geqslant \tau(x, x^0)$; thus (4.39) is correct.

Due to the fact that the information on the solution considered falls into two parts: information on the times $\tau(x, x^0)$ and that on the functions $\sigma_{-1}(x, x^0)$, $\sigma_0(x, x^0)$, the problem of defining the operator L coefficients also falls into two parts. One of them is to determine the coefficients a_{ij} by $\tau(x, x^0)$—it was considered in Chapter 3. Here the problem of defining the coefficients b_i, c by the functions $\sigma_{-1}(x, x^0)$, $\sigma_0(x, x^0)$, the coefficients a_{ij} considered known in this case, will be studied.

Consider the difficulties arising in studying the problems on the definition of b_i, c by the given functions σ_{-1}, σ_0. To calculate σ_{-1}, σ_0 make use of the equations from Section 4.1

$$\sigma_{-1}(x, x^0) = |A(x^0)|^{-1/2} \left| \frac{\partial}{\partial x} g(x, x^0) \right|^{1/2} \times \exp\left[-\frac{1}{2} \int_{\Gamma(x,x^0)} \sum_{i=1}^{3} \left(b_i(\xi) - \sum_{j=1}^{3} \frac{\partial}{\partial \xi_j} a_{ij}(\xi) \right) \tau_{\xi_i}(\xi, x^0)\, \mathrm{d}\tau \right] \tag{4.40}$$

$$\sigma_0(x, x^0) = \frac{1}{4} \sigma_{-1}(x, x^0) \frac{1}{\tau(x, x^0)} \int_{\Gamma(x,x^0)} [\sigma_{-1}(\xi, x^0)]^{-1} L_\xi \sigma_{-1}(\xi, x^0)\, \mathrm{d}\tau. \tag{4.41}$$

As the coefficients a_{ij} are known, one can find from equality (4.40)

$$\sum_{i=1}^{3} \int_{\Gamma(x,x^0)} b_i(\xi) \tau_{\xi_i}(\xi, x^0)\, \mathrm{d}\tau = \phi(x, x^0), \quad x \in S, x^0 \in S \tag{4.42}$$

$$\phi(x, x^0) = 2 \ln\left[\sigma_{-1}(x, x^0) |A(x^0)|^{1/2} \left| \frac{\partial}{\partial x} g(x, x^0) \right|^{-1/2} \right] + \sum_{i,j=1}^{3} \int_{\Gamma(x,x^0)} \left[\frac{\partial}{\partial \xi_j} a_{ij}(\xi) \right] \tau_{\xi_i}(\xi, x^0)\, \mathrm{d}\tau.$$

Along $\Gamma(x, x^0)$ the components of the vector $\nabla_\xi \tau(\xi, x^0) = p(\xi, x^0)$ can be expressed through those of the vector of the tangential $\mathrm{d}\xi/\mathrm{d}\tau$. For this purpose use should be made of the differential equation of the ray (3.75)

$$\tau_{\xi_i}(\xi, x^0) = \sum_{j=1}^{3} b_{ij}(\xi) \frac{\mathrm{d}\xi_i}{\mathrm{d}\tau}, \quad i = 1, 2, 3. \tag{4.43}$$

Substituting the expression for τ_{ξ_i} from (4.43) into (4.42), the following inequality is obtained

$$\sum_{j=1}^{3} \int_{\Gamma(x,x^0)} \bar{b}_j(\xi)\,\mathrm{d}\xi_j = \phi(x,x^0), \quad x \in S, \quad x^0 \in S \tag{4.44}$$

where

$$\bar{b}_j(\xi) = \sum_{i=1}^{3} b_i(\xi) b_{ij}(\xi). \tag{4.45}$$

Thus, one comes to the problem of determining $\bar{b}_j(x)$ inside D by the known integrals (4.44) along the geodesic line $\Gamma(x,x^0)$ connecting an arbitrary pair of points x, x^0 of the boundary of the domain D. This problem has already been studied [42] (see also [43]) for the case when the coefficients $a_{ij}(x)$ of the operator L depend only on the coordinate x_3. In this case a set of $\Gamma(x,x_0)$ is invariant with respect to transformations of the parallel transfer along the plane $x_3 = 0$. In a general case of arbitrary $a_{ij}(x)$ the problem on uniqueness of the solution of (4.44) has been investigated [161] and the results are given. It appears that (4.44) allows one to find the components

$$\mathrm{rot}\,\bar{b}, \quad \bar{b} = (\bar{b}_1, \bar{b}_2, \bar{b}_3)$$

and for $x \in S$—a projection of the vector $\bar{b}(x)$ on the plane tangential to S. These are the only quantities that can be derived from (4.44). Indeed, if in equality (4.44) one tends x to x^0 along the surface S in such a way that the unit vector $(x - x^0)/|x - x^0|$ has a limiting position l, one gets

$$\bar{b}_l(x) = (\bar{b}(x), l) = \lim_{x \to x^0} \frac{\phi(x,x^0)}{|x - x^0|}. \tag{4.46}$$

Therefore, for $x \in S$ a projection of the vector $\bar{b}(x)$ on any vector lying in the plane tangential to S and constructed at the point x, is known, which allows one to calculate the integral

$$\sum_{j=1}^{3} \int_{L(x,x^0)} \bar{b}_j(\xi)\,\mathrm{d}\xi_j$$

along any smooth curve $L(x,x^0)$ lying on the surface S and having its ends at the points x, x^0. Now form a closed curve composed of $\Gamma(x,x^0)$ and $L(x,x^0)$ and 'stretch' an arbitrary smooth surface $\Sigma(x,x^0)$ with the boundary $\Gamma(x,x^0) \cup L(x,x^0)$ on it. Using the Stokes' formula one obtains

$$\sum_{i=1}^{3} \int_{\Gamma(x,x^0) \cup L(x,x^0)} \bar{b}_i(\xi)\,\mathrm{d}\xi_i = \iint_{\Sigma(x,x^0)} (\mathrm{rot}\,\bar{b}(\xi), n)\,\mathrm{d}S. \tag{4.47}$$

Here n is the direction of the normal to the surface $\Sigma(x,x^0)$. Due to equality (4.47), setting the function $\phi(x,x^0)$ for $x, x^0 \in S$ is equivalent to setting the projection $\bar{b}(x)$, $x \in S$ onto the tangential plane and the integrals from rot $\bar{b}(x)$ over all possible surfaces $\Sigma(x,x^0)$. Therefore, the only quantities that can be derived from (4.47) are rot $\bar{b}(x)$ and the projections $\bar{b}_l(x)$, $x \in S$.

This proves that for $x, x^0 \in D$ the function $\sigma_{-1}(x,x^0)$ is defined only by the rotor of the vector $\bar{b}(x)$, $x \in D$, and by the projections $\bar{b}_l(x)$, $x \in S$. Since it will be proved below that rot $\vec{b}(x)$, $x \in D$ is uniquely defined by the function $\phi(x, x^0)$, it indicates that for $x \in D$, $x^0 \in S$ the function $\sigma_{-1}(x, x^0)$ is completely determined by its values for $x \in S$, $x^0 \in S$.

Now go over to (4.41). It can be written as follows:

$$\int_{\Gamma(x,x^0)} \left[\sum_{i=1}^{3} b_i(\xi) \frac{\partial}{\partial \xi_i} \ln \sigma_{-1}(\xi, x^0) + c(\xi) \right] d\tau = \psi(x, x^0), \quad x \in S, \quad x^0 \in S$$

$$\psi(x, x^0) = 4\sigma_0(x, x^0)\tau(x, x^0)/\sigma_{-1}(x, x^0) \qquad (4.48)$$

$$- \int_{\Gamma(x,x^0)} \sigma_{-1}(\xi, x^0) \sum_{i,j=1}^{3} a_{ij}(\xi) \frac{\partial^2}{\partial \xi_i \partial \xi_j} \sigma_{-1}(\xi, x^0)\, d\tau.$$

Since the function $\sigma_{-1}(x, x^0)$, $x \in D$, $x^0 \in S$ can be considered known as a result of solving (4.44), then $\psi(x, x^0)$ in (4.48) is also a known function. As a result, there arises the problem of determining $c(x)$, $b_i(x)$ from (4.48) under the condition that rot $\vec{b}(x)$ is also a given function. To date this complex problem has not yet been studied. If it is assumed that $b_i(x)$ are known functions, there arises the problem of defining the coefficient $c(x)$ by the integrals

$$\int_{\Gamma(x,x_0)} c(\xi)\, d\tau = \psi_1(x, x^0), \quad x \in S, \quad x^0 \in S. \qquad (4.49)$$

This problem of integral geometry has been considered in Section 3.4 for a special case when $a_{ij} = 0$, $i \neq j$, $a_{ii} = 1/n^2(x)$. For the case of arbitrary $a_{ij}(x)$ the problem of solving (4.49) has been considered [129–131, 135, 136, 148].

In a methodical sense it is convenient to consider first the problem of determining $c(x)$ from equation (4.49) and then that of determining $\vec{b}(x)$ from (4.44). Since the investigation of the former is largely based on the method described in Section 3.4, the basic notations used in that method will be used again. Assume that the surface S is described by the equation $F(x) = 0$, where $F(x) \in C^3(D)$, $\nabla F|_S = 1$, $F(x) < 0$, $x \in D$. Let

$$w(x, \xi) = \int_{\Gamma(x,\xi)} c(y)\, d\tau, \quad x \in D, \xi \in D$$

$$w(x, x^0) = \psi_1(x, x^0), \quad x \in S, \quad x^0 \in S. \qquad (4.50)$$

Since along $\Gamma(x, \xi)$

$$\frac{\partial}{\partial \tau} w(x, \xi) = \nabla_x w(x, \xi) \frac{dx}{d\tau} = \sum_{i,j=1}^{3} a_{ij}(x) w_{x_i}(x, \xi) \tau_{x_j}(x, \xi)$$

then the function $w(x, \xi)$ satisfies the equation

$$\sum_{i,j=1}^{3} a_{ij}(x) \tau_{x_j}(x, \xi) w_{x_i}(x, \xi) = c(x), \quad x \in D. \qquad (4.51)$$

For the function $w(x, \xi)$ there is an analogue of equality (3.57).

Lemma 4.2. If $a_{ij}(x) \in C^3(D)$, $i, j = 1, 2, 3$, then for any function $w(x, \xi)$, which is continuous for $x \in D$, $\xi \in D$, $x \neq \xi$ together with the derivatives up to the second order, *at $x \neq \xi$ the following identity is valid*

$$2\sum_{k,s=1}^{3} T_{ks} w_{x_s}(x,\xi)\frac{\partial}{\partial \xi_k}\left[\sum_{i,j=1}^{3} a_{ij}(x)\tau_{x_j}(x,\xi)w_{x_i}(x,\xi)\right]$$

$$\equiv \sum_{k,l=1}^{3}\left[\frac{\partial}{\partial \xi_l}F(\xi)\right]\frac{\partial}{\partial \xi_k}\left[\sum_{i,j,s=1}^{3}(-1)^{s+k+l}p_{kl}q_{kls0}\right.$$

$$\left.\times\, a_{ij}(x)\tau_{x_j}(x,\xi)w_{x_s}(x,\xi)w_{x_i}(x,\xi)\right]$$

$$+\sum_{i,s,k=1}^{3}(-1)^{i+s+k}\frac{\partial}{\partial x_i}[p_{is}q_{0kis}w_{x_s}(x,\xi)w_{\xi_k}(x,\xi)]$$

$$+\left\{\sum_{i,j=1}^{3} a_{ij}(x)w_{x_i}(x,\xi)w_{x_j}(x,\xi)\right.$$

$$\left.+\left[\sum_{i,j=1}^{3} a_{ij}(x)\tau_{x_j}(x,\xi)w_{x_i}(x,\xi)\right]^2\right\}|T| \tag{4.52}$$

where

$$p_{kl}=\begin{cases}1, & k>l,\\ 0, & k=l,\\ -1, & k<l,\end{cases}$$

and $|T|$, T_{ks}, q_{kls0} have the same sense as in Section 3.4.

Expanding the minor corresponding to T_{ks} with respect to the elements of the first column the left-hand part of identity (4.52), having previously multiplied it by 1/2, is transformed to the form

$$\sum_{k,s=1}^{3} T_{ks}w_{x_s}\frac{\partial}{\partial \xi_k}\left[\sum_{i,j=1}^{3} a_{ij}\tau_{x_j}w_{x_i}\right]$$

$$=\sum_{k,s,l=1}^{3}(-1)^{k+s+l}p_{kl}q_{kls0}w_{x_s}F_{\xi_l}\frac{\partial}{\partial \xi_k}\left[\sum_{i,j=1}^{3} a_{ij}\tau_{x_j}w_{x_i}\right]$$

$$=\sum_{k,l=1}^{3}F_{\xi_l}\frac{\partial}{\partial \xi_k}\left[\sum_{i,j,s=1}^{3}(-1)^{s+k+l}p_{kl}q_{kls0}a_{ij}\tau_{x_j}w_{x_i}w_{x_s}\right]$$

$$-\left[\sum_{i,j=1}^{3} a_{ij}\tau_{x_j}w_{x_i}\right]\left[\sum_{k,s=1}^{3} T_{ks}w_{x_s\xi_k}\right.$$

$$\left.+\sum_{s,k,l=1}^{3}(-1)^{s+k+l}p_{kl}w_{x_s}F_{\xi_l}\frac{\partial}{\partial \xi_k}q_{kls0}\right]. \tag{4.53}$$

Now see whether the following equalities

$$\sum_{k,l=1}^{3}(-1)^{s+k+l}F_{\xi_l}p_{kl}\frac{\partial}{\partial \xi_k}q_{kls0}=-2T_{0s}$$

$$=-2|T|\sum_{k=1}^{3} a_{ks}\tau_{x_k},\quad s=1,2,3. \tag{4.54}$$

are valid. Simple calculations demonstrate that

$$\sum_{k=1}^{3}(-1)^k p_{kl}\frac{\partial}{\partial\xi_k}q_{kls0}=2q_{0l0s},\quad s,l=1,2,3.$$

For instance, at $s=1$, $l=2$, one has

$$\frac{\partial}{\partial\xi_1}\begin{vmatrix}\tau_{x_2} & \tau_{x_3}\\ \tau_{x_2\xi_3} & \tau_{x_3\xi_3}\end{vmatrix}-\frac{\partial}{\partial\xi_3}\begin{vmatrix}\tau_{x_2} & \tau_{x_3}\\ \tau_{x_2\xi_1} & \tau_{x_3\xi_1}\end{vmatrix}=2\begin{vmatrix}\tau_{x_2\xi_1} & \tau_{x_3\xi_1}\\ \tau_{x_2\xi_3} & \tau_{x_3\xi_3}\end{vmatrix}=2q_{0201}.$$

Therefore

$$\sum_{k,l=1}^{3}(-1)^{s+k+l}F_{\xi_l}p_{kl}\frac{\partial}{\partial\xi_k}q_{kls0}=2\sum_{l=1}^{3}(-1)^{s+l}F_{\xi_l}q_{0ls0}=-T_{0s}.$$

The second half of equality (4.54) is verified in the following way. Multiply the sth column of the determinant of the matrix T by $\sum_{k=1}^{3}a_{ks}\tau_{x_k}$ and add to the resulting column each jth column, $1\leqslant j\leqslant 3$, $j\neq s$, previously multiplied by $\sum_{k=1}^{3}a_{kj}\tau_{x_k}$. As a result, the elements of the sth column are transformed to the form

$$\sum_{k,j=1}^{3}a_{kj}\tau_{x_k}\tau_{x_j},\ \frac{1}{2}\frac{\partial}{\partial\xi_1}\sum_{k,j=1}^{3}a_{kj}\tau_{x_k}\tau_{x_j},$$

$$\frac{1}{2}\frac{\partial}{\partial\xi_2}\sum_{k,j=1}^{3}a_{kj}\tau_{x_k}\tau_{x_j},\ \frac{1}{2}\frac{\partial}{\partial\xi_3}\sum_{k,j=1}^{3}a_{kj}\tau_{x_k}\tau_{x_j}.$$

The fact that

$$\sum_{k,j=1}^{3}a_{kj}\tau_{x_k}\tau_{x_j}=1$$

means that the first element of the sth column equals 1, the rest of the elements equal zero. Therefore

$$|T|\sum_{k=1}^{3}a_{ks}\tau_{x_k}=T_{0s}.$$

Allowing for equality (4.54), formula (4.53) takes the form

$$\sum_{k,s=1}^{3}T_{ks}w_{x_s}\frac{\partial}{\partial\xi_k}\left[\sum_{i,j=1}^{3}a_{ij}\tau_{x_j}w_{x_i}\right]=\sum_{k,l=1}^{3}F_{\xi_l}\frac{\partial}{\partial\xi_k}$$
$$\times\left[\sum_{i,j,s=1}^{3}(-1)^{s+k+l}p_{kl}q_{kls0}a_{ij}\tau_{x_j}w_{x_i}w_{x_s}\right]-\left[\sum_{i,j=1}^{3}a_{ij}\tau_{x_j}w_{x_i}\right]$$
$$\times\left[\sum_{s,k=1}^{3}T_{ks}w_{x_s\xi_k}\right]+2\left[\sum_{i,j=1}^{3}a_{ij}\tau_{x_j}w_{x_i}\right]^2|T|. \tag{4.55}$$

On the other hand

$$\sum_{k,s=1}^{3}T_{ks}w_{x_s}\frac{\partial}{\partial\xi_k}\left[\sum_{i,j=1}^{3}a_{ij}\tau_{x_j}w_{x_i}\right]=\sum_{k,s=1}^{3}T_{ks}w_{x_s}$$
$$\sum_{i,j=1}^{3}a_{ij}\tau_{x_j}w_{x_i\xi_k}+\sum_{k,s=1}^{3}T_{ks}w_{x_s}\sum_{i,j=1}^{3}a_{ij}w_{x_i}\tau_{x_j\xi_k}. \tag{4.56}$$

The second term in the above equality can be transformed using the equalities

$$\sum_{k=1}^{3} T_{ks}\tau_{x_j\xi_k} = \left(\delta_{js} - \tau_{x_j}\sum_{k=1}^{3} a_{ks}\tau_{x_k}\right)|T|, \quad s, j = 1, 2, 3, \tag{4.57}$$

where δ_{js} is the Kronecker symbol. Equalities (4.57) result from the obvious equalities

$$\sum_{k=1}^{3} T_{ks}\tau_{x_j\xi_k} + \tau_{x_j}T_{0s} = |T|\delta_{js}, \quad j, s = 1, 2, 3.$$

The left-hand part of the last formula contains a sum of the products of the elements of the jth column of the matrix T and the cofactors of the elements of the sth column of the same matrix. In line with the properties of determinants, at $j \neq s$ this expression equals zero, and at $j = s$ it coincides with the determinant of the matrix T.

Adding (4.56) and (4.55) and allowing for (4.57) one obtains the identity

$$2\sum_{k,s=1}^{3} T_{ks}w_{x_s}\frac{\partial}{\partial\xi_k}\left[\sum_{i,j=1}^{3} a_{ij}\tau_{x_j}w_{x_i}\right] \equiv \sum_{k,l=1}^{3} F_{\xi_l}\frac{\partial}{\partial\xi_k}$$

$$\times\left[\sum_{i,j,s=1}^{3}(-1)^{s+k+l}p_{kl}q_{kls0}a_{ij}\tau_{x_j}w_{x_i}w_{x_s}\right]$$

$$+\sum_{i,j=1}^{3} a_{ij}\tau_{x_j}\sum_{k,s=1}^{3} T_{ks}(w_{x_s}w_{x_i\xi_k} - w_{x_i}w_{x_s\xi_k})$$

$$+\left\{\sum_{i,j=1}^{3} a_{ij}w_{x_i}w_{x_j} + \left[\sum_{i,j=1}^{3} a_{ij}\tau_{x_j}w_{x_i}\right]^2\right\}|T|$$

that differs from the required identity (4.52) in only one term. Now make sure that

$$\sum_{i,j=1}^{3} a_{ij}\tau_{x_j}\sum_{k,s=1}^{3} T_{ks}(w_{x_s}w_{x_i\xi_k} - w_{x_i}w_{x_s\xi_k})$$

$$\equiv \sum_{i,k,s=1}^{3}(-1)^{i+s+k}\frac{\partial}{\partial x_i}(w_{x_s}w_{\xi_k}q_{0kis}p_{is}). \tag{4.58}$$

The left-hand part of this equality can be transformed by changing the order of summation

$$\sum_{i,j=1}^{3} a_{ij}\tau_{x_j}\sum_{k,s=1}^{3} T_{ks}(w_{x_s}w_{x_i\xi_k} - w_{x_i}w_{x_s\xi_k})$$

$$= \sum_{i,k,s=1}^{3} w_{x_s}w_{x_i\xi_k}\left[\sum_{j=1}^{3}(a_{ij}\tau_{x_j}T_{ks} - a_{sj}\tau_{x_j}T_{ki})\right]. \tag{4.59}$$

The expression contained in square brackets can be calculated by the procedure used when deducing (3.60), using the fact that

$$\sum_{i,j=1}^{3} a_{ij}\tau_{x_i}\tau_{x_j} = 1.$$

Using the formula

$$\left(\sum_{j=1}^{3} a_{ij}\tau_{x_j}\right)T_{ks}-\left(\sum_{j=1}^{3} a_{sj}\tau_{x_j}\right)T_{ki}=(-1)^{i+j+s}p_{is}q_{0kis}.$$

which is an analogue to formula (3.60) one can transform equality (4.59) into

$$\sum_{i,j=1}^{3} a_{ij}\tau_{x_j}\sum_{k,s=1}^{3} T_{ks}(w_{x_s}w_{x_i\xi_k}-w_{x_i}w_{x_s\xi_k})$$

$$=\sum_{i,s,k=1}^{3}(-1)^{i+s+k}\frac{\partial}{\partial x_i}(w_{x_s}w_{\xi_k}p_{is}q_{0kis})$$

$$-\sum_{s,k=1}^{3}(-1)^{s+k}w_{x_s}w_{x_k}\left[\sum_{i=1}^{3}(-1)^i\frac{\partial}{\partial x_i}(p_{is}q_{0kis})\right].$$

By way of computing one can easily verify that

$$\sum_{i=1}^{3}(-1)^i\frac{\partial}{\partial x_i}(p_{is}q_{0kis})=0,\quad k,s=1,2,3.$$

This results in identity (4.58) and thus the lemma is proved.

For the solutions of equation (4.51) the left-hand part of identity (4.52) becomes zero. The first term in the right-hand part can be written as $\nabla_\xi F\cdot\mathrm{rot}_\xi q$, where the vector q is determined by the equality

$$q=\left[\sum_{i,j=1}^{3} a_{ij}\tau_{x_j}w_{x_i}\right]\sum_{s=1}^{3}(-1)^{s+1}w_{x_s}(q_{23s0},q_{31s0},q_{12s0}).$$

Therefore (4.52) yields the following equality

$$\nabla F_\xi\cdot\mathrm{rot}_\xi q+\sum_{i,s,k=1}^{3}(-1)^{s+k+i}\frac{\partial}{\partial x_i}(w_{x_s}w_{\xi_k}q_{0kis}p_{is})$$

$$+\left\{\sum_{i,j=1}^{3} a_{ij}w_{x_i}w_{x_j}+\left[\sum_{i,j=1}^{3} a_{ij}\tau_{x_j}w_{x_i}\right]^2\right\}|T|\equiv 0,\quad x\neq\xi.$$

Multiplying both parts of this identity by $\delta(F(\xi))\,\mathrm{d}\xi\,\mathrm{d}x$, integrating over the domain resulting from the direct product of the domains $D\times D$ by way of throwing away a set of points x,ξ, obeying the inequality $|x-\xi|\leqslant\varepsilon$, and going over to the limit at $\varepsilon\to 0$, one obtains

$$\int_D \mathrm{d}x\int_S\left\{\sum_{i,j=1}^{3} a_{ij}w_{x_i}w_{x_j}+\left[\sum_{i,j=1}^{3} a_{ij}\tau_{x_j}w_{x_i}\right]^2\right\}|T|\,\mathrm{d}S_\xi$$

$$-\int_S \mathrm{d}S_\xi\int_S\Phi(w,\tau)\,\mathrm{d}S_x=0. \tag{4.60}$$

Here $\Phi(w,\tau)$ denotes determinant (3.55). In deducing this formula I have omitted the transformations totally identical to those made in Section 3.4. For the sake of total analogy let me note that in the inner integral standing in the first term of (4.60) one can go over from integrating over the surface S to that over a certain unit sphere. For this purpose at a fixed point x introduce a unit vector v of a tangential to $\Gamma(x, \xi)$ constructed at the point ξ. Let θ, ϕ be spheric coordinates of the vector v. Characterize a point ξ in the vicinity of S by the coordinates θ, ϕ, $\psi = F(\xi)$.

Since from the iconical equation one has $\nabla_x \tau(x, \xi) = vA^{-1/2}(x)$, then, according to the transformations in Section 3.4

$$|T|\,d\xi = |A(x)|^{-1/2}\,d\omega_v\,d\psi, \quad d\omega_v = \sin\theta\,d\theta\,d\phi.$$

Hence

$$|T|\,dS_\xi = |A(x)|^{-1/2}\,d\omega_v. \tag{4.61}$$

Equality (4.60) is valid if $a_{ij} \in C^2(D)$, $c(x) \in C^1(D)$. From this equality one can easily derive an estimate of stability of the solution of (4.49). Note, that for any two vectors $\alpha = (\alpha_1, \alpha_2, \alpha_3)$, $\beta = (\beta_1, \beta_2, \beta_3)$ the following inequality holds

$$\left(\sum_{i,j=1}^{3} a_{ij}\alpha_i\beta_j\right)^2 \leqslant \left(\sum_{i,j=1}^{3} a_{ij}\alpha_i\alpha_j\right)\left(\sum_{i,j=1}^{3} a_{ij}\beta_i\beta_j\right).$$

Therefore

$$c^2(x) = \left(\sum_{i,j=1}^{3} a_{ij}\tau_{x_j}w_{x_i}\right)^2 \leqslant \left(\sum_{i,j=1}^{3} a_{ij}w_{x_i}w_{x_j}\right)\left(\sum_{i,j=1}^{3} a_{ij}\tau_{x_j}\tau_{x_i}\right) = \sum_{i,j=1}^{3} a_{ij}w_{x_i}w_{x_j}.$$

Formulas (4.60) and (4.61) result in the estimate

$$\int_D c^2(x)|A(x)|^{-1/2}\,dx \leqslant \frac{1}{8\pi}\int_S dS_\xi \int_S \Phi(\psi_1, \tau)\,dS_x. \tag{4.62}$$

The same constant can assume another form if the constant μ of uniform positiveness of the matrix A is used

$$\int_D c^2(x)\,dx \leqslant \frac{\mu^{3/2}}{8\pi}\int_S dS_\xi \int_S \Phi(\psi_1, \tau)\,dS_x. \tag{4.63}$$

From estimate (4.63) it follows that the function $\psi_1 = 0$ is corresponded to by $c(x) = 0$, $x \in D$, i.e. that the solution of problem (4.49) is unique.

Now formulate the final result of the investigation carried out here.

Theorem 4.2. If the coefficients $a_{ij} \in C^2(D)$ and give rise to a Riemann metric with a regular ray behaviour in the domain D, then the solution of problem (4.49) is unique in the class of functions $c(x) \in C^1(D)$ with its stability defined by (4.62) and (4.63).

Now consider the problem associated with the solution of (4.44). The problem of uniqueness of the solutions of (4.44) has already been discussed, now, formulate an exact statement.

Theorem 4.3. Under the suppositions of Theorem 4.2 on the coefficients a_{ij}, if $b_i(x) \in C^1(D)$, then from equation (4.44) one can uniquely find rot $\bar{b}(x)$, $x \in D$ and a projection of the vector $\bar{b}$ on the tangential to S plane at every point $x \in S$.

This theorem can be considered proved if it is stated that the function $\phi(x, x^0) = 0$, $x \in S$, $x^0 \in S$, is corresponded to by rot $\bar{b} = 0$, $x \in D$. Now demonstrate that this is the case.

Consider the function

$$w(x,\xi)=\sum_{i=1}^{3}\int_{\Gamma(x,\xi)}\tilde{b}_i(x)\,\mathrm{d}x_i,\quad x\in D,\quad \xi\in D. \tag{4.64}$$

At $x\in S$, $x^0\in S$

$$w(x,x^0)=\phi(x,x^0)=0. \tag{4.65}$$

The function $w(x,\xi)$ obeys the equation

$$\sum_{i,j=1}^{3}a_{ij}(x)\tau_{x_j}(x,\xi)(w_{x_i}(x,\xi)-\tilde{b}_i(x))=0. \tag{4.66}$$

Indeed, along $\Gamma(x,\xi)$

$$\sum_{j=1}^{3}a_{ij}(x)\tau_{x_j}=\frac{\mathrm{d}x_i}{\mathrm{d}\tau}.$$

Therefore, along $\Gamma(x,\xi)$ (4.46) is equivalent to the equation

$$\mathrm{d}w=\sum_{i=1}^{3}w_{x_i}\,\mathrm{d}x_i=\sum_{i=1}^{3}\tilde{b}_i\,\mathrm{d}x_i$$

from which by way of integration formula (4.64) is derived.

Further proof of the theorem is based on the following lemma.

Lemma 4.3. In fulfilling the conditions of Lemma 4.3 for the function $w(x,\xi)$ the following identity is valid

$$\begin{aligned}
&2\sum_{k,s=1}^{3}T_{ks}(w_{x_s}-\tilde{b}_s)\frac{\partial}{\partial\xi_k}\left[\sum_{i,j=1}^{3}a_{ij}\tau_{x_j}(w_{x_i}-\tilde{b}_i)\right]\\
&\quad\equiv\sum_{k,l=1}^{3}F_{\xi_l}\frac{\partial}{\partial\xi_k}\left[\sum_{i,j,s=1}^{3}(-1)^{s+k+l}p_{kl}q_{kls0}a_{ij}\tau_{x_j}(w_{x_s}w_{x_i}-2w_{x_i}\tilde{b}_s\right.\\
&\quad\left.+\tilde{b}_s\tilde{b}_i)\right]+\sum_{i,k,s=1}^{3}\frac{\partial}{\partial x_i}[(-1)^{i+k+s}p_{is}q_{0kis}w_{x_s}w_{\xi_k}]\\
&\quad+\left\{\sum_{i,j=1}^{3}a_{ij}(w_{x_i}-\tilde{b}_i)(w_{x_j}-\tilde{b}_j)+\left[\sum_{i,j=1}^{3}a_{ij}\tau_{x_j}(w_{x_i}-\tilde{b}_i)\right]^2\right\}|T|.
\end{aligned} \tag{4.67}$$

This identity may be proved by repeating the stages of proving identity (4.52) with certain differences arising only when establishing the equalities analogous to equalities (4.53) and (4.56). Write these equalities

$$\begin{aligned}
&\sum_{k,s=1}^{3}T_{ks}(w_{x_s}-\tilde{b}_s)\frac{\partial}{\partial\xi_k}\left[\sum_{i,j=1}^{3}a_{ij}\tau_{x_j}(w_{x_i}-\tilde{b}_i)\right]\\
&\quad=\sum_{k,s,l=1}^{3}(-1)^{k+s+l}F_{\xi_l}p_{kl}q_{kls0}(w_{x_s}-\tilde{b}_s)\frac{\partial}{\partial\xi_k}\left[\sum_{i,j=1}^{3}a_{ij}\tau_{x_j}(w_{x_i}-\tilde{b}_i)\right]
\end{aligned}$$

$$= \sum_{k,l=1}^{3} F_{\xi_l} \frac{\partial}{\partial \xi_k}\left[\sum_{i,j,s=1}^{3} (-1)^{s+k+l} p_{kl} q_{klso} a_{ij} \tau_{x_j}(w_{x_i} - \tilde{b}_i)(w_{x_s} - \tilde{b}_s)\right]$$

$$- \sum_{i,j=1}^{3} a_{ij}\tau_{x_j}(w_{x_i} - \tilde{b}_i)\left[\sum_{k,s,l=1}^{3} (-1)^{s+k+l} F_{\xi_l} p_{kl} q_{klso} w_{x_s \xi_k}\right.$$

$$\left. + \sum_{k,s,l=1}^{3} (-1)^{s+k+l} F_{\xi_l} p_{kl}(w_{x_s} - \tilde{b}_s)\frac{\partial}{\partial \xi_k} q_{klso}\right]$$

$$= \sum_{k,l=1}^{3} F_{\xi_l} \frac{\partial}{\partial \xi_k}\left\{\sum_{i,j,s=1}^{3} (-1)^{s+k+l} p_{kl} q_{klso} a_{ij} \tau_{x_j}[(w_{x_i} - \tilde{b}_i)\right.$$

$$\left. \times (w_{x_s} - \tilde{b}_s) + \tilde{b}_i w_{x_s}]\right\} - \left[\sum_{i,j=1}^{3} a_{ij}\tau_{x_j} w_{x_i}\right]\left[\sum_{k,s=1}^{3} T_{ks} w_{x_s \xi_k}\right]$$

$$-\left[\sum_{i,j=1}^{3} a_{ij}\tau_{x_j}(w_{x_i} - \tilde{b}_i)\right]\left[\sum_{s,k,l=1}^{3} (-1)^{s+k+l} F_{\xi_l} p_{kl}(w_{x_s} - \tilde{b}_s)\frac{\partial}{\partial \xi_k} q_{klso}\right]$$

$$- \sum_{i,j,l,s=1}^{3} (-1)^{s+k+l} F_{\xi_l} p_{kl} w_{x_s} \tilde{b}_i a_{ij} \frac{\partial}{\partial \xi_k}(\tau_{x_j} q_{klso}),$$

$$\sum_{k,s=1}^{3} T_{ks}(w_{x_s} - \tilde{b}_s)\frac{\partial}{\partial \xi_k}\left[\sum_{i,j=1}^{3} a_{ij}\tau_{x_j}(w_{x_i} - \tilde{b}_i)\right]$$

$$= \sum_{k,s=1}^{3} T_{ks}(w_{x_s} - \tilde{b}_s) \sum_{i,j=1}^{3} a_{ij}\tau_{x_j} w_{x_i \xi_k}$$

$$+ \sum_{k,s=1}^{3} T_{ks}(w_{x_s} - \tilde{b}_s) \sum_{i,j=1}^{3} a_{ij}\tau_{x_j \xi_k}(w_{x_i} - \tilde{b}_i)$$

$$= \sum_{k,s=1}^{s} T_{ks} w_{x_s} \sum_{i,j=1}^{3} a_{ij}\tau_{x_j} w_{x_i \xi_k} + \sum_{k,s=1}^{3} T_{ks}(w_{x_s} - \tilde{b}_s)$$

$$\times \sum_{i,j=1}^{3} a_{ij}\tau_{x_j \xi_k}(w_{x_i} - \tilde{b}_i) - \sum_{k,l=1}^{3} F_{\xi_l}\frac{\partial}{\partial \xi_k}$$

$$\times \left[\sum_{i,j,s=1}^{3} (-1)^{k+s+l} p_{kl} q_{klso} a_{ij}\tau_{x_j} w_{x_i} \tilde{b}_s\right]$$

$$+ \sum_{i,s,l=1}^{3} F_{\xi_l} w_{x_i} \tilde{b}_s \sum_{k,j=1}^{3} (-1)^{s+k+l} p_{kl} a_{ij} \frac{\partial}{\partial \xi_k}(\tau_{x_j} q_{klso}).$$

Add the above equalities and use the formulas obtained earlier. As a result, a new equality is obtained

$$2\sum_{k,s=1}^{3} T_{ks}(w_{x_s} - \tilde{b}_s)\frac{\partial}{\partial \xi_k}\left[\sum_{i,j=1}^{3} a_{ij}\tau_{x_j}(w_{x_i} - \tilde{b}_i)\right] = \sum_{k,l=1}^{3} F_{\xi_l}\frac{\partial}{\partial \xi_k}$$

$$\times \left[\sum_{i,j,s=1}^{s} (-1)^{s+k+l} p_{kl} q_{klso} a_{ij}\tau_{x_j}(w_{x_s} w_{x_i} - 2w_{x_i}\tilde{b}_s + \tilde{b}_s \tilde{b}_i)\right]$$

$$+\sum_{i,s,k=1}^{3}\frac{\partial}{\partial x_i}[(-1)^{i+s+k}w_{x_s}w_{\xi_k}q_{0kis}p_{is}]$$

$$+\left\{\sum_{i,j=1}^{3}a_{ij}(w_{x_i}-\bar{b}_i)(w_{x_j}-\bar{b}_j)+\left[\sum_{i,j=1}^{3}a_{ij}\tau_{x_j}(w_{x_i}-\bar{b}_i)\right]^2\right\}|T|$$

$$+\sum_{i,l,s=1}^{3}F_{\xi_l}(w_{x_i}\bar{b}_s-w_{x_s}\bar{b}_i)\sum_{k,j=1}^{3}(-1)^{s+k+l}p_{kl}a_{ij}\frac{\partial}{\partial\xi_k}(\tau_{x_j}q_{kls0})$$

which differs from equality (4.67) only in the last term. Now prove that this term equals zero; write it as follows:

$$\sum_{i,l,s=1}^{3}F_{\xi_l}(w_{x_i}\bar{b}_s-w_{x_s}\bar{b}_i)\sum_{k,j=1}^{3}(-1)^{s+k+l}p_{kl}a_{ij}\frac{\partial}{\partial\xi_k}(\tau_{x_j}q_{kls0})$$

$$=\sum_{k,i,l,s=1}^{3}F_{\xi_l}w_{x_i}\bar{b}_s p_{kl}(-1)^{k+l}\frac{\partial}{\partial\xi_k}\left[(-1)^s\sum_{j=1}^{3}a_{ij}\tau_{x_j}q_{kls0}\right.$$

$$\left.-(-1)^i\sum_{j=1}^{3}a_{sj}\tau_{x_j}q_{kli0}\right]. \tag{4.68}$$

One can easily demonstrate the validity of the following equalities

$$(-1)\sum_{j=1}^{3}a_{ij}\tau_{x_j}q_{kls0}-(-1)^i\sum_{j=1}^{3}a_{ij}\tau_{x_j}q_{kli0}=(-1)^{s+i}p_{is}\tau_{x_n\xi_m},$$

$$i,s,k,l=1,2,3,\quad k\neq l,\quad n\neq i,s,\quad m\neq k,l. \tag{4.69}$$

The principle of checking the validity can be demonstrated for a special case. For instance, at $s=1, i=3, k=1, l=3$, one has

$$-\sum_{j=1}^{3}a_{3j}\tau_{x_j}q_{1310}+\sum_{j=1}^{3}a_{1j}\tau_{x_j}q_{1330}=-\begin{vmatrix}\tau_{x_2} & \tau_{x_3}\\ \tau_{x_2\xi_2} & \tau_{x_3\xi_2}\end{vmatrix}\sum_{j=1}^{3}a_{3j}\tau_{x_j}$$

$$+\begin{vmatrix}\tau_{x_1} & \tau_{x_2}\\ \tau_{x_1\xi_2} & \tau_{x_2\xi_2}\end{vmatrix}\sum_{j=1}^{3}a_{1j}\tau_{x_j}$$

$$=-\begin{vmatrix}\tau_{x_2} & \tau_{x_3} & \sum_{j=1}^{3}a_{3j}\tau_{x_j}\\ \tau_{x_2\xi_2} & \tau_{x_3\xi_2} & \sum_{j=1}^{3}a_{3j}\tau_{x_j}\end{vmatrix}-\begin{vmatrix}\tau_{x_2} & \tau_{x_1} & \sum_{j=1}^{3}a_{1j}\tau_{x_j}\\ \tau_{x_2\xi_2} & \tau_{x_1\xi_2} & \sum_{j=1}^{3}a_{1j}\tau_{x_j}\end{vmatrix}$$

$$=-\begin{vmatrix}\tau_{x_2} & \tau_{x_1}\sum_{j=1}^{3}a_{1j}\tau_{x_j}+\tau_{x_3}\sum_{j=1}^{3}a_{3j}\tau_{x_j}\\ \tau_{x_2\xi_2} & \tau_{x_1\xi_2}\sum_{j=1}^{3}a_{1j}\tau_{x_j}+\tau_{x_3\xi_2}\sum_{j=1}^{3}a_{3j}\tau_{x_j}\end{vmatrix}$$

$$=-\begin{vmatrix}\tau_{x_2} & \sum_{i,j=1}^{3}a_{ij}\tau_{x_i}\tau_{x_j}\\ \tau_{x_2\xi_2} & \frac{1}{2}\frac{\partial}{\partial\xi_2}\sum_{i,j=1}^{3}a_{ij}\tau_{x_i}\tau_{x_j}\end{vmatrix}=-\begin{vmatrix}\tau_{x_2} & 1\\ \tau_{x_2\xi_2} & 0\end{vmatrix}=\tau_{x_2\xi_2}.$$

Using equalities (4.69)

$$\sum_{k=1}^{3}(-1)^k p_{kl}\frac{\partial}{\partial \xi_k}\left[(-1)^s\sum_{j=1}^{3}a_{ij}\tau_{x_j}q_{kls0}-(-1)^i\sum_{j=1}^{3}a_{sj}\tau_{x_j}q_{kli0}\right]=0.$$

Therefore, the right-hand part of equality (4.68) turns to zero and thus identity (4.67) is established.

The first term in the right-hand part of identity (4.67) can be written as a scalar product

$$\nabla_\xi F\cdot\operatorname{rot}\tilde{q}$$

where

$$\tilde{q}=\sum_{i,j,s=1}^{3}(-1)^{s+1}a_{ij}\tau_{x_j}(w_{x_s}w_{x_i}-2w_{x_i}\bar{b}_s+\bar{b}_i\bar{b}_s)(q_{23s0},q_{31s0},q_{12s0}).$$

The left-hand part of the identity becomes zero for the solutions of (4.66), which results in an integral equality analogous to (4.60)

$$\int_D \mathrm{d}x\int_S\left\{\sum_{i,j=1}^{3}a_{ij}(w_{x_i}-\bar{b}_i)(w_{x_j}-\bar{b}_j)+\left[\sum_{i,j=1}^{3}a_{ij}\tau_{x_j}(w_{x_i}-\bar{b}_i)\right]^2\right\}$$

$$\times|T|\,\mathrm{d}S_\xi=\int_S \mathrm{d}S_\xi\int_S \Phi(w,\tau)\,\mathrm{d}S_x. \tag{4.70}$$

Equality (4.70) is also valid when $a_{ij}\in C^2(D)$, $\bar{b}_i\in C^1(D)$. Since $w(x,\xi)=0$, $x\in S$, $\xi\in S$, [see (4.65)] and in calculating $\Phi(w,\tau)$ the derivatives with respect to w are taken only in the directions tangential to the surface S, then the right-hand part of equality (4.70) becomes zero. Hence, $\bar{b}_i=w_{x_i}$, $i=1,2,3$. It is equivalent to the equality $\operatorname{rot}\bar{b}=0$.

Thus, (4.44) makes it possible to find rot $\bar{b}$ uniquely. The possibility to determine the projections of the vector $\bar{b}$ on the tangential plane at the points $x\in S$ has, in fact, been proved earlier.

4.3. Inverse dynamic problem: linearization method

Consider the problem

$$u_{tt}-Lu=f(x,t),\quad x\in R^3,\quad t\in R \tag{4.71}$$

$$u|_{t<0}\equiv 0. \tag{4.72}$$

If for determining the coefficients of (4.71) some information is given on the solutions $u(x,t)$ on manifolds of the time type (for instance, on a certain set of straight lines parallel to the axis t), then the problem of determining the coefficients is termed the inverse dynamic problem. This definition stresses the idea that the information considered in the problem is an oscillation process in the time of a certain set of points of the space R^3. Information of the kind can be successfully used for determining the coefficients at both higher and lower derivatives of the operator L.

General formulation of the inverse dynamic problem on determining one of the coefficients of the operator L, which is unknown inside a certain domain D limited by the surface S, can be formulated as follows. A solution to problem (4.71) and (4.72) is known at the points of S as a function of time

$$u(x,t) = g(x,t), \quad x \in S, \quad t \geqslant 0, \tag{4.73}$$

and the task is to find the unknown coefficient inside D.

In an analogous way one can formulate the problem of determining all the coefficients of the operator L which requires greater information than (4.73). Such information can be obtained by considering several problems of type (4.71) and (4.72), for instance, corresponding to various functions $f(x,t)$, with the information of type (4.73) given with respect to every problem. Another possible way is introduction of a certain parameter λ into the function f, and then $f = f(x,t,\lambda)$ and the solution of (4.71) and (4.72) is also λ-dependent. In this case information (4.73) depends on the parameter λ, an example of such parameter being the point of application of a concentrated source.

The inverse kinematic problem on determining the coefficients of the operator L is a nonlinear problem. Indeed, the solution to (4.71) and (4.72) is a certain operator

$$u = A(q,f), \qquad q = (a_{ij}, b_i, c)$$

linear with respect to the function f and nonlinear with respect to the coefficients of the operator L, i.e. to the components of the vector q which is composed of the coefficients of the operator L. Using (4.73), one obtains a nonlinear operator equation

$$A(q,f) = g. \tag{4.74}$$

In studying nonlinear equations an important role belongs to a linear equation obtained as a result of linearization. As a rule this equation reflects the overall features of the nonlinear equation and facilitates gaining insight into the problem. On the basis of this principle consider a scheme of linearization of inverse problems and discuss some problems arising in the process of investigating the obtained linear problem.

Examine the problem of determining the coefficient $c(x)$ of the operator L in the formulation (4.71)–(4.73). Assume that the coefficient $c(x)$ can be written as

$$c(x) = c_0(x) + c_1(x) \tag{4.75}$$

where the coefficient $c_0(x)$ is known and $c_1(x)$ is small by the absolute value. (Note that to substantiate linearization, when it refers to the coefficients at the operator L derivatives, the requirement of smallness of an additional coefficient should be understood as smallness by the norm containing the coefficient derivatives up to a certain order.) Thus, the inverse problem is reduced to that of determining a small addition to the function $c_0(x)$.

The essence of the method of linearization is as follows. In a formal way we introduce the parameter λ and $c(x,\lambda)$

$$c(x,\lambda) = c_0(x) + \lambda c_1(x). \tag{4.76}$$

At $c = c(x, \lambda)$ the solution to (4.71) and (4.72) is an infinite series in powers of λ

$$u(x, t, \lambda) = \sum_{n=0}^{\infty} \lambda^n u_n(x, t) \tag{4.77}$$

that is substituted into equalities (4.71) and (4.72). The equations for $u_n(x, t)$ are obtained as a result of equating the expressions containing λ^n. In this case u_0 is the solution of the problem

$$\left(\frac{\partial^2}{\partial t^2} - L_0\right) u_0 = f(x, t), \quad u_0|_{<0} \equiv 0, \tag{4.78}$$

where L_0 denotes the operator L at $c = c_0$, and at $n \geqslant 1$ the functions u_n satisfy the conditions

$$\left(\frac{\partial^2}{\partial t^2} - L_0\right) u_n = c_1 u_{n-1}, \quad u_n|_{t<0} \equiv 0, \quad n \geqslant 1. \tag{4.79}$$

As is seen from these equalities, u_0 is c_1-independent, u_1 depends on the coefficient c_1 in a linear way, while all the rest u_n, starting with $n = 2$, depend on c_1 in a nonlinear way. Therefore the problem linearization is corresponded to by breaking of the series (4.75) and its substitution with the first two terms. The fact that equality (4.75) is obtained from equality (4.76) at $\lambda = 1$ corresponds to the presentation of the solution of problem (4.71) and (4.72) in the form

$$u(x, t) = u_0(x, t) + u_1(x, t). \tag{4.80}$$

Within this approach the data of the inverse problem as those for the function $u_1(x, t)$ can be rewritten

$$u_1(x, t) = g_1(x, t) \equiv g(x, t) - u_0(x, t), \quad x \in S, \quad t \geqslant 0. \tag{4.81}$$

The function $g_1(x, t)$ can be considered known since the function $u_0(x, t)$ is known as a solution of problem (4.78).

Thus, the inverse problem in the linear approximation is reduced to the problem of defining the function $c_1(x)$ involved in the relations

$$\left(\frac{\partial^2}{\partial t^2} - L_0\right) u_1 = c_1 u_0, \quad u_1|_{t<0} \equiv 0 \tag{4.82}$$

at the known function $u_0(x, t)$ and information (4.81). As is seen, it is the problem of defining a part of a differential equation of a special type.

Similar problems arise when the role of the coefficient is played by one of the coefficients a_{ij} or b_i. For instance, for the coefficient b_i under the supposition $b_i = (b_i)_0 + (b_i)_1$, analogous to (4.75), for u_1 arises the equation

$$\left(\frac{\partial^2}{\partial t^2} - L_0\right) u_1 = (b_i)_1 \frac{\partial}{\partial x_i} u_0, \quad u_1|_{t<0} \equiv 0, \tag{4.83}$$

and the inverse problem on determining $(b_i)_1$ from relations (4.83) and (4.81).

In the case when not all the operator L coefficients are known it would be natural, as has been stated earlier, to consider several problems of type (4.71) and

(4.72) with various functions f, which means that the function $f(x, t)$ can be viewed as a vector function. Its dimensionality must coincide with the number of the unknown coefficients of the operator L, in which case the functions u, g also become vector functions. Writing each of the operator L coefficients in the form analogous to (4.75)

$$a_{ij} = (a_{ij})_0 + (a_{ij})_1, \qquad b_i = (b_i)_0 + (b_i)_1, \qquad c = c_0 + c_1$$

and denoting through L_k the operator L corresponding to the coefficients $(a_{ij})_k$, $(b_i)_k, c_k, k = 0, 1$, we get the inverse problem of determining $(a_{ij})_1, (b_i)_1, c_1$ from the relations

$$\left(\frac{\partial^2}{\partial t^2} - L_0\right)u_1 = L_1 u_0, \quad u_1|_{t<0} \equiv 0, \tag{4.84}$$

under condition (4.81). As a result, a vector variant of the inverse problem is obtained on determining the right-hand part of a special type.

Now return to problem (4.82) and (4.81) as to the most convenient object for demonstrating the method. The problem of defining $c_1(x)$ can also be treated as a problem of integral geometry. In order to prove it, solve the Cauchy problem using the formula representing the solution through the fundamental one

$$u_1(x, t) = \int_{R^4} c_1(\xi)u_0(\xi, \tau)H_0(x, t - \tau, \xi)\,d\xi\,d\tau.$$

Here H_0 corresponds to the operator L_0. Using the structure of the fundamental solution [see (4.7)] this formula can be written in the following way

$$u_1(x, t) = \frac{1}{4\pi}\int_{\tau(x,\xi)\leqslant t} c_1(\xi)u_0(\xi, t - \tau(x, \xi))\frac{d\xi}{\tau(x, \xi)}$$
$$+ \int_{\tau(x,\xi)\leqslant t} c_1(\xi)\int_0^{t-\tau(x,\xi)} u_0(\xi, \tau)v_{-1}(x, t - \tau, \xi)\,d\tau\,d\xi, \quad t \geqslant 0 \tag{4.85}$$

where v_{-1} is a regular part of the function H_0. From condition (4.81) one can derive an integral equation for determining the coefficient $c_1(x)$

$$\int_{\tau(x,\xi)\leqslant t} c_1(\xi)\rho(x, \xi, t)\,d\xi = g_1(x, t), \quad x \in S, \quad t \geqslant 0 \tag{4.86}$$

where the weight function $\rho(x, \xi, t)$ is determined by the equality

$$\rho(x, \xi, t) = u_0(\xi, t - \tau(x, \xi))/4\pi\tau(x, \xi) + \int_0^{t-\tau(x,\xi)} u_0(\xi, \tau)v_{-1}(x, t - \tau, \xi)\,d\tau.$$

The problem of solving (4.86) with respect to the function $c_1(x)$, $x \in D$, is a problem of integral geometry. Its properties are largely determined by the weight function ρ, which, in its turn, depends on the function f and the operator L_0 coefficients.

Now examine the simplest variant of (4.86). Let

$$L_0 = \Delta, \qquad f(x, t) = \delta(x - x^0, t).$$

In this case

$$\tau(x,\xi)=|x-\xi|, \qquad u_0(x,t)=\frac{1}{4\pi|x-x^0|}\delta(t-|x-x^0|), v_{-1}\equiv 0,$$

and (4.86) is reduced to the form

$$\frac{1}{(4\pi)^2}\int_{R^3} c_1(\xi)\frac{\delta(t-|x-\xi|-|\xi-x^0|)}{|x-\xi||\xi-x^0|}\,d\xi = g_1(x,t), \quad x\in S, \quad t\geqslant 0. \qquad (4.86')$$

Since the carrier of the subintegral function is concentrated on the ellipsoid $S(x,x^0,t)=\{\xi:|x-\xi|+|\xi-x^0|=t\}$, then the integral in the left-hand part of equality (4.86′) can be reduced to an integral over the surface of $S(x,x^0,t)$. The simplest way to do it is to go over to the spherical system of coordinates r,θ, and ϕ, with its pole located at the point x^0 and the polar axis running through the points x^0, x. The equation of the ellipsoid $S(x,x^0,t)$ in this system of coordinates is as follows:

$$r=(t^2-r_0^2)/2(t-r_0\cos\theta), \qquad r_0=|x-x^0|, \qquad r=|\xi-x^0|.$$

In this case $|x-\xi|=(r^2+r_0^2-2rr_0\cos\theta)^{1/2}$.

Equation (4.86′) in the new system of coordinates is transformed to the form

$$\frac{1}{(4\pi)^2}\int_{R^3}\frac{c_1(\xi)}{(r^2+r_0^2-2rr_0\cos\theta)^{1/2}}\delta[t-r-(r^2+r_0^2-2rr_0\cos\theta)^{1/2}]r\,dr\,d\omega$$
$$=g_1(x,t), \quad x\in S, t\geqslant 0.$$

Here $d\omega=\sin\theta\,d\theta\,d\phi$. To calculate the integral with respect to the variable r make use of the following property of the δ-function. Let $r=r^*$ be an ordinary zero of the differentiated function $\psi(r)$ and let this function have no other zeros in the δ-vicinity of the point r^*. Then

$$\int_{r^*-2}^{r^*+\varepsilon} f(r)\delta(\psi(r))\,dr=f(r^*)/|\psi'(r^*)|.$$

In this case

$$\psi(r)=t-r-(r^2+r_0^2-2rr_0\cos\theta)^{1/2}, \qquad r^*=(t^2-r_0^2)/2(t-r_0\cos\theta).$$

Since

$$|\psi'(r^*)|=\left[1+\frac{r-r_0\cos\theta}{(r^2+r_0^2-2rr_0\cos\theta)^{1/2}}\right]_{r=r^*}$$
$$=\frac{t-r_0\cos\theta}{|x-\xi|}=\frac{t^2-r_0^2}{2r^*|x-\xi|}, \qquad r^*=|\xi-x^0|$$

then equation (4.86′) assumes the form

$$\iint_{S(x,x^0,t)}|x^0-\xi|^2c_1(\xi)\,d\omega=8\pi^2(t^2-|x-x^0|^2)g_1(x,t), x\in S, \quad t\geqslant 0. \quad (4.87)$$

Let $S = \{x{:}x_3 = 0\}$ and x^0 is a fixed point of S. The following theorem has been stated in [162] (see also [42]).

Theorem 4.4. If the function $c_1(x)$ is even with respect to the plane S and is continuous in R^3, then it is uniquely defined from equation (4.87).

As far as any function can be represented in the form of its even and odd parts

$$c_1(x) = [c_1(x_1, x_2, x_3) + c_1(x_1, x_2, -x_3)]/2 \\ + [c_1(x_1, x_2, x_3) - c_1(x_1, x_2, -x_3)]/2,$$

then one of the consequences of the theorem is as follows: if the function $c_1(x)$ is known in the domain $x_3 < 0$, then it is uniquely defined in the domain $x_3 \geqslant 0$ by setting $g_1(x, t)$.

To prove Theorem 4.4 introduce the spherical coordinates of $x \in S$ and $\xi \in S(x, x^0, t)$

$$x = x^0 + r_0 v^0, \quad r_0 = |x - x^0|, \quad v^0 = (\cos\phi_0, \sin\phi_0, 0).$$
$$\xi = x^0 + rQ(\phi_0)v, \quad v = (\sin\theta\cos\phi, \sin\theta\sin\phi, \cos\theta) \tag{4.88}$$

$$Q(\phi_0) = \begin{Vmatrix} 0 & \sin\phi_0 & \cos\phi_0 \\ 0 & -\cos\phi_0 & \sin\phi_0 \\ 1 & 0 & 0 \end{Vmatrix}$$

as well as the ellipsoid eccentricity ε and the polar parameter p

$$\varepsilon = r_0/t, \qquad p = t(1 - \varepsilon^2)/2.$$

In this case for $\xi \in S(x, x^0, t)$

$$r = p(1 - \varepsilon\cos\theta)^{-1}. \tag{4.89}$$

Write equation (4.87) in the following way

$$\int_{|v|=1} r^2 c_1(x^0 + rQv)\,\mathrm{d}\omega = g_2(p, \varepsilon, \phi_0), \quad 0 \leqslant p < \infty, \quad 0 \leqslant \varepsilon < 1, \quad 0 \leqslant \phi \leqslant 2\pi. \tag{4.90}$$

In this expression r is defined by equality (4.89), and the function g_2 by the expression

$$g_2(p, \varepsilon, \phi) = \frac{32p^2}{1-\varepsilon^2} g_1\left(x^0 + \frac{2\varepsilon p}{1-\varepsilon^2} v_0, \frac{2p}{1-\varepsilon^2}\right).$$

Apply to both parts of equality (4.90) the operator M

$$Mg_2 \equiv p\frac{\partial}{\partial\varepsilon}\int_0^p g_2(z, \varepsilon, \phi_0)\frac{\mathrm{d}z}{z}.$$

The possibility of its use and the result of its application follow from the chain of equalities

$$
\begin{aligned}
Mg_2 &= p\frac{\partial}{\partial\varepsilon}\int_0^p \frac{dz}{z}\int_{|v|=1} r^2 c_1(x^0+rQv)|_{r=z(1+\varepsilon\cos\theta)^{-1}}\,d\omega \\
&= p\frac{\partial}{\partial\varepsilon}\int_{|v|=1}\int_0^p r^2 c_1(x^0+rQv)|_{r=z(1+\varepsilon\cos\theta)^{-1}}\frac{dz}{z}\,d\omega \\
&= p\frac{\partial}{\partial\varepsilon}\int_{|v|=1}\int_0^{p(1+\varepsilon\cos\theta)^{-1}} rc_1(x^0+rQv)\,dr\,d\omega \\
&= \int_{|v|=1} r^3\cos\theta c_1(x^0+rQv)\,d\omega.
\end{aligned}
$$

Thus, application of the operator M to equality (4.90) results in the appearance of the multiplier $r\cos\theta$ under the sign of integral. Therefore, repeated use of the operator Mk times will result in the appearance of the multiplier $(r\cos\theta)^k$

$$M^k g_2 = \int_{|v|=1} r^{k+2}\cos^k\theta c_1(x^0+rQv)\,d\omega, \quad k=0,1,2,\ldots. \tag{4.91}$$

Set in these equalities $\varepsilon=0$, then $r=p=t/2$ and the integrals over the ellipsoids $S(x,x^0,t)$ transform into those over the spheres $|\xi-x^0|=t/2$. Since in this case $r\cos\theta=(\xi_1-x_1^0)\cos\phi_0+(\xi_2-x_2^0)\sin\phi_0$, then at $\varepsilon=0$ equalities (4.91) can be written as follows:

$$\int_{|\xi-x^0|=t/2} c_1(\xi)[(\xi_1-x_1^0)\cos\phi_0+(\xi_2-x_2^0)\sin\phi_0]^k\,dS_\xi = M^k g_2|_{\varepsilon=0},$$

$$k=0,1,2,\ldots,\quad t\geqslant 0, 0\leqslant\phi_0\leqslant 2\pi. \tag{4.92}$$

Due to ϕ_0 arbitrariness, the following quantities can be obtained from here

$$\int_{|\xi-x^0|=t/2} c_1(\xi)(\xi_1-x_1^0)^n(\xi_2-x_2^0)^m\,dS_\xi = g_{nm},\quad n,m=0,1,2,\ldots. \tag{4.93}$$

On the fixed sphere $|\xi-x^0|=t/2$ a complete system of moments of the function $c_1(x)$ is formed that is even with respect to x_3, which fact proves the validity of the earlier formulated theorem.

On the sphere $|\xi-x^0|=t/2$ one can construct a system of the Fourier coefficients over the spheric functions, using moments (4.93), as well as represent $c_1(\xi)$ as a Fourier series.

It is possible to formulate the problem on determining the function $c_1(x)$ through using (4.87). Let S be a plane $x_3=0$, and x^0 is a variable point on this plane. Let us consider the case when $x=x^0$. Then the surfaces $S(x^0,x^0,t)$ are the spheres $|\xi-x^0|=t/2$ and (4.87) has the following form

$$\frac{1}{4\pi}\iint_{|\xi-x^0|=t/2} c_1(\xi)\,d\omega = 8\pi g_1(x,t),\quad x\in S, t\geqslant 0. \tag{4.94}$$

The problem here arises of determining the function $c_1(x)$ over its spherical means, when the centre of the spheres runs through many points of a fixed plane, the radius of the spheres being arbitrary.

The problem of determining (4.94) has been considered by Courant [155, p. 747] in connection with the investigation of an incorrect Cauchy problem when the data are given on the set of a time type (see also [163]). Here the result obtained in the book without giving the proof is given.

Theorem 4.5. The even with respect to the variable x_3, continuous in R^3 function $c_1(x)$ is uniquely defined by integrals (4.94).

More complex cases for (4.86) have been considered when

$$L_0 = n^2(x_3)\Delta + c_0(x), \qquad S = \{x : x_3 = 0\}$$
$$f = f(x, t, x^0) = \delta(x - x^0, t), \quad x^0 \in S$$

in [43, 164]. In this case a set of the Riemann spheres is axially symmetrical and invariant with respect to the parallel transfers along the plane S. This fact greatly simplifies the process of investigation of the problem of integral geometry (4.86).

For the case when L_0 is an arbitrary hyperbolic operator, $f = \delta(x - x^0, t)$ and under fairly general suppositions on the operator L_0 coefficients, one can study problem (4.86) in the class of functions $c_1(x)$ that can be written as

$$c_1(x) = \sum_{k=1}^{n} \phi_k(x_1, x_2)\psi_k(x_3) \tag{4.95}$$

where $\phi_k(x_1, x_2)$, $\psi_k(x_3)$ are arbitrary smooth functions, there own for every function $c_1(x)$. The function classes of type (4.95) are the classes of uniqueness for a wide range of problems of integral geometry (see [41, 165–168]).

Equation (4.86) can also be viewed as a Volterra operator equation [169]. Some results of investigating these equations from the standpoint of functional analysis are given in [41, 170, 171].

For the case when the function $f(x, t)$ is a smooth one, $f(x, 0) \neq 0$, $x \in \bar{D}$, and the operator L_0 coefficients are smooth enough and give rise to a regular set of the geodesic lines $\Gamma(x, x^0)$ inside D, the uniqueness theorem is valid for (4.86) in the class of the continuous finite functions $c_1(x)$ whose carrier is contained inside D and sufficiently small. This result was obtained in some works on integral geometry [22, 172, 173]. When the main part is constant, the same conclusion can be found in [174] (see also [41]). The next section will consider the problems discussed here but with a somewhat different approach.

4.4. General scheme of studying inverse problems for hyperbolic type equations

Consider some problems associated with uniqueness and stability of the solution of the inverse dynamic problem. These two problems appear to be closely connected with the problems of determining the right-hand part of the differential equation analogous to the problems we arrived at in Section 4.3.

Examine (4.71)–(4.73) in a more subtle formulation. Let all the coefficients of the operator L be known outside a certain closed domain $\Omega \subset D$ and be unknown in Ω. The task is to find them by the additional information (4.73). Assume that f is a vector-function and its dimensionality coincides with the number of the unknown coefficients of the operator L (in the case considered it is 10, taking into account a symmetry of the coefficients at the higher derivatives: $a_{ij} = a_{ji}$). In line with this fact the functions u, g are also vector-functions. Denote the ordered set of the operator L coefficients through $q = (a_{11}, a_{12}, a_{13}, a_{22}, a_{23}, a_{33}, b_1, b_2, b_3, c)$ and a certain set consisting of the functions $q(x)$-through Q. Also suppose that $q(x)$, $f(x, t)$ have derivatives of a sufficiently high order.

When studying the problem of stability by the inverse problem data a certain set G should be considered consisting of the functions $g(x, t)$ and for any two elements estimates should be made of a certain norm of the difference of the solutions of the inverse problems, corresponding to these elements, through the norm of the difference of the elements from G. In line with this denote any two arbitrary elements from G through $g^{(1)}$, $g^{(2)}$, and let $(u^{(1)}, q^{(1)})$, $(u^{(2)}, q^{(2)})$ be the corresponding to them solutions of (4.71)–(4.73), and

$$u^{(1)} - u^{(2)} = \tilde{u}, \qquad q^{(1)} - q^{(2)} = \tilde{q}, \qquad g^{(1)} - g^{(2)} = \tilde{g}.$$

Denote through $L^{(i)}$ the operator corresponding to $q = q^{(i)}$, $i = 1, 2$, and through $\tilde{L}$ the operator corresponding to $q = \tilde{q}$. For the functions $\tilde{u}$, $\tilde{q}$, and $\tilde{g}$ the following results

$$\tilde{u}_{tt} - L^{(1)}\tilde{u} = \tilde{q}(x)\Phi(x, t) \tag{4.96}$$

$$\tilde{u}|_{t<0} \equiv 0 \tag{4.97}$$

$$\tilde{u}(x, t) = \tilde{g}(x, t), \quad x \in S, \quad t \geqslant 0. \tag{4.98}$$

Here $\Phi(x, t)$ is a matrix defined by the equality

$$\tilde{q}(x)\Phi(x, t) = \tilde{L}u^{(2)}, \qquad \Phi(x, t) = (\Phi_{ij}, i, j = 1, \ldots, 10). \tag{4.99}$$

If problem (4.96)–(4.98) on determining $\tilde{q}(x)$ allows an estimate of its stability which is uniform with respect to the set Q, then the same stability estimate is also valid for the inverse problem (4.71)–(4.73). Uniformity of the estimate of problem (4.96)–(4.98) with respect to the set Q is necessary since the operator $L^{(1)}$ depends on $q^{(1)} \in Q$, and the matrix Φ on $q^{(2)} \in Q$. Note that in this case the stability estimate is of a conditional character since it presupposes the belonging of the inverse problem solution to the set Q.

Thus, the need of obtaining a stability estimate of (4.71)–(4.73) results in the problem of determining the right-hand part of a special kind, similar to the problem in Section 4.3 arising when linearizing the inverse problem formulation. It is one more argument in favour of the statement that the linear part of the inverse problem operator preserves the basic features of the initial problem. Note also, that the problem of uniqueness of the solution to the inverse problem is reduced to that of uniqueness of the solution to problem (4.96)–(4.98). Indeed, the fact that the function $\tilde{g} = 0$ corresponds to $\tilde{q} = 0$ means that $g^{(1)} = g^{(2)}$ corresponds to $q^{(1)} = q^{(2)}$.

When $f(x, t)$ is a generalized function, problem (4.71)–(4.73) and the corresponding problem (4.96)–(4.98) [it should be recalled that Φ is expressed through the function $u^2(x, t)$ which itself depends on $f(x, t)$] have been examined only in special classes of functions (4.95) and under simplified suppositions on the operator L structure. An exception here are the problems where the operator L coefficients depend on only one variable, in which case the inverse problem can be investigated by giving a minimum information on the solution. For instance, when the task is to determine only one coefficient, it is sufficient to set the solution of the problem only at one point $x \in S$ as a function of time [39, 88, 89, 93, 127, 175–177].

For the case of the smooth functions $f(x,t)$ satisfying certain additional assumptions, the investigation can be carried out up to the end. Here are the results obtained in [178].

Lemma 4.4. Let det $\Phi_{tt}(x,0) \neq 0$, $x \in \Omega$. *In this case the investigation of the problems of uniqueness and stability by data (4.73) of the inverse problem (4.71)–(4.73) on the set* Q *is reduced to the investigation of analogous problems for the problem of determining the vector function* $\phi(x)$, *supp* $\phi(x) \subset \Omega$, *from the relations*

$$v_{tt} - Lv = \phi(x)R(x,t) \tag{4.100}$$

$$v|_{t=0} = 0, \qquad v_t|_{t=0} = \phi(x) \tag{4.101}$$

$$v(x,t) = h(x,t), \quad x \in S, \quad t \geqslant 0, \tag{4.102}$$

where L *is a given operator;* $q \in Q$; $R(x,t)$, *and* $h(x,t)$ *are given functions;* R *is a square matrix.*

Note that

$$u^{(2)}|_{t=0} = 0, \qquad u_t^{(2)}|_{t=0} = 0, \qquad u_{tt}^{(2)}|_{t=0} = f(x,0). \tag{4.103}$$

Therefore the condition det $\Phi_{tt}(x,0) \neq 0$, $x \in \Omega$, is an easily verified condition for the function $f(x,0)$.

Differentiate equalities (4.96) and (4.98) three times over the variable t

$$\tilde{u}_{ttt} = v, \qquad \phi(x) = \tilde{q}(x)\Phi_{tt}(x,0)$$

$$R(x,t) = \Phi_{tt}^{-1}(x,0)\Phi_{ttt}(x,t), \qquad h(x,t) = \tilde{g}_{ttt}(x,t)$$

and redenote $L^{(1)}$ through L. Then, allowing for (4.103), one gets equalities (4.100)–(4.102) and the lemma is proved.

As Q, take a set of infinitely differentiated functions $q(x)$ with limited by a given constant M partial derivatives for which the constant μ of the uniformly positive operator L is fixed. Also assume that for $q \in Q$ the set of the characteristic conoids is regular in the sense that the biocharacteristics outgoing from one point cross each other at no other point. Under these assumptions and a fairly smooth function $f(x,t)$ any solution of (4.71) and (4.72) for $q \in Q$ is uniformly limited together with the derivatives in the cylinder $G_T = \{(x,t): x \in D, 0 \leqslant t \leqslant T\}$, with the corresponding constant depending only on M, T, μ and the function $f(x,t)$. In line with this fact the functions R, v can be considered uniformly limited in G_T with respect to the set Q.

Denote

$$\| \phi \|^2 = \int_\Omega |\phi(x)|^2 \, dx, \quad \| h \|_T^2 = \max_{0 \leqslant t \leqslant T} \int_S |h(x,t)|^2 \, dS$$

$$\| R \|_T^2 = \max_{(x,t) \in \Omega \times [0,T]} \sum_{i,j} |R_{ij}(x,t)|^2.$$

In these formulas dS is an element of the surface S; $R_{ij}(x,t)$ are the components of the matrix $R(x,t)$.

The two following theorems of stability of problem (4.100)–(4.102) exist.

Theorem 4.6. Let the following condition is fulfilled

$$T > \frac{1}{\mu} \operatorname{diam} D. \tag{4.104}$$

then there exists such $\delta > 0$ that when the condition

$$\operatorname{diam} \Omega < \delta, \quad \| R \|_T < \delta \tag{4.105}$$

is fulfilled, the solution of (4.100)–(4.102) uniformly with respect to Q obeys the estimate

$$\| \phi \| \leqslant C \| h_t \|_T^{1/2}. \tag{4.106}$$

Theorem 4.7. Let condition (4.104) be fulfilled; then there exists such $\delta > 0$ that under the condition

$$\operatorname{diam} D < \delta, \tag{4.107}$$

for the function $\phi(x)$ valid is estimate (4.106), uniform with respect to the set Q.

These theorems are of the same character as the small-scale theorems since they also involve limitations concerning the small size of the required function carrier. Their proof is carried out by the same scheme and based on the energetic estimates for equation (4.100).

Multiply the left- and right-hand parts of equality (4.100) by $2v_t(x,t)$ in a scalar way. The left-hand part of the resulting equality can be written as follows:

$$(2v_t, [v_{tt} - Lv]) \equiv \frac{\partial}{\partial t}\left[(v_t, v_t) + \sum_{i,j=1}^{3} a_{ij}(v_{x_i}, v_{x_j}) - c(v,v) \right]$$

$$- 2 \sum_{i,j=3}^{3} \frac{\partial}{\partial x_i}(v_t, a_{ij} v_{x_j}) - 2 \sum_{i=1}^{3} \left(b_i - \sum_{j=1}^{3} (a_{ij})_{x_j} \right)(v_t, v_{x_i}).$$

Denote through

$$I(t) = \int_{t=\text{const}} \left[(v_t, v_t) + \sum_{i,j=1}^{3} a_{ij}(v_{x_i}, v_{x_j}) - c(v,v) \right] \mathrm{d}v \tag{4.108}$$

the integral over a cross-section of the domain $G_t = D \times [0, T]$ by the plane $t = \text{const}$. Note that the quadratic form standing under the integral sign in (4.108) is positive if $-c > 0$. Assume that this condition is fulfilled. It should be recalled that it does not, in fact, limit the generality as its fulfilment can always be achieved by way of substituting a new function $\tilde{v} = ve^{\lambda t}$ for the function v, provided the numerical parameter λ is chosen in an appropriate way. However, in this case a new term containing the first derivative with respect to the variable t appears in the operator L, its presence though not being essential for the method of energetic estimates.

Integrate both parts of equality (4.100), multiplied in a scalar way by $2v_t$, over the part of the cylinder G_T that is contained between the cross-sections $t=t_1$, $t=t_2$, and $t_2>t_1$. Using the Gauss–Ostrogradsky formula

$$I(t_2)-I(t_1)=2\int_{t_1}^{t_2}\mathrm{d}\tau\int_S\left(v_\tau,\sum_{i,j=1}^{3}a_{ij}v_{x_i}\cos(n,x_j)\right)\mathrm{d}S$$
$$+2\int_{t_1}^{t_2}\mathrm{d}\tau\int_S\left[(v_\tau,\phi R)+\left(v_\tau,\sum_{i=1}^{3}\left(b_i-\sum_{j=1}^{3}(a_{ij})_{x_j}\right)v_{x_i}\right)\right]\mathrm{d}x.$$

Here n is the direction of the external normal to S. As far as $v_t|_S=h_t$, and v_{x_i} on S are uniformly limited with respect to Q then using the Cauchy–Buniakovsky inequality

$$\left|2\int_S\left(v_t,\sum_{i,j=1}^{3}a_{ij}v_{x_i}\cos(n,x_j)\right)\mathrm{d}S\right|\leqslant C_1\|h_t\|_T.$$

Besides

$$2|(v_\tau,\phi R)|\leqslant(v_\tau,v_\tau)+(\phi R,\phi R)\leqslant|v_\tau|^2+\|R\|_T^2|\phi|^2.$$

Therefore

$$2\left|\int_D(v_\tau,\phi R)\,\mathrm{d}x\right|\leqslant I(\tau)+\|R\|_T^2\|\phi\|^2.$$

In an analogous way

$$2\left|\int_D\left(v_t,\sum_{i=1}^{3}\left(b_i-\sum_{j=1}^{3}(a_{ij})_{x_j}\right)v_{x_i}\right)\mathrm{d}x\right|$$
$$\leqslant\int_D\left[|v_t|^2+\left|\sum_{i=1}^{3}\left(b_i-\sum_{j=1}^{3}(a_{ij})_{x_j}\right)v_{x_i}\right|^2\right]\mathrm{d}x\leqslant C_2I(t)$$

with a certain fairly great constant C_2, its value depending only on the estimate of b_i, $(a_{ij})_{x_j}$ in the domain D and the constant μ.

From the estimates obtained one can easily derive an estimate of the derivative from I with respect to the parameter t

$$\left|\frac{\mathrm{d}}{\mathrm{d}t}I(t)\right|\leqslant C_1\|h_1\|_T+\|R\|_T^2\|\phi\|^2+C_3I(t),\quad 0\leqslant t\leqslant T. \tag{4.109}$$

Hence

$$\frac{\mathrm{d}}{\mathrm{d}t}(I(t)\mathrm{e}^{-C_3t})\leqslant(C_1\|h_t\|_T+\|R\|_T^2\|\phi\|^2)\mathrm{e}^{-C_3t}$$

$$I(t)\leqslant[I(0)+(C_1\|h_t\|_T+\|R\|_T^2\|\phi\|^2)t]\mathrm{e}^{C_3t},\quad t\in[0,T]. \tag{4.110}$$

Estimate (4.110) is a common energetic estimate of the solution used for investigating boundary problems, but for our purposes it is not of great interest. More interesting is the estimate for $I(t)$ resulting from the estimation of the derivative $I'(t)$ from below

$$\frac{d}{dt}(I(t)e^{C_3 t}) \geqslant -(C_1 \| h_t \|_T + \| R \|_T^2 \| \phi \|^2) e^{C_3 t}.$$

Therefore

$$I(t)e^{C_3 t} \geqslant I(0) - (C_1 \| h_t \|_T + \| R \|_T^2 \| \phi \|^2) t e^{C_3 t}, \quad 0 \leqslant t \leqslant T.$$

Setting here $t = T$

$$I(0) \leqslant [I(T) + T(C_1 \| h_t \|_T + \| R \|_T^2 \| \phi \|^2)] e^{C_3 T}. \tag{4.111}$$

But in line with data (4.101) $I(0) = \| \phi \|^2$. Hence inequality (4.111) can be written as follows:

$$\| \phi \|^2 (1 - \| R \|_T^2 T e^{C_3 T}) \leqslant [I(T) + T C_1 \| h_t \|_T] e^{C_3 T}.$$

From this expression under the condition

$$\| R \|_T^2 T e^{C_3 T} < 1 \tag{4.112}$$

one can obtain an inequality for $\| \phi \|$ of type

$$\| \phi \|^2 \leqslant C_4 (I(T) + T \| h_t \|_T). \tag{4.113}$$

Condition (4.112) is certain to be fulfilled provided either the condition (4.105) for the function R with a sufficiently small δ is fulfilled, or in Theorem 4.7 T is small and coordinated with the smallness of δ in such a way that (4.104) is valid. Thus, the theorem conditions make it possible to go over to inequality (4.113) with a universal constant C_4 which is uniform with respect to the set Q.

In order to estimate $I(T)$ represent the solutions of (4.100) with the Cauchy data (4.101) as $v = v_1 + v_2$, where v_1 is the solution of the nonuniform equation (4.100) with the zero Cauchy data, and v_2 is the solution of a uniform equation corresponding to (4.100) with data (4.101). If the integrals of (4.108) are denoted through $I_1(T)$, $I_2(T)$ at $t = T$, corresponding to the functions v_1 and v_2, then it is obvious that

$$I(T) \leqslant 2[I_1(T) + I_2(T)].$$

The integral $I_1(T)$ can be also estimated through the method of energetic estimates of the solution to the Cauchy problem. For this purpose one should construct a dome-like domain limited from above with a plane-like surface containing the upper base of the cylinder G_T and from below with the plane $t = 0$ (Fig. 4.2).

The energetic estimates for the cross-sections of this domain by the planes $t = \text{const.}$ are of type (4.110), with the term containing h_t omitted. In this case the constant C_3 changes since the cross-sections of the dome-like domain by the planes $t = \text{const.}$ contain the corresponding cross-sections of the domain G_T. An estimate for $I_1(T)$ is as follows:

$$I_1(T) \leqslant \| R \|_T^2 \| \phi \|^2 e C_3^{\prime T}. \tag{4.114}$$

The fact that $I_1(0) = 0$ is due to the zero Cauchy data is allowed for.

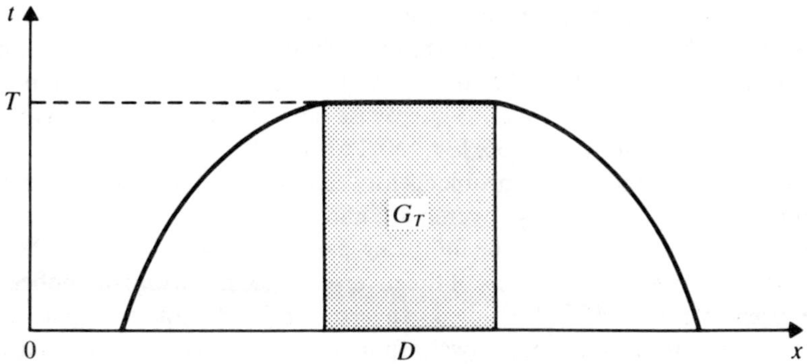

Figure 4.2.

In order to estimate $I_2(T)$ make use of the representation of the solution V_2 through the fundamental solution $H(x, t-\tau, \xi)$

$$v(x,t) = \int_{\Omega} H(x,t,\xi)\phi(\xi)\,d\xi. \tag{4.115}$$

For $x \in D$, $\xi \in \Omega$ and the t values close to T, the function $H(x,t,\xi)$ is a common regular function since the conoid $t = T - \tau(x,\xi)$ where the singular part of the function is concentrated crosses in the case of (4.104), the plane $t = 0$ outside the domain D: the surface $\tau(x,\xi) = T$ contains the domain D for any $\xi \in \Omega$. Therefore, in the discussed domain of the derivative changes the function $H(x,t,\xi)$ is limited together with its first-order partial derivatives in a uniform way with respect to Q. Using formula (4.115) to estimate the function V_2 and its derivatives contained in the integral $I_2(T)$, one gets

$$I_2(T) \leqslant C_6 \|\phi\|^2$$

where

$$C_6 = \int_D \left\{ \int_{\Omega|} (H_t^2 - cH^2)\,d\xi + \sum_{i,j=1}^{3} a_{ij}(x) \left(\int_{\Omega} H_{x_i}^2\,d\xi \right)^{1/2} \times \left(\int_{\Omega} H_{x_j}^2\,d\xi \right)^{1/2} \right\}_{t=T} dx \leqslant C_7 (\operatorname{diam} D \operatorname{diam} \Omega)^3.$$

It follows from the estimates obtained for $I_2(T)$ and $I_2(T)$ that

$$I(T) \geqslant C_8 \|\phi\|^2,$$

where $C_8 \to 0$ if $\|R\|_T \to 0$, $\operatorname{diam} \Omega \to 0$ or if $T \to 0$ $\operatorname{diam} D \to 0$. In this case estimate (4.106) results from inequality (4.113), and from estimate (4.106) follows the theorem of uniqueness: if $h = 0$, then $\phi = 0$, $x \in \Omega$.

The limitation associated with the dimension smallness of the function $\phi(x)$ carrier proves to be a result of the method used. In a number of papers [179–181] another method of investigating problem (4.96)–(4.98) has been suggested in which case the requirement of the smallness of the function $\tilde{q}(x)$ carrier can be eliminated in solving the problem of the solution uniqueness. This method also makes it possible to obtain some stability estimates of the solution. By way of concluding this section, I will point out that the methods of investigation of the inverse problems in the class of piecewise analytical functions have been developed by Anikonov (see [35, 138, 150, 182–186]).

Investigation of the inverse problems for second-order hyperbolic type linear equations has been extensively covered in a number of papers. Alongside with the works mentioned above, note the following papers: [34, 40, 41, 89–93, 128, 187–234], where various formulations of the inverse problems arising in applications have been considered and studied. Inverse problems for the quasi-linear hyperbolic equations have been dwelt upon in [235–241]. Note also that there are a number of investigations associated with the problems arising in studying the inverse problems. There are the works examining the problems of integral geometry [137, 171, 242–247], as well as the Cauchy problem for the hyperbolic equations with the data on a time-like surface [248–253].

Chapter 5

Inverse problems for first-order linear hyperbolic systems

First-order hyperbolic systems of equations describe many physical processes associated with wave propagation. For instance, there exist systems of equations for acoustics, electromagnetic oscillations, and those of dynamic equations of the theory of elasticity. As a rule, under certain additional assumptions one can derive from them second-order equations. Therefore, it seems obvious to examine the inverse problems on determining the coefficients of the system of equations in terms of the system itself, this approach being most convenient as far as the algorithm is concerned (in the sense of constructing a closed system of the inverse problem equations).

This chapter deals with certain general questions of the inverse problem theory for first-order systems, followed by investigation of the inverse problem formulation for the system of equations of the theory of elasticity and that of thermodynamics.

5.1. Systems of equations with a single spatial variable

Systems of this kind arise when studying one-dimensional motions. Recall some basic facts concerning the theory of first-order linear hyperbolic systems (see [49, 155]).

Examine a system of n equations

$$\frac{\partial u}{\partial t} + A\frac{\partial u}{\partial x} + Bu = f \tag{5.1}$$

with respect to the function $u(x, t)$ with the components $u_1, u_2, \ldots, u_n$. Here A, B are the quadratic matrices of dimensionality $n \times n$; $f = f(x, t)$ is a vector with the components $f_1, f_2, \ldots, f_n$. System (5.1) is hyperbolic, according to Petrovskij I, if all the roots of the characteristic equation

$$|A - \lambda E| = 0 \tag{5.2}$$

are real and different. In the case of a symmetrical matrix A, system (5.1) is referred to as a symmetrical hyperbolic system [254]. Then the roots of equation (5.2) are real, but there may be divisible ones among them. In both cases for the smooth matrix $A = A(x, t)$ there exists such a nonsingular matrix $T = T(x, t)$, that

$$T^{-1}AT = \Lambda. \tag{5.3}$$

where Λ is a diagonal matrix, with its diagonal composed of the matrix A

eigenvalues

$$\Lambda = \left\| \begin{matrix} \lambda_1 & 0 & \cdots & 0 \\ 0 & \lambda_2 & \cdots & 0 \\ \cdots & \cdots & \cdots & \cdots \\ 0 & 0 & \cdots & \lambda_n \end{matrix} \right\|.$$

Formula (5.3) results in the following equality

$$AT = T\Lambda \tag{5.4}$$

which means that the kth column of the matrix T is an eigenvalue of the matrix A, corresponding to the eigenvalue λ_k.

System (5.1) can be reduced to a classical form through the matrix T, for which purpose one should replace the required function by the function

$$v = Tu.$$

After this substitution multiply the left-hand part of (5.1) by T^{-1} and get

$$\frac{\partial v}{\partial t} + \Lambda \frac{\partial v}{\partial x} + Cv = F \tag{5.5}$$

where

$$C = T^{-1}(T_t + AT_x + BT), \quad F = T^{-1}f.$$

System (5.5) is convenient since it falls and appears to be connected only through the lower terms. The components v_i, $i = 1, 2, \ldots, n$, of the function v are called the Riemann invariants of (5.1) and they remain constant along the characteristics of (5.1) in the case when $C = 0$, $F = 0$.

Let (5.1) or its equivalent (5.5) be considered in the domain $D = \{(x, t): 0 \leqslant x \leqslant H, t > 0\}$. Then, in order to isolate one of the system solutions, one must set the initial conditions

$$u|_{t=0} = \phi, \qquad v|_{t=0} = \psi = T^{-1}\phi, \quad 0 \leqslant x \leqslant H \tag{5.6}$$

and the boundary conditions at $x = 0$, $x = H$. For the resulting problem to be correct, setting the boundary conditions should be brought into accord with the eigenvalues of the matrix A. Let, for the sake of definiteness, λ_k retain the sign in the domain D and

$$\lambda_k > 0, k = 1, 2, \ldots, s, \qquad \lambda_k < 0, k = s+1, \ldots, n, \quad 0 \leqslant s \leqslant n. \tag{5.7}$$

In this case on the left boundary $x = 0$ one can set s boundary conditions that must be solved with respect to the first s Riemann invariants

$$v_i(0, t) + \sum_{j=s+1}^{n} \alpha_{ij}(t) v_j(0, t) = g_i(t), \quad t \geqslant 0, \quad i = 1, 2, \ldots, s. \tag{5.8}$$

In an analogous way, on the right-hand boundary $x = H$ the boundary conditions can be set in the following way

$$v_i(H, t) + \sum_{j=1}^{s} \alpha_{ij}(t) v_j(H, t) = g_i(t), \quad t \geqslant 0, \quad i = s+1, \ldots, n. \tag{5.9}$$

In the above $\alpha_{ij}(t)$, $g_i(t)$ are known functions.

The boundary problem (5.5)–(5.9) formulated above has a unique solution, with its smoothness essentially depending on smoothness of the functions included in the initial and boundary conditions, as well as on the system coefficients. Besides, for a smooth solution to be obtained for the functions F, ψ, α_{ij}, and g_i, the conditions of agreement must be met at the angular points of the domain D.

Now go over to the formulation proper and to studying the inverse problems for hyperbolic systems. For simplicity I will limit the following to the case when the matrix A is known, $\alpha_{ij} = 0$, and the inverse problem is to determine the matrix B. In this case the matrices T, Λ are also known and the problem of determining the matrix B is equivalent to that of determining the matrix $C = (c_{ij}, i, j = 1, \ldots, n)$ in (5.5). It is in terms of this equation that the inverse problem will be formulated. Consider n problems of type (5.5)–(5.9) each with its own set of the functions F, ψ, and g_i, but with one and the same matrices Λ, C. All the functions corresponding to the problem numbered l, $1 \leqslant l \leqslant n$, will be marked with a superior l. Therefore

$$\left(E\frac{\partial}{\partial t} + \Lambda\frac{\partial}{\partial x} + C\right)v^l = F^l, \quad v^l|_{t=0} = \psi^l \tag{5.10}$$

$$\begin{aligned} &v_i^l(0,t) = g_i^l(t), \quad i = 1, 2, \ldots, s, \\ &v_i^l(H,t) = g_i^l(t), \quad i = s+1, \ldots, n, \\ &\qquad l = 1, 2, \ldots, n, v^l(x,0) = \psi^l(x). \end{aligned}$$

The task is to find the matrix $C = C(x)$, $x \in [0, H]$, provided the following information concerning the solutions of problem (5.10) is known

$$\begin{aligned} &v_i^l(0,t) = h_i^l(t), \quad i = s+1, \ldots, n \\ &v_i^l(H,t) = h_i^l(t), \quad i = 1, 2, \ldots, s, \quad l = 1, 2, \ldots, n. \end{aligned} \tag{5.11}$$

When examining this problem assume that the matrix Λ is constant [though it is insignificant while the fulfilment of conditions (5.7) is of importance] and that the following conditions of the agreement between the initial and boundary data are met

$$\begin{aligned} &\psi_i^l(0) = g_i^l(0), \quad i = 1, 2, \ldots, s, \\ &\psi_i^l(H) = g_i^l(0), \quad i = s+1, \ldots, n, \quad l = 1, 2, \ldots, n. \end{aligned} \tag{5.12}$$

Here ψ_i^l are the vector ψ^l components.

Lemma 5.1. Let the matrix C be continuous on the intercept $[0, H]$, the functions $F^l \in C(D)$, $\psi^l \in C[0, H]$, $g_i^l \in C[0, \infty)$. Then at every l the solution to problem (5.10) and (5.12) is continuous in the domain D.

Now prove this lemma with the view of clarifying the principle of reducing problem (5.10) to a certain system of integral equations of Volterra type.

Consider the characteristics of system (5.10) which are the lines along which the following equality holds

$$\mathrm{d}x/\mathrm{d}t = \lambda_i, \quad i = 1, 2, \ldots, n.$$

Under the suppositions assumed they are the straight lines $x - \lambda_i t = \text{const}$. Along each of the lines the expression

$$\left(\frac{\partial}{\partial t} + \lambda_i \frac{\partial}{\partial x}\right) v_i^l$$

is a derivative taken in the direction of the characteristic. It allows one to integrate every component of equality (5.10) along the corresponding characteristic from an arbitrary point (x, t) lying inside D to the point of the characteristic's crossing with the boundary of the domain D, and at the points of the D boundary to employ the initial and boundary conditions depending on where the point of crossing is located. Now describe it in detail, beginning with denoting the domain D boundary through Γ, after singling out its component parts Γ_0, Γ_1, and Γ_2, so as $\Gamma = \Gamma_0 \cup \Gamma_1 \cup \Gamma_2$

$$\Gamma_0 = \{(x, t): 0 \leqslant x \leqslant H, t = 0\}$$
$$\Gamma_1 = \{(x, t): x = 0, t > 0\}$$
$$\Gamma_2 = \{(x, t): x = H, t > 0\}.$$

Let (x, t) be an arbitrary point of the domain D. Draw the characteristic L_i corresponding to λ_i through the point (x, t) on the plane of the variables ξ and τ, and then continue it up to the crossing with the boundary Γ in the domain $\tau \leqslant t$, the point of crossing denoted through (x_Γ^i, t_Γ^i). In this case $x_\Gamma^i = x_\Gamma^i(x, t)$, $t_\Gamma^i = t_\Gamma^i(x, t)$. For the values $i = 1, 2, \ldots, s$ the point (x_Γ^i, t_Γ^i) lies on Γ_0 or Γ_1, while for the values $i = s + 1, \ldots, n$, on Γ_0 or Γ_2.

Integrate the ith component of equality (5.10) along the characteristic L_i from (x_Γ^i, t_Γ^i) to the point (x, t). In this case one gets

$$v_i^l(x, t) = v_i^l(x_\Gamma^i, t_\Gamma^i) + \int_{t_\Gamma^i}^{t} \left[F_i^l(\xi, \tau) + \sum_{i,j=1}^{n} c_{ij}(\xi) v_j^l(\xi, \tau) \right]_{\xi = x + \lambda_i(\tau - t)} d\tau,$$

$$i, l = 1, 2, \ldots, n. \quad (5.13)$$

Through adding to the system of equalities the initial and boundary conditions from relations (5.10), a closed system of integral Volterra equations of the second kind with continuous kernels and free terms is obtained. This system of equations is understood in the following way: if the point $(x_\Gamma^i, t_\Gamma^i) \in \Gamma_0$, then the values of $v_i^l(x_\Gamma^i, t_\Gamma^i)$ should be calculated using the initial conditions, while if $(x_\Gamma^i, t_\Gamma^i) \in \Gamma_1 \cup \Gamma_2$, using the boundary conditions.

Let

$$\lambda_0 = \max_{1 \leqslant i \leqslant n} |\lambda_i|$$

(x_0, t_0) is an arbitrary point of the domain D. The system of equations (5.13) together with the initial and boundary conditions is closed in any domain

$$G(x_0, t_0) = \{(x, t): (x, t) \in D, \tau \leqslant t_0 - |x - x_0| / \lambda_0\}$$

and defines a continuous function $v^l(x, t)$ in it. Therefore, the lemma is proved.

To investigate the inverse problem it is convenient to introduce the functions

$$w^l(x,t) = \frac{\partial}{\partial t} v^l(x,t), \quad l = 1, 2, \ldots, n.$$

In order to study their properties differentiate (5.10) and the boundary conditions over t, and find the w^l value at $t = 0$ using (5.10) and the initial conditions. As a result each function w^l is a solution to the problem

$$\left(E\frac{\partial}{\partial t} + \Lambda\frac{\partial}{\partial x} + C\right) w^l = F_t^l$$

$$w^l|_{t=0} = F^l(x,0) - \Lambda\frac{\mathrm{d}}{\mathrm{d}x}\psi^l(x) - C(x)\psi^l(x)$$

$$w_i^l(0,t) = \frac{\mathrm{d}}{\mathrm{d}t} g_i^l(t), \quad i = 1, 2, \ldots, s \tag{5.14}$$

$$w_i^l(H,t) = \frac{\mathrm{d}}{\mathrm{d}t} g_i^l(t), \quad i = s+1, \ldots, n.$$

Problem (5.14) is of the same kind as (5.10). In the same way of integrating along the characteristics one can write integral relations for the components w_i^l. They are as follows:

$$w_i^l(x,t) = w_i^l(x_\Gamma^i, t_\Gamma^i) + \int_{t_\Gamma^i}^{t} \left[\frac{\partial}{\partial \tau} F_i^l(\xi,\tau) - \sum_{j=1}^{n} c_{ij}(\xi) w_j^l(\xi,\tau)\right]_{\xi = x + \lambda_i(\tau - t)} \mathrm{d}\tau,$$

$$(x,t) \in D, \quad i = 1, 2, \ldots, n \tag{5.15}$$

and in the totality with the initial and boundary conditions define a closed system of equations in any domain $G(x_0, t_0)$.

Nevertheless, there is a distinction between (5.14) and (5.10). In (5.10) the conditions of conformity between the initial and boundary conditions at the angular points of the domain are fulfilled and thus the solution is kept continuous in the domain D. In (5.14) to demand that the conformity conditions be met would be illogical from the standpoint of the inverse problem, since these conditions would involve the values of the matrix C elements at the points $x = 0$, $x = H$. In the case when the conformity conditions are not met w_i^l, $i = 1, 2, \ldots, s$ are discontinuous at the point $(0,0)$ and w_i^l, $i = s+1, \ldots, n$ are discontinuous at the point $(H, 0)$, these discontinuities propagating along the characteristics inside the domain D. As a result, even at infinitely smooth functions F^l, ψ^l each of the functions w_i^l, $i = 1, 2, \ldots, s$ will be discontinuous along the characteristic L_i outgoing from the point $(0,0)$, and each of the functions w_i^l, $i = s+1, \ldots, n$ along the characteristic L_i outgoing from the point $(H,0)$ to the inside of the domain D. Therefore, each component w_i^l of the function w^l appears to be discontinuous along the characteristic L_i outgoing either from the point $(0,0)$ for $i = 1, 2, \ldots, s$, or from the point $(H, 0)$ for $i = s+1, \ldots, n$. The following fairly obvious lemma is valid.

Lemma 5.2. Let the matrix C be continuous on the intercept $[0, H]$, the functions $F^l \in C^1(D)$, $\psi^l \in C^1[0, H]$, $g_i^l \in C^1(0, \infty)$. In this case the solution of (5.14) is a piecewise-continuous function in the domain D, each of the functions $w_i^l(x, t)$ having a finite discontinuity along the characteristic L_i, outgoing from one of the angular points of the domain D.

To solve the inverse problem (5.10) and (5.11), when fulfilling the conditions of Lemma 5.2, it is necessary that the functions $h_i^l(t)$ continuous on the segment $[0, \infty)$, satisfy the conditions of conformity with the initial data

$$h_i^l(0) = \psi_i^l(0), \quad i = s+1, \ldots, n, \quad l = 1, 2, \ldots, n$$
$$h_i^l(0) = \psi_i^l(H), \quad i = 1, 2, \ldots, s, \quad l = 1, 2, \ldots, n$$

and have piecewise-continuous derivatives on $[0, \infty)$, in which case for $(d/dt)h_i^l(t)$ finite discontinuities are allowed only at the points $t = H/|\lambda_i|$.

For determining the matrix C elements on the intercept $[0, H]$ one shall need only a part of information (5.11), i.e. it is sufficient to consider each of the functions $h_i^l(t)$ known only on the intercept $[0, H/|\lambda_i|]$. In conformity with the consequence discussed, the function $h_i^l(t) \in C^1[0, H/|\lambda_i|]$.

Denote through Ψ the matrix formed by the columns of ψ^l: $\Psi = (\psi^1, \psi^2, \ldots, \psi^n)$.

Theorem 5.1. If the conditions of Lemma 5.2 and the necessary condition for the inverse problem data are met and, besides

$$\det \Psi(x) \neq 0, \quad x \in [0, H] \tag{5.16}$$

then there exists such $H^* > 0$, that for $H \in (0, H^*)$ a solution to the inverse problem (5.10) and (5.11) exists, is unique and is determined by setting $h_i^l(t)$ for $t \in [0, H/|\lambda_i|]$.

To prove the theorem, consider an arbitrary point $(x, 0) \in \Gamma_0$ and draw the characteristic line L_i through it until its crossing with the boundary $\Gamma_1 \cup \Gamma_2$. Integrating the ith component of (5.14) and using the initial conditions and data (5.14), one obtains

$$\sum_{j=1}^{n} c_{ij}(x)\psi_j^l(x) = F_i^l(x, 0) - \lambda_i \frac{d}{dx}\psi_i^l(x) - \frac{d}{dt}h_i^l(t_i(x))$$
$$+ \int_0^{t_i(x)} \left[\frac{\partial}{\partial \tau}F_i^l(\xi, \tau) - \sum_{j=1}^{n} c_{ij}(\xi) w_j^l(\xi, \tau)\right]_{\xi = x + \lambda_i \tau} d\tau,$$
$$x \in [0, H], \quad i, l = 1, 2, \ldots, n. \tag{5.17}$$

Here

$$t_i(x) = \frac{1}{|\lambda_i|}\begin{cases} H - x, & i = 1, 2, \ldots, s \\ x, & i = s+1, \ldots, n. \end{cases}$$

Let

$$\mu = \min_{1 \leqslant i \leqslant n} |\lambda_i|$$
$$\Pi = \{(x, t): 0 \leqslant x \leqslant H, 0 \leqslant t \leqslant H/\mu\}.$$

The system of equations (5.15) and (5.17), modified with the initial and boundary conditions from (5.14) is a closed system of integral equations of the second kind with respect to the known functions c_{ij}, w_i^l. Indeed, due to condition (5.16) the system of equations (5.17) can be solved with respect to $c_{ij}(x)$. Therefore, in order to reduce the whole system (5.15) and (5.17) to a normal form it is sufficient to replace in the initial data (5.14) the matrix C elements with their expressions from the solved system of equations (5.17). For a small parameter this system contains an interval of integration that does not exceed H/μ, which makes it possible to apply the principle of contracted mapping at small H. One can easily verify that each iteration results in continuous in the domain Π functions $c_{ij}(x)$ and piecewise-continuous functions w_i^l with discontinuities along the corresponding characteristics outgoing from the point $(0,0)$, $(H,0)$. As the scheme of using the principle of contracted mapping has been considered in detail in Chapter 2, I refer the reader there for more detailed clarifications of the proof.

In the case if $\lambda_i < 0, i = 1, 2, \ldots, n$, the right-hand boundary $x = H$ can be moved back to infinity, by setting $H = \infty$, in which case the Cauchy problem can serve as a direct problem, and the additional data can be given at $x = 0$. For the inverse problem formulated in such a way valid is the theorem of uniqueness of the solution as a whole, which is analogous to Theorem 2.2.

Theorem 5.1 is of a small-scale character. It is of interest to check whether the supposition on the smallness of H is necessary indeed, or its use is necessitated by the method used.

5.2. Inverse problems using focused sources of wave generation

Here the inverse problem in the formulation considered in Section 5.1 is considered under the assumption that the data of the direct problem (5.5)–(5.9) are generalized functions concentrated on the boundaries of the domain D, i.e. $v^l(x,t)$, $l = 1, 2, \ldots, n$, are the solutions of the following boundary problems

$$\left(E\frac{\partial}{\partial t} + \Lambda\frac{\partial}{\partial x} + C\right)v^l = 0, \quad (x,t) \in D$$

$$\begin{aligned} &v^l|_{t<0} = 0 \qquad (5.18)\\ &v_i^l(0,t) = \delta(t)\delta_i, \quad i = 1, 2, \ldots, s\\ &v_i^l(H,t) = \delta(t)\delta_{il}, \quad i = s+1, \ldots, n, \quad l = 1, 2, \ldots, n. \end{aligned}$$

Here δ_{ij} is the Kronecker symbol. The task is to find the matrix $C = (c_{ij}, i, j = 1, 2, \ldots, n)$ on the intercept $[0, H]$, provided the following information is available on the solutions to (5.18)

$$\begin{aligned} &v_i^l(0,t) = h_i^l(t), \quad i = s+1, \ldots, n \qquad (5.19)\\ &v_i^l(H,t) = h_i^l(t), \quad i = 1, 2, \ldots, s, \quad l = 1, 2, \ldots, n. \end{aligned}$$

In investigating this problem it is assumed that all λ_k are different.

Transition from distributed data to those with localized carriers is very important in an applied sense since in practice one always deals with effects concentrated within a comparatively limited domain of the space. At the same time the inverse problems with focused sources of wave generation possess their own peculiarities, as has been shown when considering the problems in Chapters 2 and 4. One of the peculiarities lies in the fact that when the data are focused on the boundary of a physical domain occupied by a body (in the case considered at the points $x=0$, $x=H$), one is unable to find all the coefficients of the differential equation if this equation is sufficiently general (see Section 2.7). As will be shown below, in the inverse problem (5.18)–(5.19) one is also unable to find all the elements of the matrix C, but it is possible, however, to find $n(n-1)$ nonlinear combinations between the elements of this matrix. The fact that it is impossible to find all the matrix C elements results from the following construction. Introduce new unknown functions $\bar{v}_i^l(x,t)$, having determined them by the formulas

$$\bar{v}_i^l(x,t)=v_i^l(x,t)p_i(x), \quad i,l=1,2,\dots,n,$$

$$p_i(x)=\exp\left[\frac{1}{\lambda_i}\int_0^x c_{ii}(\xi)\,\mathrm{d}\xi\right], \quad i=1,2,\dots,n. \tag{5.20}$$

In terms of the newly introduced functions relations (5.18)–(5.19) can be written as follows:

$$\begin{aligned}
&\left(E\frac{\partial}{\partial t}+\Lambda\frac{\partial}{\partial x}+\bar{C}\right)\bar{v}^l=0\\
&\quad \bar{v}^l|_{t<0}=0\\
&\quad \bar{v}_i^l(0,t)=\delta(t)\delta_{il}, \quad i=1,2,\dots,s\\
&\quad \bar{v}_i^l(H,t)=\delta(t)\delta_{il}p_i(H), \quad i=s+1,\dots,n, \quad l=1,2,\dots,n\\
&\quad \bar{v}_i^l(0,t)=h_i^l(t), \quad i=s+1,\dots,n\\
&\quad \bar{v}_i^l(H,t)=p_i(H)h_i^l(t), \quad i=1,2,\dots,s, \quad l=1,2,\dots\ n.
\end{aligned} \tag{5.21}$$

Here $\bar{C}$ is the matrix with the elements $\bar{c}_{ij}$, $i,j=1,2,\dots,n$, defined by

$$\bar{c}_{ii}=0, \qquad \bar{c}_{ij}=c_{ij}p_i/p_j, \quad i\neq j, \quad i,j=1,2,\dots,n. \tag{5.22}$$

The system of equalities (5.21) and (5.22) includes $n(n-1)$ combinations of the matrix C elements (the elements $\bar{c}_{ij}$ for $i\neq j$), as well as n numbers $p(H)$, $i=1,2,\dots,n$, which are the functionals of the elements c_{ii} of the matrix C. Therefore, one can only find $\bar{c}_{ij}$ and $p_i(H)$ in the inverse problem. I shall demonstrate below that in fact they can be found, and uniquely, provided the length of the intercept $[0,H]$ is sufficiently small.

First study a structure of the solution of the direct problem (5.21).

Lemma 5.3. If $c_{ij}\in C[0,H]$, then the solution to problem (5.21) can be written as follows:

$$\bar{v}_i^l(x,t)=\delta_{il}\delta\left(t-\frac{x}{\lambda_i}\right)+w_i^l(x,t), \quad i=1,2,\dots,s \tag{5.23}$$

$$\bar{v}_i^l(x,t)=p_i(H)\delta_{il}\delta\left(t-\frac{x-H}{\lambda_i}\right)+w_i^l(x,t),$$

$$i = s+1, \ldots, n, l = 1, 2, \ldots, n$$

where $w_i^l(x,t)$ are piecewise-continuous functions in the domain D.

To prove this lemma replace $\bar{v}_i^l$ in (5.21) with expressions from (5.23). Then for each function $w^l = (w_i^l,\ i = 1, 2, \ldots, n)$ the following is obtained

$$\frac{\partial}{\partial t} w_i^l + \lambda_i \frac{\partial}{\partial x} w_i^l + \sum_{j=1}^{n} \bar{c}_{ij} w_j^l + \sum_{j=1}^{s} \bar{c}_{ij} \delta_{jl} \delta\left(t - \frac{x}{\lambda_j}\right)$$

$$+ \sum_{j=s+1}^{n} \bar{c}_{ij} p_j(H) \delta_{jl} \delta\left(t - \frac{x-H}{\lambda_j}\right) = 0 \tag{5.24}$$

$$\begin{aligned} w_i^l|_{t<0} &= 0, \quad i = 1, 2, \ldots, n \\ w_i^l(0,t) &= 0, \quad i = 1, 2, \ldots, s \\ w_i^l(H,t) &= 0, \quad i = s+1, \ldots, n, \quad l = 1, 2, \ldots, n. \end{aligned}$$

Integrating the system of equalities along the characteristics, problem (5.24) is reduced to the system of integral equations

$$w_i^l(x,t) + \int_{t_\Gamma^i(x,t)}^{t} \sum_{j=1}^{n} c_{ij}(\xi) w_j^l(\xi,\tau)|_{\xi = x + \lambda_i(\tau - t)} \,\mathrm{d}\tau$$

$$+ F_i^l(x,t) = 0, \quad (x,t) \in D, \quad i, l = 1, 2, \ldots, n \tag{5.25}$$

where

$$t_\Gamma^i(x,t) = \begin{cases} \max[0, t - x/\lambda_i], & i = 1, 2, \ldots, s \\ \max[0, t + (H-x)/\lambda_i], & i = s+1, \ldots, n \end{cases}$$

$$F_i^l(x,t) = \frac{\lambda_l}{|\lambda_l - \lambda_i|} \bar{c}_{il} \left(\frac{\lambda_l (x - \lambda_i t)}{\lambda_l - \lambda_i} \right)$$

$$\times \begin{cases} 0, x < \lambda_i t, & i \leqslant s, \lambda_l \geqslant \lambda_i \\ 0, x > \lambda_l t, & i \leqslant s, \lambda_l \geqslant \lambda_i \\ 1, \lambda_i t < x < \lambda_l t, & i \leqslant s, \lambda_l \geqslant \lambda_i \\ 0, x > \lambda_i t, & i \leqslant s, \lambda_l < \lambda_i \\ 0, x < \lambda_l t, & i \leqslant s, \lambda_l < \lambda_i \\ 1, \lambda_l t < x < \lambda_i t, & i \leqslant s, \lambda_l < \lambda_i \\ 0, x > \lambda_l t, & i > s \\ 0, x > H + \lambda_i (t - H/\lambda_l), & i > s \\ 1, x < H + \lambda_i (t - H/\lambda_l), x < \lambda_l t, & i > s \end{cases}$$

$$l = 1, 2, \ldots, s$$

$$F_i^l(x,t) = \frac{|\lambda_l| p_l(H)}{|\lambda_l - \lambda_i|} \bar{c}_{il} \left(\frac{\lambda_l (x - \lambda_i t) - \lambda_i H}{\lambda_l - \lambda_i} \right)$$

$$\times\begin{cases} 0, x < H + \lambda_l t, & i \leqslant s \\ 0, x < \lambda_i(t + H/\lambda_l), & i \leqslant s \\ 1, x > \lambda_i(t + H/\lambda_l), x > H + \lambda_l t, & i \leqslant s \\ 0, x > H + \lambda_i t, & i > s, \quad |\lambda_l| \geqslant |\lambda_i| \\ 0, x < H + \lambda_l t, & i > s, \quad |\lambda_l| \geqslant |\lambda_i| \\ 1, H + \lambda_l t < x < H + \lambda_i t, & i > s, \quad |\lambda_l| \geqslant |\lambda_i| \\ 0, x < H + \lambda_i t, & i > s, \quad |\lambda_l| < |\lambda_i| \\ 0, x > H + \lambda_l t, & i > s, \quad |\lambda_l| < |\lambda_i| \\ 1, H + \lambda_i t < x < H + \lambda_l t, & i > s, \quad |\lambda_l| < |\lambda_i| \end{cases}$$

$$l = s + 1, \ldots, n.$$

The formulas for calculating $F_i^l(x, t)$ look quite cumbersome, but their structure is easily understood using additional graphics. For the case of $l \leqslant s$ there are three possible typical situations associated with different positions of the characteristic L_l, where the delta-function is concentrated, and the characteristic L_i along which the integration takes place (Fig. 5.1). In this case the domain D is subdivided into three subdomains and, depending on the fact into which of them the point (x, t) gets, $F_i^l(x, t)$ is calculated. In this case $F_i^l(x, t)$ is nonzero only when the characteristic L_i, passing through the point (x, t) crosses the characteristic L_l, the same being true for $l > s$.

The system of integral equations (5.25) has continuous kernels and piecewise-continuous free terms. Its solution exists, that can be easily proved by using the method of successive approximations, and is also piecewise-continuous, the lines of finite discontinuities of the function w_i^l coinciding in this case with the lines where the functions F_i^l get discontinuous.

It follows from the lemma that for the inverse problem to be solvable the functions $h(t)$, which are the inverse problem data, must have the following structure

$$h_i^l(t) = p_i(H)\delta_{il}\delta(t + H/\lambda_i) + \theta(t)\bar{h}_i^l(t), \quad i = s + 1, \ldots, n$$

$$h_i^l(t) = \frac{1}{p_i(H)}\delta_{il}\delta(t - H/\lambda_i) + \theta(t)\bar{h}_i^l(t) \tag{5.26}$$

$$i = 1, 2, \ldots, s, l = 1, 2, \ldots, n$$

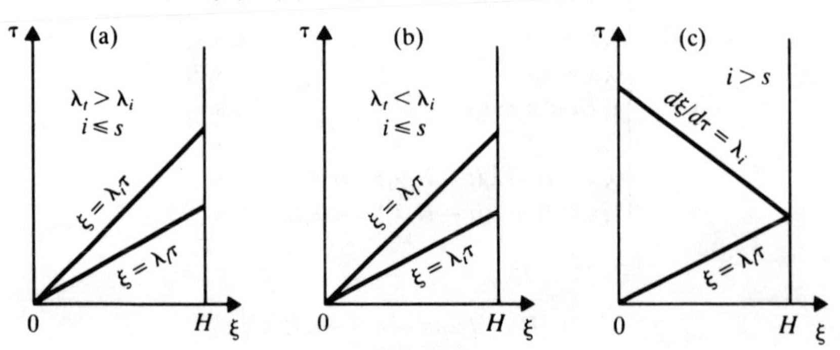

Figure 5.1.

where $\bar{h}_i^l(t)$ are piecewise-continuous functions, in which case for $i = 1, 2, \dots, s$ the $\bar{h}_i^l(t)$ discontinuities are possible only at the points $t = H/\lambda_i$, $t = H/\lambda_l$, if $l \leqslant s$, or at the point $t = H/\lambda_i - H/\lambda_l$ if $l > s$, while for $i > s$; at the point $t = H/\lambda_l$; H/λ_i, if $l \leqslant s$, or at the points $t = H/|\lambda_i|$, $t = H/|\lambda_l|$, if $l > s$.

Now analyse the inverse problem; (5.26) demonstrates that at $i, l = 1, 2, \dots, n$ the inverse problem data make it possible to find $p_i(H)$ as coefficients at the singular parts of $h_i^l(t)$. It should be pointed out that since $\bar{h}_i^l(t)$ is going to be used to determine the coefficients $\bar{c}_{ij}$ that not all the functions $\bar{h}_i^l(t)$ will be needed: $\bar{h}_i^l(t)$ at $i = l = 1, 2, \dots, n$ prove to be useless, while at $i \neq l$ one only uses a part of the domain of setting of the remaining $\bar{h}_i^l(t)$. That is, it is sufficient to consider $\bar{h}_i^l(t)$ known on certain finite lengths of the axis t, the position of these lengths depending on the indices i, l. For $l \leqslant s$ and $i \leqslant s$ the role of such a length is played by a set of points on the axis t between the points $t = H/\lambda_i$, $t = H/\lambda_l$, while for $i > s$ it is the length $[0, H/\lambda_l - H/\lambda_i]$. If $l > s$ one needs to know $\bar{h}_i^l(t)$ at $i \leqslant s$ on the length $[0, H/\lambda_i - H/\lambda_l)$, and when $i > s$—for the t values included between $H/|\lambda_i|$, $H/|\lambda_l|$. Bearing this in mind, limit the inverse problem formulation and henceforth consider given only those functions $\bar{h}_i^l(t)$ and only on those sets that are needed. It should be recalled that $\bar{h}_i^l(t)$ are continuous functions on the sets used.

Theorem 5.2. If $c_{ij} \in C[0, H]$ and the necessary conditions for $h_i^l(t)$ are met, then there exists such $H^* > 0$ that for $H \in (0, H^*)$ there is a unique solution of the inverse problems (5.20) and (5.21).

Let $l \leqslant s$, $i \leqslant s$, $i \neq l$. Consider the characteristic L_l outgoing from the point $(0, 0)$ into the inside of the domain D. Let $(x, t) \in D$ be an arbitrary point of this characteristic and, hence, $t = x/\lambda_l$, $x \in [0, H]$. Draw the characteristic L_i through

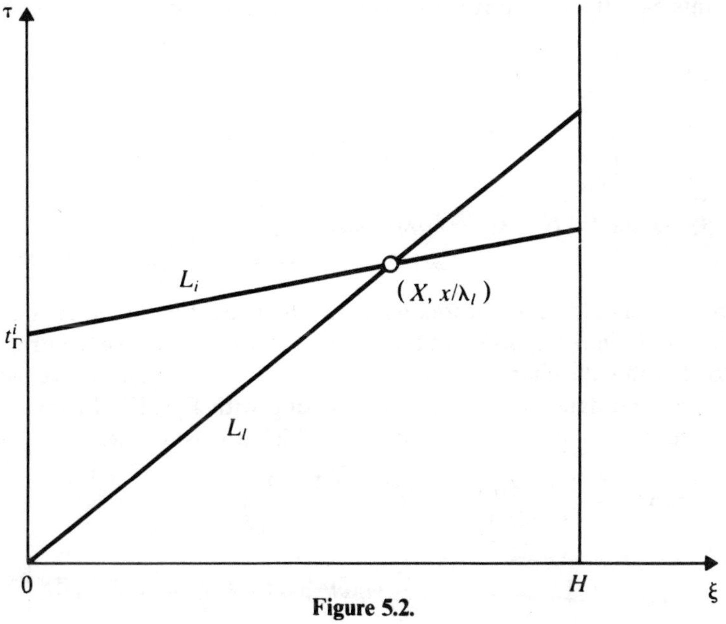

Figure 5.2.

the point $(x, x/\lambda_l)$ up to its crossing the points of the boundary Γ, integrate the ith equation of system (5.24) along this characteristic and use the inverse problem data on the boundary Γ_2 (Fig. 5.2). Then one gets the relations

$$\frac{h_i^l}{\lambda_i}\left(H - x\left(1 - \frac{\lambda_i}{\lambda_l}\right)\right) + \frac{\lambda}{|\lambda_l - \lambda_i|}\bar{c}_{il}(x)$$
$$+ \int_{t_\Gamma^i(x)}^{[H - x(1-\lambda_i/\lambda_l)]/\lambda_i} \sum_{j=1}^{n} c_{ij}(\xi) w_j^l(\xi, \tau)|_{\xi = \lambda_i\tau + x(1-\lambda_i/\lambda_l)} \, d\tau = 0,$$
$$x \in [0, H], \quad i, l = 1, 2, \ldots, s, \quad i \neq l, \quad (5.27)$$

where

$$t_\Gamma^i(x) = \max\left[0, x\left(\frac{1}{\lambda_l} - \frac{1}{\lambda_i}\right)\right].$$

From equalities (5.27) one obtains

$$\bar{c}_{il}(x) = -\frac{|\lambda_l - \lambda_i|}{\lambda_l}\left[h_i^l\left(\frac{H}{\lambda_i} - x\left(\frac{1}{\lambda_i} - \frac{1}{\lambda_l}\right)\right)\right.$$
$$\left. + \int_{t_\Gamma^i(x)}^{[H - x(1-\lambda_i/\lambda_l)]/\lambda_i} \sum_{j=1}^{n} \bar{c}_{ij}(\xi) w_j^l(\xi, \tau)|_{\xi = \lambda_i\tau + x(1-\lambda_i/\lambda_l)} \, d\tau\right],$$
$$x \in [0, H], \quad i, l = 1, 2, \ldots, s, \quad i \neq l. \quad (5.28)$$

For the case when $l \leqslant s, i > s$ I shall carry out the same transformations, except for one point: the inverse problem data will be used on a portion of the boundary Γ_1. In this case the relations for $\bar{c}_{il}(x)$ will be as follows:

$$\bar{c}_{il}(x) = -\frac{|\lambda_l - \lambda_i|}{\lambda_l}\left[h_i^l\left(-\frac{x}{\lambda_i}\left(1 - \frac{\lambda_i}{\lambda_l}\right)\right)\right.$$
$$\left. + \int_{t_\Gamma^i(x)}^{-x(1-\lambda_i/\lambda_l)/\lambda_i} \sum_{j=1}^{n} \bar{c}_{ij}(\xi) w_j^l(\xi, \tau)|_{\xi = \lambda_i\tau + x(1-\lambda_i/\lambda_l)} \, d\tau\right]$$
$$t_\Gamma^i = \max\{0, [H - x(1 - \lambda_i/\lambda_l)]/\lambda_i\},$$
$$l = 1, 2, \ldots, s; \quad i = s+1, \ldots, n, \quad x \in [0, H]. \quad (5.29)$$

If $l > s$, one should act in the following way: to let out the characteristic L_l from the point $(H, 0)$ into the inside of the domain D, to consider an arbitrary point $(x, t) \in L_l$, to draw the characteristic L_i through it at $i \neq l$, and to make use of the inverse problem data at the points of crossing with $\Gamma_1 \cup \Gamma_2$. In this case the required equations for $\bar{c}_{il}(x)$ are obtained, which are as follows:

$$\bar{c}_{il}(x) = -\frac{|\lambda_l - \lambda_i|}{|\lambda_l| p_l(H)}\left[h_i^l\left((H - x)\left(\frac{1}{\lambda_i} - \frac{1}{\lambda_l}\right)\right)\right.$$
$$\left. + \int_{t_\Gamma^i(x)}^{(H-x)(\lambda_i^{-1} - \lambda_l^{-1})} \sum_{j=1}^{n} \bar{c}_{ij}(\xi) w_j^l(\xi, \tau)|_{\xi = \lambda_i(\tau - (x - H/\lambda_l)) + x} \, d\tau\right]$$

$$t_\Gamma^i(x) = \max\left[0, -\frac{x}{\lambda_i} + \frac{x-H}{\lambda_l}\right],$$

$$i = 1, 2, \ldots, s, \quad l = s+1, \ldots, n, \quad x \in [0, H] \quad (5.30)$$

$$\bar{c}_{il}(x) = -\frac{|\lambda_l - \lambda_i|}{|\lambda_l| p_l(H)}\left[\bar{h}_i^l\left(-\frac{x}{\lambda_i} + \frac{x-H}{\lambda_l}\right)\right.$$

$$\left. + \int_{t_\Gamma^i(x)}^{-x/\lambda_i + (x-H)/\lambda_l} \sum_{j=1}^{n} \bar{c}_{ij}(\xi) w_j^l(\xi, \tau)|_{\xi = \lambda_i(\tau - (x - H/\lambda_l)) + x} \, d\tau\right]$$

$$t_\Gamma^i(x) = \max\left[0, (H-x)\left(\frac{1}{\lambda_i} - \frac{1}{\lambda_l}\right)\right],$$

$$i, l = s+1, \ldots, n, \quad i \neq l, \quad x \in [0, H]. \quad (5.30')$$

The system of equalities (5.25) and (5.28)–(5.30′) is a closed system of equations in the domain Π with respect to $\bar{c}_{il}$, w_i^l. At small H the principle of contracted mapping can be applied to the system, which means that the lemma is proved.

Therefore, in the inverse problem (5.18) and (5.19) one can find only the combinations of the matrix C elements that are determined by $\bar{c}_{ij}$, $i \neq j$, and the functionals $p_i(H)$, $i = 1, 2, \ldots, n$.

The problem discussed can result from a more complex one-dimensional inverse problem in a multidimensional space. Here is its formulation: let D_0 be a domain of the space R^{m+1} of the variables $x, y, y = (y_1, \ldots, y_m)$ which can be written as a layer of the H width

$$D_0 = \{(x, y): 0 \leqslant x \leqslant H, y \in R^m\}.$$

In the domain $D = \{(x, y, t): (x, y) \in D_0, t \in R\}$ consider the system of equations

$$\left(E\frac{\partial}{\partial t} + \Lambda\frac{\partial}{\partial x} + \sum_{k=1}^{m} B_k \frac{\partial}{\partial y_k} + Q\right) v = 0 \quad (5.31)$$

with the symmetrical matrices $B_k = (b_{ij}^k, i, j = 1, 2, \ldots, n)$ and an arbitrary matrix $Q = (q_{ij}, i, j = 1, 2, \ldots, n)$. Assume that the matrices B_k, Q depend only on the variable x, and that the matrix Λ is diagonal and constant, all its diagonal elements λ_i being different. Formulate the problem of defining the matrices B_k, Q. For this purpose consider n boundary problems for (5.31)

$$\left(E\frac{\partial}{\partial t} + \Lambda\frac{\partial}{\partial x} + \sum_{k=1}^{m} B_k \frac{\partial}{\partial y_k} + Q\right) v^l = 0, \quad v^l|_{t<0} \equiv 0$$

$$v_i^l(0, y, t) = \delta_{il}\delta(y, t), \quad i = 1, 2, \ldots, s$$

$$v_i^l(H, y, t) = \delta_{il}\delta(y, t), \quad i = s+1, \ldots, n, \quad l = 1, 2, \ldots, n. \quad (5.32)$$

Let the components of the solution to problem (5.32) be known at $x = 0$ and $x = H$

$$v_i^l(0, y, t) = h_i^l(y, t), \quad i = s+1, \ldots, n$$

$$v_i^l(H, y, t) = h_i^l(y, t), \quad i = 1, 2, \ldots, s, \quad l = 1, 2, \ldots, n. \quad (5.33)$$

The task is to find the matrices B_k, Q.

Denote

$$p_{i0}(x) = \exp\left[\frac{1}{\lambda_i}\int_0^x q_{ii}(\xi)\,d\xi\right], p_{ik}(x) = \frac{1}{\lambda_i}\int_0^x b_{ii}^k(\xi)\,d\xi,$$

$$i = 1, 2, \ldots, n, \quad k = 1, 2, \ldots, m.$$

Theorem 5.3. If $B_k, Q \in C[0, H]$, then there exist such $H^* > 0$, that for any $H \in (0, H^*)$ the data (5.33) uniquely define on the intercept $[0, H]$ the functions

$$\begin{aligned} \bar{q}_{ij}(x) &= q_{ij}(x)p_{i0}(x)/p_{j0}(x) \\ \bar{b}_{ij}^k(x) &= b_{ij}^k(x)p_{i0}(x)/p_{j0}(x) \\ d_{ij}^k(x) &= \frac{1}{\lambda_i}b_{ii}^k(x) - \frac{1}{\lambda_j}b_{jj}^k(x), \\ i, j &= 1, 2, \ldots, n, \quad i \neq j, \quad k = 1, 2, \ldots, m \end{aligned} \tag{5.34}$$

and the functionals

$$p_{i0}(H), p_{ik}(H), \quad i = 1, 2, \ldots, n, \quad k = 1, 2, \ldots, m. \tag{5.35}$$

To prove the theorem make use of the Fourier transform of problem (5.32) and (5.33) with respect to the variable y. Such a transform is allowed since for every t the solution of problem (5.32) is finite. In terms of the Fourier images the inverse problem can be written as follows:

$$\begin{aligned} &\left[E\frac{\partial}{\partial t} + \Lambda\frac{\partial}{\partial x} + C(x, \nu)\right]\tilde{v}^l = 0, \quad \tilde{v}^l|_{t<0} \equiv 0 \\ &\tilde{v}_i^l(0, \nu, t) = \delta_{il}\delta(t), \quad t = 1, 2, \ldots, s \\ &\tilde{v}_i^l(H, \nu, t) = \delta_{il}\delta(t), \quad i = s+1, \ldots, n, \quad l = 1, 2, \ldots, n \\ &C(x, \nu) = Q(x) - i\sum_{k=1}^m \nu_k B_k(x), \quad \nu = (\nu_1, \ldots, \nu_m) \end{aligned} \tag{5.36}$$

$$\begin{aligned} \tilde{v}_i(0, \nu, t) &= \tilde{h}_i^l(\nu, t), \quad i = s+1, \ldots, n \\ \tilde{v}_i(H, \nu, t) &= \tilde{h}_i^l(\nu, t), \quad i = 1, 2, \ldots, s, \quad l = 1, 2, \ldots, n. \end{aligned} \tag{5.37}$$

In these formulas ν denotes the parameter of the Fourier transform.

At every fixed ν problems (5.36) and (5.37) coincide with problems (5.18) and (5.19). Consider the ν values lying in a sphere centred at the origin of coordinates and having the radius $\delta > 0$. In conformity with Theorem 5.2, in this case one can find such $H^* = H^*(\delta)$, that in the inverse problems (5.36) and (5.37) for every $|\nu| \leqslant \delta$ on $[0, H]$ at $H \in (0, H^*)$ one can uniquely define the functions

$$\bar{c}_{ij}(x, \nu) = c_{ij}(x, \nu)p_i(x, \nu)/p_j(x, \nu), \quad i, j = 1, 2, \ldots, n, \quad i \neq j$$

where

$$p_i(x, \nu) = \exp\left[\frac{1}{\lambda_i}\int_0^x c_{ii}(\xi, \nu)\,d\xi\right], \quad i = 1, 2, \ldots, n,$$

and the functionals $p_i(H,\nu)$, $i=1,2,\ldots,n$.

Since in the assumed denotations one has

$$p_i(x,\nu)=p_{io}(x)\exp\left[-i\sum_{k=1}^{m}\nu_k p_{ik}\right]$$

$$\bar{c}_{ij}(x,\nu)=\left[\bar{q}_{ij}(x)-i\sum_{k=1}^{m}\nu_k\bar{b}_{ij}^k(x)\right]$$

$$\times\exp\left\{-i\sum_{k=1}^{m}\nu_k[p_{ik}(x)-p_{jk}(x)]\right\}$$

then the functionals $p_i(H,\nu)$, $|\nu|\leqslant\delta$ determine $p_{io}(H)$, $p_{ik}(H)$, and $\bar{c}_{ij}(x,\nu)$, $|\nu|\leqslant\delta$ enables one to determine on $[0,H]$ $\bar{q}_{ij}(x)$, $\bar{b}_{ij}^k(x)$ and $p_{ik}(x)-p_{jk}(x)$ for $i\neq j$. Therefore, Theorem 5.3 is proved.

5.3. Problem of determining the right-hand part of a hyperbolic system

For the equation

$$\left(E\frac{\partial}{\partial t}+\Lambda\frac{\partial}{\partial x}+C(x)\right)v=\Phi(x,t)f(x) \tag{5.38}$$

consider a problem of determining the function $f=(f_1,\ldots,f_n)$ at known matrices Λ, $C(x)$, $\Phi(x,t)$, $\Phi=(\Phi_{ij},\ i,\ j=1,2,\ldots,n)$. As to the elements of the constant diagonal matrix Λ, assume that condition (5.7) is met. Let equation (5.38) be considered in the domain

$$D=\{(x,t):0\leqslant x\leqslant H,t\geqslant 0\}.$$

The task is to find $f(x)$ on the intercept $[0,H]$ provided a trace of the continuously differentiated solution of (5.38)

$$v|_\Gamma=\psi(x,t). \tag{5.39}$$

is known on the boundary Γ of the domain D.

In formulating the problem in such a way, one does not formulate a direct problem for (5.38), though it could be easily done by considering a part of the data from (5.39) as the direct problem data.

Theorem 5.4. If $C(x)\in C[0,H]$, $\Phi(x,t)$, $(\partial/\partial t)\,\Phi(x,t)\in C(D)$ and

$$\det\Phi(x,0)\neq 0,\quad x\in[0,H] \tag{5.40}$$

then there exists such $H^*>0$, that for any $H\in(0,H^*)$ the data (5.39) uniquely define $f(x)\in C[0,H]$.

In proving the theorem assume that $\Phi(x,0)=E$. It is reduced to a general case due to it fulfilling condition (5.40) by replacing Φ by $\bar{\Phi}=\Phi(x,t)\Phi^{-1}(x,0)$ and $f(x)$ by $\bar{f}(x)=\Phi(x,0)\,f(x)$, in which case $\Phi f=\bar{\Phi}\bar{f}$, $\bar{\Phi}(x,0)=E$.

Since the above formulated problem is linear, it is sufficient to prove that the function $\psi(x,t)=0$ is corresponded to by a unique solution from $C[0,H]$ equal to zero. Therefore assume $\psi(x,t)=0$. Consider an additional function

$$w(x,t)=\frac{\partial}{\partial t}v(x,t).$$

This function is a solution to the equation

$$\left(E\frac{\partial}{\partial t}+\Lambda\frac{\partial}{\partial x}+C\right)w=\Phi_t(x,t)f(x) \tag{5.41}$$

and its trace on Γ is calculated with the formula

$$w|_{\Gamma_1\cap\Gamma_2}=0, \qquad w|_{\Gamma_0}=f(x). \tag{5.42}$$

For Γ_k, $k=0,1,2$, the notations of the preceding paragraphs are preserved here. In calculating the trace w on Γ use has been made of (5.38) and the condition $v|_\Gamma=0$, $\Phi(x,0)=E$.

Now consider a rectangle

$$\Pi=\{(x,t):0\leqslant x\leqslant H, 0\leqslant t\leqslant H/\mu\}, \quad \mu=\min_{1\leqslant i\leqslant n}|\lambda_i|.$$

Let (x,t) be an arbitrary point of this rectangle, through which the characteristic L_i is drawn up to its crossing with $\Gamma_1\cup\Gamma_2$, denoting through (x^i_Γ, t^i_Γ) the point of crossing (for the case when there are two points of crossing the one for which $t^i_\Gamma\leqslant t$ is chosen). Integrating the ith component of equality (5.41) along the characteristic L_i from the point (x^i_Γ, t^i_Γ) to the point (x,t), one gets

$$w_i(x,t)=\int_{t^i_\Gamma(x,t)}^{t}[\Phi_\tau(\xi,\tau)f(\xi)-C(\xi)w(\xi,\tau)]_i|_{\xi=x+\lambda_i(\tau-t)}\,d\tau, \quad i=1,2,\ldots,n. \tag{5.43}$$

On the other hand, choosing an arbitrary point $(x,0)\in\Gamma_0$ and integrating the ith component of system (5.41) along the characteristic L_i from the point $(x,0)$ to the crossing point with the boundary $\Gamma_1\cup\Gamma_2$, one obtains

$$f_i(x)=-\int_0^{t^i_\Gamma(x,0)}[\Phi_\tau(\xi,\tau)f(\xi)-C(\xi)w(\xi,\tau)]_i|_{\xi=x+\lambda_i(\tau-t)}\,d\tau, \quad i=1,2,\ldots,n. \tag{5.44}$$

Within the rectangle Π the system of equations (5.43) and (5.44) is a closed system of uniform integral equations.

Denote

$$\|w\|=\max_{1\leqslant i\leqslant n}\max_{(x,t)\in\Pi}|w_i(x,t)|$$

$$\|\Phi_t\|=\max_{1\leqslant i\leqslant n}\sum_{j=1}^{n}\max_{(x,t)\in\Pi}\left|\frac{\partial}{\partial t}\Phi_{ij}(x,t)\right|.$$

Equalities (5.43) and (5.44) result in the estimate

$$\max[\|f\|,\|w\|]\leqslant H/\mu[\|\Phi_t\|\,\|f\|+\|C\|\,\|w\|]. \tag{5.45}$$

From this estimate under the condition

$$H/\mu(\|C\| + \|\Phi_t\|) < 1 \tag{5.46}$$

one has only a zero solution $\|w\| = \|f\| = 0$. Thus, under condition (5.46) $f(x) = 0$, $x \in [0, H]$, and the theorem is proved.

It should be recalled that in proving the theorem use has only been made of that part of information (5.39) which corresponds to the values $t \in [0, H/\mu]$. Thus, in formulating the inverse problem it is sufficient to consider the function $\psi(x, t)$ given for $t \in [0, H/\mu]$.

Now consider a hyperbolic system with a great number of spatial variables of the kind

$$\left(E\frac{\partial}{\partial t} + \Lambda\frac{\partial}{\partial x} + \sum_{k=1}^{m} B_k\frac{\partial}{\partial y_k} + Q\right)v = \Phi(x, t)f(x, y) \tag{5.47}$$

and the matrices B_k and Q depending only on the variable x. Let B_k be symmetrical matrices and $f(x, y)$ be a finite continuous function in the domain $D_0 = \{(x, y): 0 \leqslant x \leqslant H,\ y \in R^m\}$.

The task is to find $f(x, y)$ if on the boundary Γ of the domain $D = \{(x, y, t): (x, y) \in D_0, t \geqslant 0\}$ known is the trace of the continuously differentiated solution of system (5.47) with continuous matrices B_k, Q, Φ, and Φ_t

$$v|_\Gamma = \psi(x, y, t). \tag{5.48}$$

Since the function $f(x, y)$ is finite and system (5.47) is hyperbolic, then the solution $v(x, y, t)$ is also finite at every fixed t. Therefore, one can apply the Fourier transform with respect to the variable y to the system of equalities (5.47) and (5.48), in which case the inverse problem is reduced to a series of one-dimensional problems which depend on the parameter $\nu = (\nu_1, \ldots, \nu_m)$ of the Fourier transform

$$\left(E\frac{\partial}{\partial t} + \Lambda\frac{\partial}{\partial x} + C(x, \nu)\right)\tilde{v}(x, \nu, t) = \Phi(x, t)\tilde{f}(x, \nu) \tag{5.49}$$

$$\tilde{v}|_{\tilde{\Gamma}} = \tilde{\psi}(x, \nu, t). \tag{5.50}$$

Here $\tilde{\Gamma}$ denotes a projection of the set Γ onto the space of the variables x, t

$$C(x, \nu) = Q(x) - i\sum_{k=1}^{m} \nu_k B_k(x).$$

Thus, the initial problem is reduced to the one studied earlier at every fixed ν. It should be recalled that the Fourier image of a finite function is an integral analytical function and, hence, $\tilde{f}(x, \nu)$ is an integral function with respect to ν at every $x \in [0, H]$, i.e. it is determined by its values in a sphere of any smallness $|\nu| \leqslant \delta$, $\delta > 0$. Therefore, when fulfilling the conditions [see (5.46)]

$$H/\mu(\|C(x, \nu)\| + \|\Phi_t(x, t)\|) < 1, \quad |\nu| \leqslant \delta, \tag{5.51}$$

the solution of the inverse problem (5.47) and (5.48) is unique. Since one can choose δ to be as small as desired, then condition (5.51) is equivalent to the condition

$$H/\mu(\|Q\| + \|\Phi_t\|) < 1. \tag{5.52}$$

Thus, if H is chosen from condition (5.52), then the function $f(x, y)$ is uniquely defined by the data (5.48).

In this section we have finished getting acquainted with the general theory of inverse problems for first-order linear hyperbolic systems and now go over to studying some concrete and extremely important, in an applied sense, inverse problems for a system of equations of the theory of elasticity and a system of equations for electrodynamics. Of course, the problems discussed above do not fully cover the state of the general theory of inverse problems for the systems of the first order. A survey of the results obtained in this field can be found in [41] and some publications on particular results are also available [255–269].

5.4. Lamb one-dimensional inverse problem

Alekseev [188] considered for the first time a dynamic problem for a system of equations of the theory of elasticity (see also [187]). Under the supposition that the medium density and the Lamé elastic parameters depend only on one coordinate and through a special choice of the functionals, the problem was reduced from solving the system and using the Fourier–Bessel transform to several spectral Sturm–Liouville inverse problems. Later the same problem was investigated by Blagoveshchenskij [270] who considered it in a nonstationary formulation and also reduced it to several inverse problems for second-order equations. Practical application problems of the inverse dynamic problem of seismology and the associated problem of creating numerical algorithms of its solution have been investigated by a number of authors [189, 190, 192, 203, 216, 271–274].

Consider here a one-dimensional formulation of the inverse problem, writing the system of equations of the theory of elasticity in terms of strains, particle velocities as a system of first-order equations.

Let $x = (x_1, x_2, x_3)$ be a point of the domain

$$D = \{x : x_3 \geqslant 0\}.$$

Denote through σ_{ij} a projection onto the axis x_i of the strain affecting a small area with its normal parallel to the axis x_i, analogous notations taken for the components ε_{ij} of the strain tensor. The Hooke's law for an isotropic elastic body can be written as follows:

$$\begin{aligned} \varepsilon_{ij} &= a\sigma_{ij} - b(\sigma_{11} + \sigma_{22} + \sigma_{33})\delta_{ij}, \quad i, j = 1, 2, 3 \\ a &= 1/2\mu, \qquad b = \lambda/2\mu(3\lambda + 2\mu). \end{aligned} \tag{5.53}$$

Here λ and μ are the Lamé coefficients.

Let, then, u_i be a projection onto the axis x_i of the vector of a particle velocity and ρ the medium density. In this case equations of motion of an elastic body particles in the absence of external forces are as follows (see, for instance, [275]):

$$\rho\frac{\partial u_i}{\partial t}-\sum_{j=1}^{3}\frac{\partial}{\partial x_j}\sigma_{ij}=0,\quad i=1,2,3. \tag{5.54}$$

Components of the strain tensor are connected with the particle velocities u_i through the relations

$$2\frac{\partial}{\partial t}\varepsilon_{ij}=\frac{\partial u_i}{\partial x_j}+\frac{\partial u_j}{\partial x_i},\quad i,j=1,2,3. \tag{5.55}$$

If now in equalities (5.55) one replaces ε_{ij} by their expressions obtained from (5.53), the resulting equations together with (5.54) will form a system of first-order equations with respect to u_i, σ_{ij}. Allowing for the symmetry of the strain tensor $\sigma_{ij}=\sigma_{ji}$, this system can be written as that with respect to a nine-dimensional vector-column U with the components $u_1, u_2, u_3, \sigma_{11}, \sigma_{12}, \sigma_{13}, \sigma_{22}, \sigma_{23}, \sigma_{33}$. The vector U components will be sometimes also denoted by U_i, $i=1,2,\dots,9$. A system of equations of the theory of elasticity is a symmetrical hyperbolic one

$$\left(A\frac{\partial}{\partial t}-\sum_{k=1}^{3}B_k\frac{\partial}{\partial x_k}\right)U=0. \tag{5.56}$$

Here A is a symmetrical matrix of the type

$$A=\begin{Vmatrix} p_1 & 0 & 0\\ 0 & p_2 & p_3\\ 0 & p_3^* & p_4\end{Vmatrix}$$

$$p_1=\begin{Vmatrix}\rho & 0 & 0\\ 0 & \rho & 0\\ 0 & 0 & \rho\end{Vmatrix},\quad p_2=\begin{Vmatrix} a-b & 0 & -b\\ 0 & 2a & 0\\ -b & 0 & a-b\end{Vmatrix}$$

$$p_3=\begin{Vmatrix} 0 & 0 & -b\\ 0 & 0 & 0\\ 0 & 0 & -b\end{Vmatrix},\quad p_4=\begin{Vmatrix} 2a & 0 & 0\\ 0 & 2a & 0\\ 0 & 0 & a-b\end{Vmatrix}.$$

The symbol p_3^* denotes a matrix transposed with respect to the matrix p_3. The elements b_{ij}^k, $i,j=1,2,\dots,9$ of the symmetrical matrices B_k are either zeros or units, i.e.

$$b_{14}^1=b_{25}^1=b_{37}^1=b_{41}^1=b_{52}^1=b_{73}^1=1$$
$$b_{15}^2=b_{26}^2=b_{38}^2=b_{51}^2=b_{62}^2=b_{83}^2=1$$
$$b_{17}^3=b_{28}^3=b_{39}^3=b_{71}^3=b_{82}^3=b_{93}^3=1$$

and all the other $b_{ij}^k=0$.

The matrix A is positive if ρ,λ,μ are positive. Henceforth consider these conditions met.

For the system of equations (5.56) the Lamb problem is correct in order to find a generalized solution to (5.56) satisfying the initial conditions

$$U|_{t<0}\equiv 0 \tag{5.57}$$

and the boundary conditions of the type

$$\sigma_{3j}|_{x_3=0} = f_j\delta(x_1 - x_1^0, x_2 - x_2^0, t), \quad j = 1, 2, 3. \tag{5.58}$$

Here f_j are components of the numerical vector f, and $(x_1^0, x_2^0, 0)$ is a fixed point of the domain D boundary. If the Lamb problem is solved for three linearly independent vectors f and for the arbitrary point $(x_1^0, x_2^0, 0) = x^0$, then one obtains the Green's tensor, using which one can write a solution of the boundary problem with arbitrarily set strains on the plane $x_3 = 0$ as an explicit formula.

Consider a problem inverse to the Lamb problem, the essence of which is in determining the parameters ρ, λ, and μ, characterizing an elastic medium, through the first and third components of the vector U, known on the plane $x_3 = 0$, i.e. through the particle velocities u_i, $i = 1, 3$ given as functions of a point of the plane $x_3 = 0$ and the time t

$$u_i|_{x_3=0} = g_i(x_i, x_2, t), \quad i = 1, 3. \tag{5.59}$$

In this section a one-dimensional formulation of the Lamb inverse problem will be discussed, i.e. it will be assumed that ρ, λ, and μ depend only on the variable x_3, the point x^0 in this case assumed fixed. The next section will be devoted to the three-dimensional Lamb inverse problem within a linear approach, in which case it is necessary that the point x^0 run through all the set of points of the plane $x_3 = 0$, with the functions g_i in (5.59) also depending on x_1^0, x_2^0.

So, begin by analysing the one-dimensional Lamb inverse problem. Assume that in (5.58)

$$f_1 = 1, \qquad f_2 = 0, \qquad f_3 = 1. \tag{5.60}$$

Use the Fourier transform

$$\tilde{U}(\nu, x_3, t) = \iint_{R^2} U(x_1, x_2, x_3, t) \exp\left[i \sum_{k=1}^{2} \nu_k (x_k - x_k^0) \right] dx_1\, dx_2$$

$$\nu = (\nu_1, \nu_2).$$

In terms of this transform problem (5.56)–(5.59) can be written as follows:

$$\left(A\frac{\partial}{\partial t} - B_3 \frac{\partial}{\partial x_3} + i \sum_{k=1}^{2} \nu_k B_k \right) \tilde{U} = 0 \tag{5.61}$$

$$\tilde{U}|_{t<0} \equiv 0, \quad \tilde{\sigma}_{3j}|_{x_3=0} = f_j \delta(t), \quad j = 1, 2, 3$$
$$\tilde{u}_i|_{x_3=0} = \tilde{g}_i(\nu, t), \quad i = 1, 3. \tag{5.62}$$

Theorem 5.5. If λ, μ, and ρ are positive, belong to the space $C^1(0, \infty)$ and

$$\lambda(x_3) \neq 2\mu(x_3), \quad x_3 \in [0, \infty)$$

then the data (5.59) uniquely define λ, μ, and ρ on the segment $[0, \infty)$. Moreover, in order to uniquely define λ, μ, and ρ it is sufficient to set the following three functions of one variable

$$\tilde{g}_1(0, t), \quad \tilde{g}_3(0, t), \quad \frac{\partial}{\partial \nu_1} \tilde{g}_1(\nu, t)|_{\nu=0}.$$

Therefore, information (5.59) is excessive, and selecting from it the information allowing one to uniquely define the required parameters of the medium is realized through the Fourier transform. The proof of the theorem is carried out in several steps: first system (5.61) is written in a classical form and then the investigation of the inverse problem is subdivided into particular inverse problems. In one of them two combinations out of three required coefficients are obtained through the functions $\tilde{g}_1(0,t)$, $\tilde{g}_3(0,t)$, in the other the third combination is found through the function

$$\frac{\partial}{\partial v_1}\tilde{g}_1(v,t)|_{v=0}.$$

In totality these combinations make it possible to find all the medium parameters required.

Now go over from the variable x_3 to the variable

$$z=\int_0^{x_3}\rho(\xi)\,\mathrm{d}\xi. \tag{5.63}$$

Since $\rho>0$, then z is a monotonic function of the variable x_3. Hence, there exists the inverse function $x_3=g(z)$. Replace x_3 by $g(z)$ in all the functions involved in equalities (5.61) and (5.62) and preserve the former notations for them. In this case the term $B_3(\partial/\partial x_3)$ in system (5.61) is replaced by the expression $\rho B_3(\partial/\partial z)$.

Now reduce system (5.61) to a classical form. For this purpose find the matrix T as a solution of the matrix equation

$$ATK+\rho B_3T=0 \tag{5.64}$$

with the condition of normalization

$$T^*AT=\rho E. \tag{5.65}$$

In these formulas H is a unit matrix, and K is a diagonal matrix, with the eigenvalues $k_i, i=1,2,\ldots,9$ of the matrix $-\rho A^{-1}B_3$ standing on its diagonal. The k_i values are the roots of the characteristic equation

$$\det(kA+\rho B_3)=0 \tag{5.66}$$

and the columns of the matrix T are the eigenvectors of the matrix $-\rho A^{-1}B_3$. It can be easily verified that (5.66) can be reduced to the form $k^3[k^2-(\rho v_p)^2][k^2-(\rho v_s)^2]^2=0$ where

$$v_p=\surd((\lambda+2\mu)/\rho), \qquad v_s=\surd(\mu/\rho)$$

define velocities of the transverse and longitudinal seismic waves. Now index the characteristic numbers k_i allowing for their multiplicity in the following way

$$k_1=-k_p \quad k_2=k_3=-k_s, \quad k_4=k_5=k_6=0, \quad k_7=k_8=k_s,$$
$$k_9=k_p, \quad k_p=\rho v_p, \quad k_s=\rho v_s.$$

In this case the matrix T satisfying conditions (5.64) and (5.65) has the following form

$$T=\left\|\begin{matrix} 0 & 0 & q_1 & 0 & 0 & 0 & q_1 & 0 & 0\\ 0 & q_1 & 0 & 0 & 0 & 0 & 0 & q_1 & 0\\ q_1 & 0 & 0 & 0 & 0 & 0 & 0 & 0 & q_1\\ q_2 & 0 & 0 & q_3 & 0 & q_4 & 0 & 0 & -q_2\\ 0 & 0 & 0 & 0 & q_5 & 0 & 0 & 0 & 0\\ q_2 & 0 & 0 & 0 & 0 & q_6 & 0 & 0 & -q_2\\ 0 & 0 & q_7 & 0 & 0 & 0 & -q_7 & 0 & 0\\ 0 & q_7 & 0 & 0 & 0 & 0 & 0 & -q_7 & 0\\ q_8 & 0 & 0 & 0 & 0 & 0 & 0 & 0 & -q_8 \end{matrix}\right\|$$

$$q_1=\frac{1}{\sqrt{2}},\qquad q_2=\frac{k_p^2-2k_s^2}{\sqrt{(2)}k_p},\qquad q_3=\sqrt{\left(\frac{3k_p^2-4k_s^2}{k_p^2-k_s^2}k_s\right)}$$

$$q_4=\frac{k_p^2-2k_s^2}{\sqrt{(k_p^2-k_s^2)}}\frac{k_s}{k_p},\qquad q_5=k_s,\qquad q_6=2\sqrt{(k_p^2-k_s^2)}\frac{k_s}{k_p}$$

$$q_7=\frac{k_s}{\sqrt{2}},\qquad q_8=\frac{k_p}{\sqrt{2}}.$$

Now introduce the function V through the equality $\tilde{U}=TV$. Using conditions (5.64) and (5.65) of choosing the matrix T, the following problem for the function V is obtained

$$\left(E\frac{\partial}{\partial t}+K\frac{\partial}{\partial z}+C\right)V=0$$

$$V|_{t<0}\equiv 0,\qquad (TV)_{6+j}|_{z=0}=f_j\delta(t),\quad j=1,2,3. \tag{5.67}$$

Here the matrix C is defined by the formula

$$C=i\frac{1}{\rho}T^*(v_1B_1+v_2B_2)T-T^*B_3\frac{\partial}{\partial z}T.$$

Direct calculations result in the following structure of the matrix C

$$C=\left\|\begin{matrix} c_1 & c_2v_2 & c_2v_1 & 0 & 0 & 0 & c_3v_1 & c_3v_2 & -c_1\\ c_2v_2 & c_4 & 0 & 0 & c_5v_1 & c_6v_2 & 0 & -c_4 & -c_3v_2\\ c_2v_1 & 0 & c_4 & c_7v_1 & c_5v_2 & c_8v_1 & -c_4 & 0 & -c_3v_1\\ 0 & 0 & c_7v_1 & 0 & 0 & 0 & c_7v_1 & 0 & 0\\ 0 & c_5v_1 & c_5v_2 & 0 & 0 & 0 & c_5v_2 & c_5v_1 & 0\\ 0 & c_6v_2 & c_8v_1 & 0 & 0 & 0 & c_8v_1 & c_6v_2 & 0\\ c_3v_1 & 0 & c_4 & c_7v_1 & c_5v_2 & c_8v_1 & -c_4 & 0 & -c_2v_1\\ c_3v_2 & c_4 & 0 & 0 & c_5v_1 & c_6v_2 & 0 & -c_4 & -c_2v_2\\ c_1 & -c_3v_2 & -c_3v_1 & 0 & 0 & 0 & -c_2v_1 & -c_2v_2 & -c_1 \end{matrix}\right\|$$

$$c_1=-\frac{1}{2}\frac{d}{dz}k_p,\qquad c_2=i\frac{1}{\rho}\frac{k_p^2-2k_s^2+k_pk_s}{2k_p}$$

$$c_3=i\frac{1}{\rho}\frac{k_p^2-2k_s^2-k_pk_s}{2k_p},\qquad c_4=-\frac{1}{2}\frac{\mathrm{d}}{\mathrm{d}z}k_s$$

$$c_5 = i\frac{1}{\rho}\frac{k_s}{\sqrt{2}}, \qquad c_6 = i\frac{1}{\rho}\sqrt{(2(k_p^2 - k_s^2))}\frac{k_s}{k_p}$$

$$c_7 = i\frac{1}{\rho}\sqrt{\left(\frac{3k_p^2 - 4k_s^2}{2}\right)}\frac{k_s}{k_p}, \qquad c_8 = i\frac{1}{\rho}(k_p^2 - 2k_s^2)\frac{k_s}{\sqrt{(2(k_p^2 - k_s^2))}k_p}.$$

In terms of the function V the data (5.62) are as follows:

$$(TV)_i = \tilde{g}_i(v, t), \quad i = 1, 3. \tag{5.68}$$

Set that the parameter $v = 0$ in equalities (5.67). In this case problem (5.67) in the components of the function $W = V|_{v=0}$, $W = (W_i,\ i = 1, 2, \ldots, 9)$ falls into three independent problems which, with equalities (5.60) accounted for, are as follows:

$$\left(\frac{\partial}{\partial t} - k_p\frac{\partial}{\partial z}\right)W_1 + c_1(W_1 - W_9) = 0$$

$$\left(\frac{\partial}{\partial t} + k_p\frac{\partial}{\partial z}\right)W_9 + c_1(W_1 - W_9) = 0 \tag{5.69}$$

$$W_1|_{t<0} = W_9|_{t<0} \equiv 0, \qquad (W_1 - W_9)_{z=0} = \delta(t)/q_8(0)$$

$$\left(\frac{\partial}{\partial t} - k_s\frac{\partial}{\partial z}\right)W_2 + c_4(W_2 - W_8) = 0$$

$$\left(\frac{\partial}{\partial t} + k_s\frac{\partial}{\partial z}\right)W_8 + c_4(W_2 - W_8) = 0 \tag{5.70}$$

$$W_2|_{t<0} = W_8|_{t<0} \equiv 0, \qquad (W_2 - W_8)_{z=0} = 0$$

$$\left(\frac{\partial}{\partial t} - k_s\frac{\partial}{\partial z}\right)W_3 + c_4(W_3 - W_7) = 0$$

$$\left(\frac{\partial}{\partial t} + k_s\frac{\partial}{\partial z}\right)W_7 + c_4(W_3 - W_7) = 0 \tag{5.71}$$

$$W_3|_{t<0} = W_7|_{t<0} \equiv 0, \qquad (W_3 - W_7)_{z=0} = \delta(t)/q_7(0).$$

It follows from equalities (5.70) that $W_2 = W_8 = 0$. Having written the equations for the components W_4, W_5, W_6, one can state that $W_4 = W_5 = W_6 = 0$. Using information (5.68) one gets

$$(W_1 + W_9)_{z=0} = \sqrt{(2)}\tilde{g}_3(0, t), \tag{5.72}$$

$$(W_3 + W_7)_{z=0} = \sqrt{(2)}\tilde{g}_1(0, t). \tag{5.73}$$

As a result, two inverse problems are obtained of the same type on determining k_p, k_s from equalities (5.69), (5.72), and (5.73). Now investigate the first of them, beginning from going over in equalities (5.69) from the variable z to the variable

$$s = \int_0^z \frac{d\xi}{k_p(\xi)}, \quad z = z(s).$$

In this case the system of equalities (5.69) has the following form

$$\left(\frac{\partial}{\partial t}-\frac{\partial}{\partial s}\right)W_1+\phi(s)(W_1-W_9)=0$$

$$\left(\frac{\partial}{\partial t}+\frac{\partial}{\partial s}\right)W_9+\phi(s)(W_1-W_9)=0$$

$$W_1|_{t<0}=W_9|_{t<0}\equiv 0,\qquad (W_1-W_9)_{s=0}=\frac{\sqrt{2}}{k_p(0)}\delta(t) \tag{5.74}$$

$$\phi(s)=-\frac{1}{2}\frac{\mathrm{d}}{\mathrm{d}z}\ln k_p(z(s)).$$

Now introduce the new unknown functions W_{10} and W_{90}, having isolated the singular part of the solution of problem (5.74)

$$W_1(s,t)=W_{10}(s,t)$$

$$W_9(s,t)=-\frac{\sqrt{2}}{k_p(0)}\exp\left[\int_0^s\phi(\xi)\,\mathrm{d}\xi\right]\delta(t-s)+W_{90}(s,t).$$

For the functions W_{10} and W_{90} one has the system of equalities

$$\left(\frac{\partial}{\partial t}-\frac{\partial}{\partial s}\right)W_{10}+\phi(s)(W_{10}-W_{90})+\frac{\sqrt{2}}{k_p(0)}\phi(s)$$

$$\times\exp\left[\int_0^s\phi(\xi)\,\mathrm{d}\xi\right]\delta(t-s)=0 \tag{5.75}$$

$$\left(\frac{\partial}{\partial t}+\frac{\partial}{\partial s}\right)W_{90}+\phi(s)(W_{10}-W_{90})=0$$

$$W_{10}|_{t<0}=W_{90}|_{t<0}\equiv 0,\qquad (W_{10}-W_{90})_{z=0}=0.$$

The solution of problem (5.75) equals zero for the values $t<s$. By integrating over the characteristics at $t\geqslant s$ the following system of integral equations is obtained

$$W_{10}(s,t)+\int_{(s+t)/2}^{t}\phi(\xi)[W_{10}(\xi,\tau)-W_{90}(\xi,\tau)]_{\xi=t+s-\tau}\,\mathrm{d}\tau$$

$$+\frac{1}{\sqrt{(2)}k_p(0)}\phi\left(\frac{s+t}{2}\right)\exp\left[\int_0^{(s+t)/2}\phi(\xi)\,\mathrm{d}\xi\right]=0$$

$$W_{90}(s,t)+\int_{t-s}^{t}\phi(\xi)[W_{10}(\xi,\tau)-W_{90}(\xi,\tau)]_{\xi=\tau+s-t}\,\mathrm{d}\tau \tag{5.76}$$

$$+\int_{(t-s)/2}^{t-s}\phi(\xi)[W_{10}(\xi,\tau)-W_{90}(\xi,\tau)]_{\xi=t-s-\tau}\,\mathrm{d}\tau$$

$$+\frac{1}{\sqrt{(2)}k_{p(0)}}\phi\left(\frac{t-s}{2}\right)\exp\left[\int_0^{(t-s)/2}\phi(\xi)\,\mathrm{d}\xi\right]=0,\quad t\geqslant s\geqslant 0.$$

These equations demonstrate that W_{10} and W_{90} are continuous in the domain $t \geqslant s \geqslant 0$.

As a result, the function $\tilde{g}_3(0,t)$ must be of the form

$$\tilde{g}_3(0,t) = -\frac{1}{k_p(0)}\delta(t) + \bar{g}_3(t)\theta(t)$$

where $\bar{g}_3(t)$ is a continuous function. Therefore, knowing the function $\tilde{g}_3(0,t)$ it can be found by its singular part $k_p(0)$. A regular part of the function $\tilde{g}_3(0,t)$ allows one to obtain an additional equation for finding $\phi(s)$. Indeed, setting $s=0$ in equations (5.76) and adding them results in

$$\sqrt{(2)}\bar{g}_3(t) + \frac{\sqrt{2}}{k_p(0)}\phi(t/2)\exp\left(\int_0^{t/2}\phi(\xi)\,\mathrm{d}\xi\right)$$
$$+2\int_{t/2}^{t}\phi(\xi)[W_{10}(\xi,\tau) - W_{90}(\xi,\tau)]_{\xi=t-\tau}\,\mathrm{d}\tau = 0.$$

Hence

$$\phi(t/2) = -\exp\left[-\int_0^{t/2}\phi(\xi)\,\mathrm{d}\xi\right]$$
$$\times\left\{k_p(0)\bar{g}_3(t) + \sqrt{(2)}k_p(0)\int_{t/2}^{t}\phi(\xi)[W_{10}(\xi,\tau) - W_{90}(\xi,\tau)]_{\xi=t-\tau}\,\mathrm{d}\tau\right\}. \tag{5.77}$$

The system of equations (5.76) and (5.77) is closed in any triangular domain $D_T = \{(s,t): 0 \leqslant s \leqslant t \leqslant T\}$. Its solution can be one and only for any T. At small T one can easily prove that the solution also exists under the supposition of $\bar{g}_3(t)$ continuity. If the function $\phi(s)$ is known one can easily find the function $k_p(z)$ as well.

As a function of the variable s it can be calculated by the formula

$$k_p(z(s)) = k_p(0)\exp[-2\phi(s)].$$

A correlation between the variables z and s can be established through the equality

$$z = z(s) = \int_0^s k_p(z(s))\,\mathrm{d}s = k_p(0)\int_0^s \exp[-2\phi(s)]\,\mathrm{d}s.$$

Thus, one can uniquely find $k_p(z)$ by the function $\tilde{g}_3(0,t)$ and, in an analogous way, $k_s(z)$ is uniquely defined by the function $\tilde{g}_1(0,t)$.

The functions $k_p(z)$, $k_s(z)$ make it possible to find the products $\lambda\rho$, $\mu\rho$ as functions of the variable z, related to x_3 through formula (5.63). For all the parameters λ, μ, and ρ to be completely determined, the task remains to find the function $\rho(z) = \rho(x_3(z))$, which will result in defining λ, μ, and ρ, as functions of the variable z. A correlation can be found between the variables x_3, z through knowing the function $\rho(z)$. Indeed, differentiating equality (5.63) over z, one gets

$$1 = \rho(x_3(z))\,\mathrm{d}x_3/\mathrm{d}z.$$

Hence

$$x_3 = \int_0^z \frac{dz}{\rho(x_3(z))}.$$

Now demonstrate that setting the function

$$\frac{\partial}{\partial v_1}\tilde{g}_1(v,t)|_{v=0}$$

allows one to define the function $\rho(z)$ uniquely.

Denote

$$Y = (Y_j, j = 1, 2, \ldots, 9) \equiv \frac{\partial}{\partial v_1} V|_{v=0}.$$

Differentiating the system of equalities (5.67) and (5.68) over the parameter v_1 and setting $v = 0$ in it, one gets

$$\left(\frac{\partial}{\partial t} - k_s\frac{\partial}{\partial z}\right)Y_3 + c_4(Y_3 - Y_7) + c_2 W_1 - c_3 W_9 = 0$$

$$\left(\frac{\partial}{\partial t} + k_s\frac{\partial}{\partial z}\right)Y_7 + c_4(Y_3 - Y_7) + c_3 W_1 - c_2 W_9 = 0 \qquad (5.78)$$

$$Y_3|_{t<0} = Y_7|_{t<0} \equiv 0, (Y_3 - Y_7)_{z=0} = 0$$

$$(Y_3 + Y_7)_{z=0} = \sqrt{(2)}\frac{\partial}{\partial v_1}\tilde{g}_1(v,t)|_{v=0} = g_{11}(t). \qquad (5.79)$$

In (5.78), W_1 and W_9 are the solution to problem (5.69). Since k_p, k_s have been obtained earlier, then the functions c_4, W_1, W_9 can be considered known here. However, the function $\rho(z)$ involved in the expressions for the coefficients c_2 and c_3 is unknown. The coefficients are now transformed to the form $c_2 = c_{20}/\rho$ and $c_3 = c_{30}/\rho$, where

$$c_{20} = i(k_p^2 - 2k_s^2 + k_p k_s)/2k_p, \qquad c_{30} = i(k_p^2 - 2k_s^2 - k_p k_s)/2k_p$$

are the known functions. Note that c_{30} becomes zero when $k_p = 2k_s$, the latter being possible only if the equality $\lambda = 2\mu$ holds, which is excluded by the theorem conditions. Therefore, under the theorem conditions $c_{30} \neq 0$.

Denote

$$\tau_p(z) = \int_0^z \frac{dz}{k_p(z)}, \qquad \tau_s(z) = \int_0^z \frac{dz}{k_s(z)}.$$

Denote the functions inverse to the functions $t = \tau_s(z)$, $t = \tau_p(z)$ through $z = \tau_s^{-1}(t)$, $z = \tau_p^{-1}(t)$. Making use of the introduced functions one can rewrite the equation for the characteristics involved in (5.69) and (5.78) as follows:

$$\tau = t \pm (\tau_p(\xi) - \tau_p(z)), \qquad \tau = t \pm (\tau_s(\xi) - \tau_s(z)).$$

Isolate the singular components of the solution W_1, W_9 of problem (5.69). As far as problem (5.69) is concerned, the following representation is valid

$$W_1(z,t) = W_{10}(z,t)\theta(t-\tau_p(z)) \tag{5.80}$$

$$W_9(z,t) = -\sqrt{\left(\frac{2}{k_p(0)k_p(z)}\right)}\delta(t-\tau_p(z)) + W_{90}(z,t)\theta(t-\tau_p(z))$$

where $W_{10}(z,t)$, $W_{90}(z,t)$ are continuous in the domain $t \geqslant \tau_p(z)$ functions. It can be verified by way of substituting the expressions for W_1, W_9 from (5.80) into (5.69).

Now write a system of integral equations for problem (5.78). From the point (0, 0) the following characteristics $t=\tau_p(z)$, $t=\tau_s(z)$, are removed. The domain $D_0=\{(z,t)\colon z\geqslant 0,\ t\geqslant 0\}$ is divided into three parts $D_1=\{(z,t)\colon 0\leqslant t<\tau_p(z)\}$, $D_2=\{(z,t)\colon \tau_p(z)\leqslant t<\tau_s(z)\}$, and $D_3=\{(z,t)\colon \tau_s(z)\leqslant t\}$. In the domain D_1 $Y_3=Y_7=0$. In $D_2\cup D_3$ the solution of system (5.78) is other than zero. Substitute the expressions for W_1 and W_9 from (5.80) into (5.78) and integrate the system of equalities along the characteristics using the initial and boundary conditions. Figure 5.3(a) and (b) shows the scheme of integration for the case when the point (z,t) is located in the domain D_2 or in the domain D_3.

As a result of integration, the following system of integral equations is obtained

$$\begin{aligned} Y_3(z,t) &+ \int_{\tau_p(\xi_1^*)}^{t} \{c_4(\xi)[Y_3(\xi,\tau)-Y_7(\xi,\tau)] \\ &+ c_2(\xi)W_{10}(\xi,\tau) - c_3(\xi)W_{90}(\xi,\tau)\}_{\xi=\tau_s^{-1}(t-\tau+\tau_s(z))}\,\mathrm{d}\tau \\ &+ \frac{c_{30}(\xi_1^*)}{\rho(\xi_1^*)}\frac{\sqrt{(2k_p(\xi_1^*))}}{\sqrt{(k_p(0))}[k_p(\xi_1^*)+k_s(\xi_1^*)]} = 0 \\ Y_7(z,t) &+ \int_{\tau_p(\xi_2^*)}^{t} \{c_4(\xi)[Y_3(\xi,\tau)-Y_7(\xi,\tau)] \\ &+ c_3(\xi)W_{10}(\xi,\tau) - c_2(\xi)W_{90}(\xi,\tau)\}_{\xi=\tau_s^{-1}(\tau-t+\tau_s(z))}\,\mathrm{d}\tau \\ &+ \frac{c_{20}(\xi_2^*)}{\rho(\xi_2^*)}\frac{\sqrt{(2k_p(\xi_2^*))}}{\sqrt{(k_p(0))}[k_p(\xi_2^*)-k_s(\xi_2^*)]} = 0, \quad (z,t)\in D_2. \end{aligned} \tag{5.81}$$

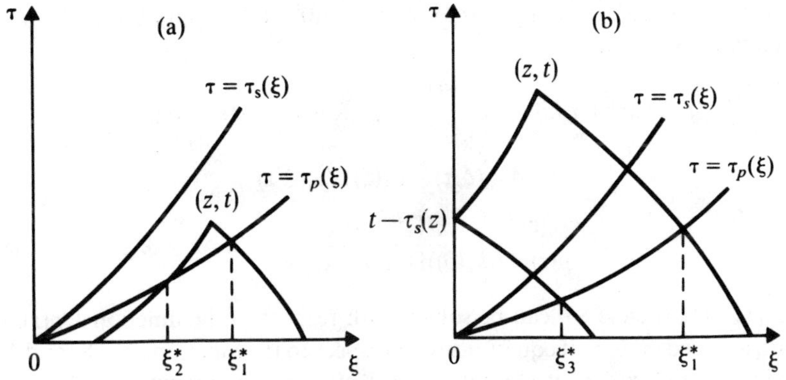

Figure 5.3.

Here ξ_1^* and ξ_2^* are the roots of the equations $\xi_1^* = \xi_1^*(z, t)$: $\tau_p(\xi) + \tau_s(\xi) - t - \tau_s(z) = 0$ and $\xi_2^* = \xi_2^*(z, t)$: $\tau_p(\xi) - \tau_s(\xi) - t + \tau_s(z) = 0$.

An analogous system of equations arises in the domain D_3 as well

$$
\begin{aligned}
&Y_3(z,t) + \int_{\tau_s(\xi_1^*)}^{t} \{c_4(\xi)[Y_3(\xi,\tau) - Y_7(\xi,\tau)] \\
&\quad + c_2(\xi)W_{10}(\xi,\tau) - c_3(\xi)W_{90}(\xi,\tau)\}_{\xi=\tau_s^{-1}(t-\tau+\tau_s(z))}\,\mathrm{d}\tau \\
&\quad + \frac{c_{30}(\xi_1^*)}{\rho(\xi_1^*)} \frac{\sqrt{(2k_p(\xi_1^*))}}{\sqrt{(k_p(0))}[k_p(\xi_1^*) + k_s(\xi_1^*)]} = 0 \\
&Y_7(z,t) + \int_{t-\tau_s(z)}^{t} \{c_4(\xi)[Y_3(\xi,\tau) - Y_7(\xi,\tau)] \\
&\quad + c_3(\xi)W_{10}(\xi,\tau) - c_2(\xi)W_{90}(\xi,\tau)\}_{\xi=\tau_s^{-1}(\tau-t+\tau_s(z))}\,\mathrm{d}\tau \\
&\quad + \int_{\tau_p(\xi_3^*)}^{t-\tau_p(z)} \{c_4(\xi)[Y_3(\xi,\tau) - Y_7(\xi,\tau)] + c_2(\xi)W_{10}(\xi,\tau) \\
&\quad - c_3(\xi)W_{90}(\xi,\tau)\}_{\xi=\tau_s^{-1}(t-\tau-\tau_s(z))}\,\mathrm{d}\tau \\
&\quad + \frac{c_{30}(\xi_3^*)}{\rho(\xi_3^*)} \frac{\sqrt{(2k_p(\xi_3^*))}}{\sqrt{(k_p(0))}[k_p(\xi_3^*) + k_s(\xi_3^*)]} = 0, \quad (z,t)\in D_3
\end{aligned}
\tag{5.82}
$$

here $\xi_3^* = \xi_3^*(z, t)$ is the root of the equation $\tau_p(\xi) + \tau_s(\xi) - t + \tau_s(z) = 0$.

The system of equations (5.81) and (5.82) shows that the functions Y_3 and Y_7 are piecewise-continuous in the domain D_0 and continuous in each of the domains D_k, $k = 1, 2, 3$ (note that the function Y_3 is continuous on the general boundary of the domains D_2, D_3). Therefore, the function $g_{11}(t)$ in (5.79) must be also continuous. Since, due to the boundary condition, $Y_3|_{z=0} = Y_7|_{z=0}$, then (5.79) can be written in the following way

$$Y_3|_{z=0} = \tfrac{1}{2} g_{11}(t).$$

Choose a point $(z, \tau_p(z))$ and integrate the first of the equations of system (5.78) along the passing through it characteristic $\tau + \tau_s(\xi) = \tau_p(z) + \tau_s(z)$ from the point of the crossing with the axis $t = 0$ to the point $(0, \tau_p(z) + \tau_s(z))$. Using the inverse problem data, one gets

$$
\begin{aligned}
&\tfrac{1}{2} g_{11}(\tau_p(z) + \tau_s(z)) + \int_{\tau_p(z)}^{\tau_p(z)+\tau_s(z)} \{c_4(\xi)[Y_3(\xi,\tau) - Y_7(\xi,\tau)] \\
&\quad + c_2(\xi)W_{10}(\xi,\tau) - c_3(\xi)W_{90}(\xi,\tau)\}_{\xi=\tau_s^{-1}(\tau_p(z)+\tau_s(z)-\tau)}\,\mathrm{d}\tau \\
&\quad + \frac{c_{30}(z)}{\rho(z)} \frac{\sqrt{(2k_p(z))}}{\sqrt{(k_p(0))}[k_p(z) + k_s(z)]} = 0, \quad z \geqslant 0.
\end{aligned}
\tag{5.83}
$$

As $c_{30}(z) \neq 0$, then (5.83) can be solved with respect to the function $1/\rho(z)$. As a result, a closed system of equations with respect to the functions Y_3, Y_7 and $1/\rho$ is obtained, which makes it possible to find $\rho(z)$ in a unique way.

Thus, Theorem 5.5 is proved.

Remark. In determining $\rho(z)$ instead of the function

$$\frac{\partial}{\partial v_1}\tilde{g}_1(v,t)|_{v=0}$$

one could use the function $\tilde{g}_1(v,t)$ at a fixed value of the parameter $v=(v_1,v_2)$, $v_1 \neq 0$. In this case investigation of the problem becomes more cumbersome since it requires the use of the total system of equations (5.67), the basic scheme of this investigation, however, remaining the same.

An inverse problem for the system of equations of the theory of elasticity in anisotropic media with a sufficiently general Hook's law has been reported in [276].

5.5. Lamb three-dimensional inverse problem within a linear approach

Here, in line with [277], the problem of determining λ, μ, and ρ as functions of the point $x=(x_1, x_2, x_3)$ in the domain $D=\{x:x_3 \geqslant 0\}$ under the supposition that each of these functions can be presented as a known function, which is only x_3-dependent, and an unknown small addition to it depending on all the three variables x_1, x_2, and x_3 will be considered. Taking into account the form of presentation of the matrix A of system (5.56), it is convenient to consider an equivalent system of the functions ρ, λ, and μ, instead of the functions ρ, a, and b.

Let the functions ρ, a, and b in the domain D be written as follows:

$$\begin{aligned}
&\rho(x)=\rho_0(x_3)+\rho_1(x), \qquad && a(x)=a_0(x_3)+a_1(x)\\
&b(x)=b_0(x_3)+b_1(x) && \\
&a_0=1/2\mu_0, && b_0=\lambda_0/2\mu_0(3\lambda_0+2\mu_0)
\end{aligned}$$

where the functions ρ_0, a_0, and b_0 are known, and ρ_1, a_1, and b_1 are small and finite in D. Henceforth it will be considered that ρ_k, a_k, b_k, $k=0,1$, are twice continuously differentiable in the domain D; $\rho_0, a_0, b_0, \rho, a$, and b are positive and uniformly isolated from zero.

Now consider the problem of determining ρ_1, a_1, and b_1, within a linear approach. In this case the starting point will be the Lamb problem (5.56)–(5.58) and (5.60) considered under the assumption that the parameter $x^0=(x_1^0, x_2^0, 0)$ runs through all the set of points of the domain D boundary. In order to determine ρ_1, a_1, and b_1 the first and third components of the Lamb problem solution will be considered to be set as functions of the points x, x^0 of the domain D boundary and of the time t

$$u_i|_{x_3=0}=g_i(x_1,x_2,t,x_1^0,x_2^0), \quad i=1,3. \tag{5.59'}$$

By writing the Lamb problem solution as

$$U=U^0+U^1$$

where U^0 is the solution of this problem at $\rho=\rho_0$, $a=a_0$, $b=b_0$, the following problem for U^1 in the linear approach is obtained

$$\left(A_0\frac{\partial}{\partial t}-\sum_{k=1}^{3}B_k\frac{\partial}{\partial x_k}\right)U^1=-A_1\frac{\partial}{\partial t}U^0$$

$$U^1|_{t<0}\equiv 0,\qquad U_j^1|_{x_3=0}=0,\qquad j=7,8,9. \tag{5.84}$$

Here A_0 and A_1 are the matrices obtained from the matrix A by way of subscribing the functions ρ, a, and b with the indices 0, 1, respectively; U_j^1 are the components of the vector U^1.

Theorem 5.6. If for any $h>0$ the condition

$$\lambda_0\neq 2\mu_0,\qquad x_3\in[0,h]$$

is met and the data (5.59′) are set in the band

$$-\infty<x_1,x_2,x_1^0,x_2^0<\infty,\qquad 0\leqslant t\leqslant 2\int_0^h\sqrt{\left(\frac{\rho_0(z)}{\mu_0(z)}\right)}\,\mathrm{d}z$$

then these data uniquely define the inverse problem solution in the domain

$$D_h=\{x:0\leqslant x_3\leqslant h\}.$$

In proving this theorem only part of the information (5.59′) is used; this is discussed further below.

Since ρ_0, a_0, and b_0 depend only on the variable x_3, then the solution U^0 of the Lamb problem depends on the variables x_1, x_2, x_1^0, x_2^0 in a special way

$$U^0=U^0(x_1-x_1^0,x_2-x_2^0,x_3,t). \tag{5.85}$$

Apply to both parts of (5.84) and (5.59′) the Fourier transform

$$\tilde U^1(\nu,x_3,t,\kappa)=\int_{R^4}U^1(x,t,x^0)$$
$$\times\exp\left\{i\sum_{k=1}^{2}[(\kappa_k-\nu_k)x_k^0+\nu_kx_k]\right\}\mathrm{d}x_1\,\mathrm{d}x_2\,\mathrm{d}x_1^0\,\mathrm{d}x_2^0,$$
$$\nu=(\nu_1,\nu_2),\quad \kappa=(\kappa_1,\kappa_2).$$

Using formula (5.85), one has

$$\left(A_0\frac{\partial}{\partial t}-B_3\frac{\partial}{\partial x_3}+i\sum_{k=1}^{2}\nu_kB_k\right)\tilde U^1(\nu,x_3,t,\kappa)=\tilde A_1(\kappa,x_3)\frac{\partial}{\partial t}\tilde U^0(\nu-\kappa,x_3,t) \tag{5.86}$$

$$\tilde U^1|_{t<0}\equiv 0,\qquad \tilde U_j^1|_{x_3=0}=0,\quad j=7,8,9$$
$$\tilde U_j^1|_{x_3=0}=\tilde g_j(\nu,t,\kappa),\quad j=1,3. \tag{5.87}$$

At every fixed κ value the problem of determining the matrix $\tilde A_1(\kappa,x_3)$, corresponding to the functions $\tilde\rho_1(\kappa,x_3)$, $\tilde a_1(\kappa,x_3)$, $\tilde b_1(\kappa,x_3)$ is one-dimensional. Investigate it by the scheme suggested in Section 5.5.

First of all, transform system (5.86) to a classical form through using the matrix T corresponding to $\lambda=\lambda_0$, $\mu=\mu_0$, and $\rho=\rho_0$, which has been defined earlier. For this purpose introduce the functions V^0 and V^1 through the equalities

$= TV^0$ and $\tilde{U}^1 = TV^1$ and the variable z through formula (5.63), having set $\rho = \rho_0$.

In terms of these functions (5.86) and (5.87) assume the form

$$\left(E\frac{\partial}{\partial t} + K\frac{\partial}{\partial z} + C\right)V^1 = -R(\kappa, z)\frac{\partial}{\partial t}V^0(v - \kappa, z, t)$$

$$V^1|_{t<0} \equiv 0, \qquad (TV^1)_j|_{z=0} = 0, \quad j = 7, 8, 9 \tag{5.88}$$

$$(TV^1)_j|_{z=0} = \tilde{g}_j(v, t, \kappa), \quad j = 1, 3. \tag{5.89}$$

Here

$$R(\kappa, z) = \frac{1}{\rho_0} T^* \tilde{A}_1(\kappa, z) T.$$

The matrix $R(\kappa, z)$ is of the following structure

$$R = \frac{1}{\rho_0} \times \begin{Vmatrix} r_1 & 0 & 0 & r_2 q_3 & 0 & r_2(q_4+q_6) & 0 & 0 & r_3 \\ 0 & r_4 & 0 & 0 & 0 & 0 & 0 & r_5 & 0 \\ 0 & 0 & r_4 & 0 & 0 & 0 & r_5 & 0 & 0 \\ r_2 q_3 & 0 & 0 & r_6 & 0 & r_7 & 0 & 0 & -r_2 q_3 \\ 0 & 0 & 0 & 0 & r_8 & 0 & 0 & 0 & 0 \\ r_2(q_4+q_6) & 0 & 0 & r_7 & 0 & r_9 & 0 & 0 & -r_2(q_4+q_6) \\ 0 & 0 & r_5 & 0 & 0 & 0 & r_4 & 0 & 0 \\ 0 & r_5 & 0 & 0 & 0 & 0 & 0 & r_4 & 0 \\ r_3 & 0 & 0 & -r_2 q_3 & 0 & -r_2(q_4+q_6) & 0 & 0 & r_1 \end{Vmatrix}$$

$$r_1 = \tilde{\rho}_1 q_1^2 + \tilde{a}_1(2q_2^2 + q_8^2) - \tilde{b}_1(2q_2 + q_8)^2, \quad r_2 = \tilde{a}_1 q_2 - \tilde{b}_1(q_8 + 2q_2)$$
$$r_3 = \tilde{\rho}_1 q_1^2 - \tilde{a}_1(2q_2^2 + q_8^2) + \tilde{b}_1(2q_2 + q_8)^2, \quad r_4 = \tilde{\rho}_1 q_1^2 + 2\tilde{a}_1 q_7^2$$
$$r_5 = \tilde{\rho}_1 q_1^2 - 2\tilde{a}_1 q_7^2, \quad r_6 = (\tilde{a}_1 - \tilde{b}_1)q_3^2, \quad r_7 = [\tilde{a}_1 q_4 - \tilde{b}_1(q_4 + q_6)]q_3$$
$$r_8 = 2\tilde{a}_1 q_5^2, \quad r_9 = (\tilde{a}_1 - \tilde{b}_1)(q_4^2 + q_6^2) - 2\tilde{b}_1 q_4 q_6.$$

Demonstrate that the functions

$$\tilde{g}_1(0, l, \kappa), \quad \tilde{g}_3(0, t, \kappa), \quad \frac{\partial}{\partial v_1}\tilde{g}_1(v, t, \kappa)|_{v=0} \tag{5.90}$$

allow one to define $\tilde{\rho}_1$, $\tilde{a}_1$, and $\tilde{b}_1$ in a unique way.

The functions $\tilde{\rho}_1$, $\tilde{a}_1$, and $\tilde{b}_1$, being Fourier images of the finite functions, are the integral analytical functions of the parameter κ. Therefore they are uniquely determined by their derivatives

$$D_\kappa^\alpha \tilde{\rho}_1|_{\kappa=0} = \rho_1^\alpha(z), \qquad D_\kappa^\alpha \tilde{a}_1|_{\kappa=0} = a_1^\alpha(z)$$
$$D_\kappa^\alpha \tilde{b}_1|_{\kappa=0} = b_1^\alpha(z), \qquad \alpha = (\alpha_1, \alpha_2), |\alpha| \geqslant 0.$$

Using the mathematical induction method it can be proved that $\rho_1^\alpha(z)$, $a_1^\alpha(z)$, and $b_1^\alpha(z)$ are uniquely defined by functions (5.90). From the standpoint of calculations, it is convenient to prove that $\rho_1^\alpha = a_1^\alpha = b_1^\alpha = 0$ if functions (5.90) equal zero. Therefore, assume that

$$\tilde{g}_1|_{v=0} = \tilde{g}_3|_{v=0} = \frac{\partial}{\partial v_1}\tilde{g}_1|_{v=0} = 0. \tag{5.91}$$

The functions V^0 and V^1 are also integral functions of the Fourier transform parameters. Denote

$$D_\kappa^\alpha V^1(v, z, t, \kappa)|_{\kappa=v=0} = W^{1\alpha}(z, t)$$

$$D_\kappa^\alpha V^0(\kappa, z, t)|_{\kappa=0} = W^{0\alpha}(z, t)$$

$$\frac{\partial}{\partial v_1} D_\kappa^\alpha V^1(v, z, t, \kappa)|_{\kappa=v=0} = Y^{1\alpha}(z, t)$$

$$\frac{\partial}{\partial v_1} D_\kappa^\alpha V^0(v-\kappa, z, t)|_{\kappa=v=0} = Y^{0\alpha}(z, t)$$

$$D_\kappa^\alpha R|_{\kappa=0} = R^\alpha(z), \quad C|_{v=0} = C^0, \quad \frac{\partial}{\partial v_1} C|_{v=0} = C^1.$$

In the first step in the mathematical induction method assume that $\alpha = 0$. From equalities (5.88), (5.89), and (5.91) one gets

$$\begin{gathered}\left(E\frac{\partial}{\partial t} + K\frac{\partial}{\partial z} + C^0\right)W^{10} = -R^0\frac{\partial}{\partial t}W^{00} \\ W^{10}|_{t<0} \equiv 0, (TW^{10})_j|_{z=0} = 0, \quad j = 7, 8, 9 \\ (TW^{10})_j|_{z=0} = 0, \quad j = 1, 3\end{gathered} \tag{5.92}$$

$$\begin{gathered}\left(E\frac{\partial}{\partial t} + K\frac{\partial}{\partial z} + C^0\right)Y^{10} + C^1W^{10} = -R^0\frac{\partial}{\partial t}Y^{00} \\ Y^{10}|_{t=0} \equiv 0, \qquad (TY^{10})_j|_{z=0} = 0, \quad j = 7, 8, 9, \\ (TY^{10})_j|_{z=0} = 0, \quad j = 1.\end{gathered} \tag{5.93}$$

Each of the systems (5.92) and (5.93) falls into a number of mutually independent subsystems. If those subsystems are written out that are analogous to systems (5.69), (5.71), and (5.78), one will get three inverse problems interconnected only through the coefficients ρ_1^0, a_1^0, and b_1^0

$$\begin{gathered}\left(\frac{\partial}{\partial t} - k_p\frac{\partial}{\partial z}\right)W_1^{10} + c_1(W_1^{10} - W_9^{10}) = -\frac{1}{\rho_0}\left(r_1^0\frac{\partial}{\partial t}W_1^{00} + r_3^0\frac{\partial}{\partial t}W_9^{00}\right) \\ \left(\frac{\partial}{\partial t} + k_p\frac{\partial}{\partial z}\right)W_9^{10} + c_1(W_1^{10} - W_9^{10}) = -\frac{1}{\rho_0}\left(r_3^0\frac{\partial}{\partial t}W_1^{00} + r_1^0\frac{\partial}{\partial t}W_9^{00}\right) \\ W_1^{10}|_{t<0} = W_9^{10}|_{t<0} \equiv 0, \qquad W_1^{10}|_{z=0} = W_9^{10}|_{z=0} = 0\end{gathered} \tag{5.94}$$

$$\left(\frac{\partial}{\partial t} - k_s\frac{\partial}{\partial z}\right)W_3^{10} + c_4(W_3^{10} - W_7^{10}) = -\frac{1}{\rho_0}\left(r_4^0\frac{\partial}{\partial t}W_3^{00} + r_5^0\frac{\partial}{\partial t}W_7^{00}\right) \tag{5.95}$$

$$\left(\frac{\partial}{\partial t}+k_s\frac{\partial}{\partial t}\right)W_7^{10}+c_4(W_3^{10}-W_7^{10})=-\frac{1}{\rho_0}\left(r_5^0\frac{\partial}{\partial t}W_3^{00}+r_4^0\frac{\partial}{\partial t}W_7^{00}\right)$$

$$W_3^{10}|_{t<0}=W_7^{10}|_{t<0}=0,\qquad W_3^{10}|_{z=0}=W_7^{10}|_{z=0}=0$$

$$\left(\frac{\partial}{\partial t}-k_s\frac{\partial}{\partial z}\right)Y_3^{10}+c_4(Y_3^{10}-Y_7^{10})+c_2W_1^{10}-c_3W_9^{10}$$

$$=-\frac{1}{\rho_0}\left[r_4^0\frac{\partial}{\partial t}Y_3^{00}+r_5^0\frac{\partial}{\partial t}Y_7^{00}\right],$$

$$\left(\frac{\partial}{\partial t}+k_s\frac{\partial}{\partial z}\right)Y_7^{10}+c_4(Y_3^{10}-Y_7^{10})$$

$$+c_3W_1^{10}-c_2W_9^{10}=-\frac{1}{\rho_0}\left[r_5^0\frac{\partial}{\partial t}Y_3^{00}+r_4^0\frac{\partial}{\partial t}Y_7^{00}\right]$$

$$Y_3^{10}|_{t<0}=Y_7^{10}|_{t<0}=0,\qquad Y_3^{10}|_{z=0}=Y_7^{10}|_{z=0}=0. \tag{5.96}$$

In these equations r_k^0 denotes the values of the elements r_k involved in the matrix R, at $\kappa=0$

$$k_p=\sqrt{(\rho_0(\lambda_0+2\mu_0))},\qquad k_s=\sqrt{(\rho_0\mu_0)}.$$

According to the studies carried out in Section 5.4 the functions W_1^{00}, W_3^{00}, W_7^{00}, and W_9^{00} are of the following structure

$$W_1^{00}(z,t)=W_{10}^{00}(z,t)\theta(t-\tau_p(z))$$

$$W_3^{00}(z,t)=W_{30}^{00}(z,t)\theta(t-\tau_s(z))$$

$$W_7^{00}(z,t)=-\sqrt{\left(\frac{2}{k_s(0)k_s(z)}\right)}\delta(t-\tau_s(z))+W_{70}^{00}(z,t)\theta(t-\tau_s(z))$$

$$W_9^{00}(z,t)=-\sqrt{\left(\frac{2}{k_p(0)k_p(z)}\right)}\delta(t-\tau_p(z))+W_{90}^{00}(z,t)\theta(t-\tau_p(z)).$$

Under the assumptions suggested in Section 5.4 ρ_0, a_0, and $b_0\in C^2(0,\infty)$. Therefore, the functions $W_{k0}^{00}(z,t)$ at $k=1$, 3, 7, 9 are continuous together with their first derivatives in the domains where they are other than zero. In line with this fact the derivatives $(\partial/\partial t)W_k^{00}$, $k=1,3,7,9$, have the following structure

$$\frac{\partial}{\partial t}W_1^{00}(z,t)=d_1(z)\delta(t-\tau_p(z))+\theta(t-\tau_p(z))\frac{\partial}{\partial t}W_{10}^{00}(z,t)$$

$$\frac{\partial}{\partial t}W_9^{00}(z,t)=-\sqrt{\left(\frac{2}{k_p(0)k_p(z)}\right)}\delta'(t-\tau_p(z))$$

$$+d_2(z)\delta(t-\tau_p(z))+\theta(t-\tau_p(z))\frac{\partial}{\partial t}W_{90}^{00}(z,t)$$

$$\frac{\partial}{\partial t}W_3^{00}(z,t)=d_3(z)\delta(t-\tau_s(z))+\theta(t-\tau_s(z))\frac{\partial}{\partial t}W_{30}^{00}(z,t) \tag{5.97}$$

$$\frac{\partial}{\partial t} W_7^{00}(z,t) = -\sqrt{\left(\frac{2}{k_s(0)k_s(z)}\right)}\delta'(t-\tau_s(z))$$

$$+ d_4(z)\delta(t-\tau_s(z)) + \theta(t-\tau_s(z))\frac{\partial}{\partial t} W_{70}^{00}(z,t)$$

$$d_1(z) = W_{10}^{00}(z, \tau_p(z)), \qquad d_2(z) = W_{90}^{00}(z, \tau_p(z))$$
$$d_3(z) = W_{30}^{00}(z, \tau_s(z)), \qquad d_4(z) = W_{70}^{00}(z, \tau_s(z)).$$

The values of the functions $d_k(z)$ at $k = 1, 2, 3, 4$ can be easily calculated using the integral equations obtained in the preceding paragraph, in which case one should recalculate the variable s into the variable z. A concrete form of $d_k(z)$ does not play an important role; these calculations will be omitted.

Now write out the structure of the functions $(\partial/\partial t) Y_k^{00}$, $k = 3, 7$, involved in (5.96). Since Y_3^{00} and Y_7^{00} are piecewise-continuous functions and are discontinuous along the characteristics $t = \tau_p(z)$, $t = \tau_s(z)$, Y_3^{00} being continuous while going over the characteristic $t = \tau_s(z)$, then their derivatives with respect to time are of the following structure

$$\frac{\partial}{\partial t} Y_3^{00} = d_5(z)\delta(t-\tau_p(z)) + \theta(t-\tau_p(z))\frac{\partial}{\partial t} Y_{30}^{00}(z,t) \tag{5.98}$$

$$\frac{\partial}{\partial t} Y_7^{00} = d_6(z)\delta(t-\tau_p(z)) + d_7\delta(t-\tau_s(z)) + \theta(t-\tau_p(z))\frac{\partial}{\partial t} Y_{70}^{00}(z,t)$$

where the coefficients d_5, d_6, and d_7 are continuous functions of the variable z, that are equal to the jumps of the functions Y_3^{00} and Y_7^{00} at the corresponding characteristics

$$d_5(z) = -\sqrt{\left(\frac{2k_p(z)}{k_p(0)}\right)}\frac{c_3(z)}{k_p(z)+k_s(z)}$$

$$d_6(z) = -\sqrt{\left(\frac{2k_p(z)}{k_p(0)}\right)}\frac{c_2(z)}{k_p(z)-k_s(z)} \tag{5.99}$$

$(\partial/\partial t) Y_{30}^{00}$ and $(\partial/\partial t) Y_{70}^{00}$ are the piecewise-continuous functions in the domain D. The coefficient $d_7(z)$ can be readily calculated but since its concrete expression is of no importance it will not be determined.

The analysis carried out above, enables one to write out the structure of the functions W_k^{10}, $k = 1, 3, 7, 9$; Y_k^{10}, $k = 1, 4, 6, 9$. Begin with the functions W_1^{10} and W_9^{10}, writing them in the following way

$$W_1^{10}(z,t) = \alpha_1(z)\delta(t-\tau_p(z)) + W_{10}^{10}(z,t) \tag{5.100}$$
$$W_9^{10}(z,t) = \alpha_2(z)\delta'(t-\tau_p(z)) + \alpha_3(z)\delta(t-\tau_p(z)) + W_{90}^{10}(z,t)$$

with the indefinite coefficients α_1, α_2, and α_3. Substituting the expressions for W_1^{10} and W_9^{10} into (5.94) and equating the coefficients at singularities, the equation for determining α_1, α_2, and α_3 is obtained

$$2\alpha_1 - c_1\alpha_2 = \frac{1}{\rho_0(z)}\sqrt{\left(\frac{2}{k_p(0)k_p(z)}\right)} r_3^0(z), \quad \alpha_1(0) = 0$$

$$k_p\alpha_2'(z) - c_1\alpha_2 = \frac{1}{\rho_0(z)}\sqrt{\left(\frac{2}{k_p(0)k_p(z)}\right)}r_1^0(z), \quad \alpha_2(0) = 0 \tag{5.101}$$

$$k_p\alpha_3'(z) + c_1(\alpha_1 - \alpha_3) = -\frac{1}{\rho_0(z)}(r_3^0 d_1 + r_1^0 d_2), \quad \alpha_3(0) = 0$$

and the system of equations for determining the functions W_{10}^{10} and W_{90}^{10}

$$\left(\frac{\partial}{\partial t} - k_p\frac{\partial}{\partial z}\right)W_{10}^{10} + c_1(W_{10}^{10} - W_{90}^{10})$$
$$= \gamma_1(z)\delta(t - \tau_p(z)) + r_1^0\Phi_1(z,t) + r_3^0\Phi_2(z,t) \tag{5.102}$$
$$\left(\frac{\partial}{\partial t} + k_p\frac{\partial}{\partial z}\right)W_{90}^{10} + c_1(W_{10}^{10} - W_{90}^{10}) = r_3^0\Phi_1(z,t) + r_1^0\Phi_2(z,t)$$
$$W_{10}^{10}|_{t<0} = W_{90}^{10}|_{t<0} = 0, \qquad W_{10}^{10}|_{z=0} = W_{90}^{10}|_{z=0} = 0$$

where

$$\gamma_1(z) = k_p\alpha_1'(z) - c_1(\alpha_1 - \alpha_3) - \frac{1}{\rho_0}(r_1^0 d_1 + r_3^0 d_2)$$
$$\Phi_1(z,t) = -\frac{1}{\rho_0}\theta(t - \tau_p(z))\frac{\partial}{\partial t}W_{10}^{00}(z,t) \tag{5.103}$$
$$\Phi_2(z,t) = -\frac{1}{\rho_0}\theta(t - \tau_p(z))\frac{\partial}{\partial t}W_{90}^{10}(z,t).$$

From (5.101) one finds in a successive way

$$\alpha_2(z) = \sqrt{\left(\frac{2}{k_p(0)k_p(z)}\right)}\int_0^z \frac{r_1^0(\xi)}{\rho_0(\xi)k_p(\xi)}\,\mathrm{d}\xi$$
$$\alpha_1(z) = -\tfrac{1}{4}k_p'(z)\alpha_2(z) + \frac{1}{2\rho_0}\sqrt{\left(\frac{2}{k_p(0)k_p(z)}\right)}r_3^0(z) \tag{5.104}$$
$$\alpha_3(z) = \frac{1}{\sqrt{(k_p(z))}}\int_0^z\left[\tfrac{1}{2}k_p'(\xi)\alpha_1(\xi) - \frac{1}{\rho_0}(r_3^0 d_1 + r_1^0 d_2)\right]\frac{d\xi}{\sqrt{(k_p(\xi))}}.$$

From (5.102) by way of integrating along the characteristic (Fig. 5.4) one obtains a system of integral equations for W_{10}^{10} and W_{90}^{10}

$$W_{10}^{10}(z,t) - \int_0^z k_p(\xi)\{c_1(\xi)[W_{10}^{10}(\xi,\tau) - W_{90}^{10}(\xi,\tau)]$$
$$- r_1^0(\xi)\Phi_1(\xi,\tau) - r_3^0(\xi)\Phi_2(\xi,\tau)\}_{\tau = t + \tau_p(z) - \tau_p(\xi)}\,\mathrm{d}\xi = 0$$
$$W_{90}^{10}(z,t) + \int_0^z k_p(\xi)\{c_1(\xi)[W_{10}^{10}(\xi,\tau) - W_{90}^{10}(\xi,\tau)] \tag{5.105}$$
$$- r_3^0(\xi)\Phi_1(\xi,\tau) - r_1^0(\xi)\Phi_2(\xi,\tau)\}_{\tau = t - \tau_p(z) + \tau_p(\xi)}\,\mathrm{d}\xi = 0, \quad t \geqslant \tau_p(z).$$

System (5.105) shows that the functions W_{10}^{10} and W_{90}^{10} are continuous in the domain $t \geqslant \tau_p(z)$. At $t < \tau_p(z)$ they equal zero.

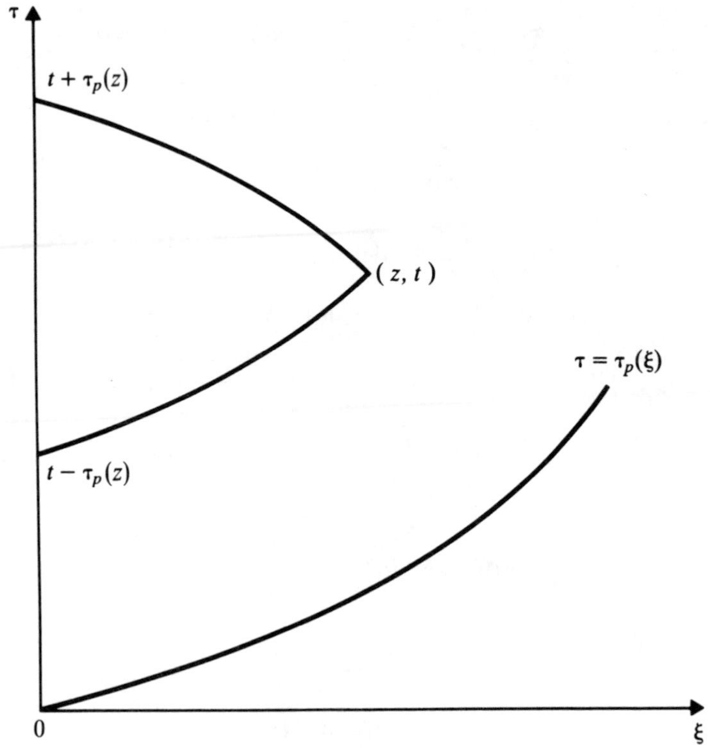

Figure 5.4.

Now consider an arbitrary point $(z, \tau_p(z))$. Integrating the first of equations (5.102) along the characteristic passing through $(z, \tau_p(z))$ from the point of its crossing with the axis $\xi = 0$ to the point of its crossing with $\tau = 0$, one finds an equation for $\gamma_1(z)$, which is as follows:

$$\gamma_1(z) = 2\int_0^z k_p(\xi)\{c_1(\xi)[W_{10}^{10}(\xi,\tau) - W_{90}^{10}(\xi,\tau)] - r_1^0(\xi)\Phi_1(\xi,\tau) - r_3^0(\xi)\Phi_2(\xi,\tau)\}_{\tau = 2\tau_p(z)-\tau_p(\xi)}\,\mathrm{d}\xi, \quad z \geqslant 0. \tag{5.106}$$

Using the first of equalities (5.103) one can construct using the function $\gamma_1(z)$ the equation for determining $\alpha_1(z)$

$$\alpha_1(z) = \int_0^z \frac{1}{k_p(\xi)}\left\{\gamma_1(\xi) + c_1(\xi)[\alpha_1(\xi) - \alpha_3(\xi)] + \frac{1}{\rho_0(\xi)}[r_1^0(\xi)d_1(\xi) + r_3^0(\xi)d_2(\xi)]\right\}\mathrm{d}\xi. \tag{5.107}$$

For the functions $\alpha_2(z)$ and $\alpha_3(z)$ one has the integral equations (5.104). Rewrite the second of equations (5.104) as that for determining the function $r_3^0(z)$

$$r_3^0(z) = \rho_0(z)\sqrt{\left(\frac{k_p(0)k_p(z)}{8}\right)}[4\alpha_1(z) + k_p'(z)\alpha_2(z)]. \tag{5.108}$$

The system of relations (5.95) is identical to system (5.94). Therefore, total analogy of equalities (5.100)–(5.108) can be obtained for it. Write the functions W_3^{10} and W_7^{10} as follows:

$$\begin{aligned} &W_3^{10}(z,t) = \alpha_4(z)\delta(t - \tau_s(z)) + W_{30}^{10}(z,t) \\ &W_7^{10}(z,t) = \alpha_5(z)\delta'(t - \tau_s(z)) + \alpha_6(z)\delta(t - \tau_s(z)) + W_{70}^{10}(z,t). \end{aligned} \tag{5.109}$$

The equations for determining α_4, α_5, and α_6 will have the following form

$$\begin{aligned} 2\alpha_4 - c_4\alpha_5 &= \frac{1}{\rho_0(z)}\sqrt{\left(\frac{2}{k_s(0)k_s(z)}\right)}r_5^0(z) \\ k_s\alpha_5'(z) - c_4\alpha_5(z) &= \frac{1}{\rho_0(z)}\sqrt{\left(\frac{2}{k_s(0)k_s(z)}\right)}r_4^0(z) \\ k_s\alpha_6'(z) + c_4(\alpha_4 - \alpha_6) &= -\frac{1}{\rho_0(z)}[r_4^0(z)d_3(z) + r_5^0(z)d_4(z)], \\ \alpha_4(0) = \alpha_5(0) &= \alpha_6(0) = 0, \end{aligned} \tag{5.110}$$

and for the functions W_{30}^{10} and W_{70}^{10} the equalities

$$\begin{aligned} &\left(\frac{\partial}{\partial t} - k_s\frac{\partial}{\partial z}\right)W_{30}^{10} + c_4(W_{30}^{10} - W_{70}^{10}) \\ &\quad = \gamma_2(z)\delta(t - \tau_s(z)) + r_4^0(z)\Phi_3(z,t) + r_5^0(z)\Phi_4(z,t), \\ &\left(\frac{\partial}{\partial t} + k_s\frac{\partial}{\partial z}\right)W_{70}^{10} + c_4(W_{30}^{10} - W_{70}^{10}) = r_5^0(z)\Phi_3(z,t) + r_4^0(z)\Phi_4(z,t) \\ &W_{30}^{10}|_{t<0} = W_{70}^{10}|_{t<0} = 0, \qquad W_{30}^{10}|_{z=0} = W_{70}^{10}|_{z=0} = 0 \end{aligned} \tag{5.111}$$

are obtained, where

$$\begin{aligned} \gamma_2(z) &= k_s\alpha_4'(z) - c_4(\alpha_4 - \alpha_6) - \frac{1}{\rho_0}(r_4^0 d_3 + r_5^0 d_4) \\ \Phi_3(z,t) &= -\frac{1}{\rho_0}\theta(t - \tau_s(z))\frac{\partial}{\partial t}W_{30}^{00}(z,t) \\ \Phi_4(z,t) &= -\frac{1}{\rho_0}\theta(t - \tau_s(z))\frac{\partial}{\partial t}W_{70}^{00}(z,t). \end{aligned} \tag{5.112}$$

From equalities (5.11) the system of integral equations for W_{30}^{10} and W_{70}^{10} is obtained

$$\begin{aligned} W_{30}^{10}(z,t) - \int_0^z &[k_s(\xi)]^{-1}\{c_4(\xi)[W_{30}^{10}(\xi,\tau) - W_{70}^{(10)}(\xi,\tau)] \\ &- r_4^0(\xi)\Phi_3(\xi,\tau) - r_5^0(\xi)\Phi_4(\xi,\tau)\}_{\tau = t + \tau_s(z) - \tau_s(\xi)}\,d\xi = 0 \end{aligned}$$

$$W_{70}^{10}(z,t)+\int_0^z [k_s(\xi)]^{-1}\{c_4(\xi)[W_{30}^{10}(\xi,\tau)-W_{70}^{10}(\xi,\tau)]$$
$$-r_5^0(\xi)\Phi_3(\xi,\tau)-r_4^0(\xi)\Phi_4(\xi,\tau)\}_{\tau=t-\tau_s(z)+\tau_s(\xi)}\,\mathrm{d}\xi=0,\quad t\geqslant\tau_s(z), \tag{5.113}$$

and the equation for determining $\gamma_2(z)$

$$\gamma_2(z)=2\int_0^z [k_s(\xi)]^{-1}\{c_4(\xi)[W_{30}^{10}(\xi,\tau)-W_{70}^{10}(\xi,\tau)]$$
$$-r_4^0(\xi)\Phi_3(\xi,\tau)-r_5^0(\xi)\Phi_4(\xi,\tau)\}_{\tau=2\tau_s(z)-\tau_s(\xi)}\,\mathrm{d}\xi,\quad z\geqslant 0. \tag{5.114}$$

The first of equalities (5.112) makes it possible to get the equation for $\alpha_4(z)$

$$\alpha_4(z)=\int_0^z \frac{1}{k_s(\xi)}\Big\{\gamma_2(\xi)+c_4(\xi)[\alpha_4(\xi)-\alpha_6(\xi)]$$
$$+\frac{1}{\rho_0(\xi)}[r_4^0(\xi)d_3(\xi)+r_5^0(\xi)d_4(\xi)]\Big\}\,\mathrm{d}\xi. \tag{5.115}$$

From equalities (5.110)

$$\alpha_5(z)=\sqrt{\left(\frac{2}{k_s(0)k_s(z)}\right)}\int_0^z \frac{r_4^0(\xi)}{\rho_0(\xi)k_s(\xi)}\,\mathrm{d}\xi$$

$$\alpha_6(z)=\frac{1}{k_s(z)}\int_0^z \Big\{\tfrac{1}{2}k_s'(\xi)\alpha_4(\xi)-[r_4^0(\xi)d_3(\xi)+r_5^0(\xi)d_4(\xi)]\frac{1}{\rho_0(\xi)}\Big\}\frac{\mathrm{d}\xi}{k_s(\xi)}$$

$$r_5^0(z)=\rho_0(z)\sqrt{\left(\frac{k_s(0)k_s(z)}{8}\right)}[4\alpha_4(z)+k_s(z)\alpha_5(z)]. \tag{5.116}$$

Now go over to the system of equalities (5.96). Express Y_3^{10} and Y_7^{10} in the following way

$$Y_3^{10}(z,t)=\alpha_7(z)\delta(t-\tau_p(z))+\alpha_8(z)\delta(t-\tau_s(z))+Y_{30}^{10}(z,t)$$
$$Y_7^{10}(z,t)=\alpha_9(z)\delta(t-\tau_p(z))+\alpha_{10}(z)\delta(t-\tau_s(z))+Y_{70}^{10}(z,t). \tag{5.117}$$

Substituting these expressions into (5.96) and equating the coefficients at $\delta'(t-\tau_p(z))$ and $\delta'(t-\tau_s(z))$ in the first of them, and the coefficients at $\delta'(t-\tau_p(z))$ and $\delta(t-\tau_s(z))$ in the second, one gets the equations for determining α_k

$$\left(1+\frac{k_s}{k_p}\right)\alpha_7-c_3\alpha_2=0,\quad 2\alpha_8=0$$

$$\left(1-\frac{k_s}{k_p}\right)\alpha_9-c_2\alpha_2=0 \tag{5.118}$$

$$k_s\alpha_{10}'(z)+c_4(\alpha_8-\alpha_{10})=-\frac{1}{\rho_0}r_4^0d_7,\quad \alpha_{10}(0)=-\alpha_9(0).$$

For the functions Y_{30}^{10} and Y_{70}^{10} one obtains the equalities

$$\left(\frac{\partial}{\partial t} - k_s \frac{\partial}{\partial z}\right) Y_{30}^{10} + c_4(Y_{30}^{10} - Y_{70}^{10}) + c_2 W_{10}^{10} - c_3 W_{90}^{10}$$
$$= \gamma_3(z)\delta(t - \tau_p(z)) + \gamma_4(z)\delta(t - \tau_s(z))$$
$$+ r_4^0(z)\Phi_5(z,t) + r_5^0(z)\Phi_6(z,t), \tag{5.119}$$
$$\left(\frac{\partial}{\partial t} + k_s \frac{\partial}{\partial z}\right) Y_{70}^{10} + c_4(Y_{30}^{10} - Y_{70}^{10}) + c_3 W_{10}^{10} - c_2 W_{90}^{10}$$
$$= \gamma_5(z)\delta(t - \tau_p(z)) + r_5^0(z)\Phi_5(z,t) + r_4^0(z)\Phi_6(z,t)$$
$$Y_{30}^{10}|_{t<0} = Y_{70}^{10}|_{t<0} = 0, \qquad Y_{30}^{10}|_{z=0} = Y_{70}^{10}|_{z=0} = 0.$$

Here

$$\gamma_3(z) = k_s\alpha_7' - c_4(\alpha_7 - \alpha_9) - c_2\alpha_1 + c_3\alpha_3 - \frac{1}{\rho_0}(r_4^0 d_5 + r_5^0 d_6)$$
$$\gamma_4(z) = c_4\alpha_{10} - \frac{1}{\rho_0} r_5^0 d_7 \tag{5.120}$$
$$\gamma_5(z) = -k_s\alpha_9' - c_4(\alpha_7 - \alpha_9) - c_3\alpha_1 + c_2\alpha_3 - \frac{1}{\rho_0}(r_5^0 d_5 + r_4^0 d_6)$$
$$\Phi_5(z,t) = -\frac{1}{\rho_0}\frac{\partial}{\partial t} Y_{30}^{00}(z,t)$$
$$\Phi_6(z,t) = -\frac{1}{\rho_0}\frac{\partial}{\partial t} Y_{70}^{00}(z,t).$$

From equalities (5.119) one finds the integral equations for Y_{30}^{10} and Y_{70}^{10}. These equations are written in the domains D_2 and D_3 in different ways: $D_2 = \{(z,t): \tau_p(z) \leqslant t < \tau_s(z)\}$ and $D_3 = \{(z,t): t \geqslant \tau_s(z)\}$. In line with the scheme of integration, shown in Fig. 5.5a and b, one gets

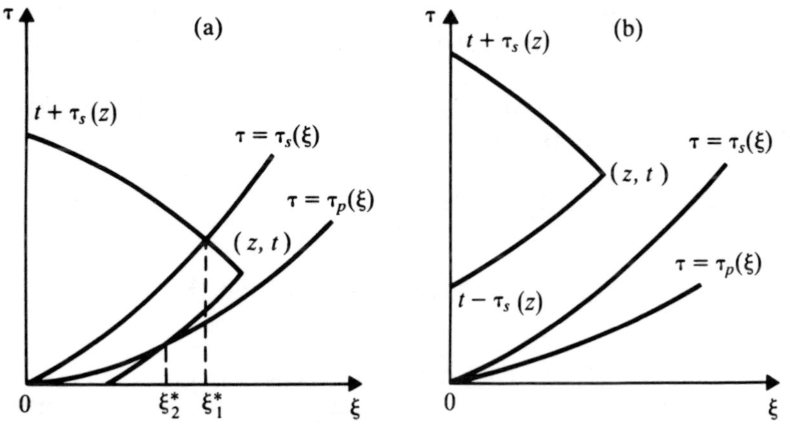

Figure 5.5.

$$Y_{30}^{10}(z,t) - \int_0^z [k_s(\xi)]^{-1}\{c_4(\xi)[Y_{30}^{10}(\xi,\tau) - Y_{70}^{10}(\xi,\tau)] + c_2(\xi)W_{10}^{10}(\xi,\tau)$$
$$- c_3(\xi)W_{90}^{10}(\xi,\tau)\}_{\tau=t+\tau_s(z)-\tau_s(\xi)}\,d\xi = \tfrac{1}{2}\gamma_4(\xi_1^*)$$
$$- \int_0^z [k_s(\xi)]^{-1}[r_4^0(\xi)\Phi_5(\xi,\tau) + r_5^0(\xi)\Phi_6(\xi,\tau)]_{\tau=t+\tau_s(z)-\tau_s(\xi)}\,d\xi$$

$$Y_{70}^{10}(z,t) + \int_{\xi_2^*}^z [k_s(\xi)]^{-1}\{c_4(\xi)[Y_{30}^{10}(\xi,\tau) - Y_{70}^{10}(\xi,\tau)]$$
$$+ c_3(\xi)W_{10}^{10}(\xi,\tau) - c_2(\xi)W_{90}^{10}(\xi,\tau)\}_{\tau=t-\tau_s(z)+\tau_s(\xi)}\,d\xi$$
$$= \gamma_5(\xi_2^*)\frac{k_p(\xi_2^*)}{k_p(\xi_2^*) - k_s(\xi_2^*)} + \int_{\xi_2^*}^z [k_s(\xi)]^{-1}[r_4^0(\xi)\Phi_5(\xi,\tau)$$
$$+ r_5^0\Phi_6(\xi,\tau)]_{\tau=t-\tau_s(z)+\tau_s(\xi)}\,d\xi, \quad (z,t)\in D_2 \tag{5.121}$$

$$Y_{30}^{10}(z,t) - \int_0^z [k_s(\xi)]^{-1}\{c_4(\xi)[Y_{30}^{10}(\xi,\tau) - Y_{70}^{10}(\xi,\tau)] + c_2(\xi)W_{10}^{10}(\xi,\tau)$$
$$- c_3(\xi)W_{90}^{10}(\xi,\tau)\}_{\tau=t+\tau_s(z)-\tau_s(\xi)}\,d\xi \tag{5.122}$$
$$= \int_0^z [k_s(\xi)]^{-1}[r_4^0(\xi)\Phi_5(\xi,\tau) + r_5^0(\xi)\Phi_6(\xi,\tau)]_{\tau=t+\tau_s(z)-\tau_s(\xi)}\,d\xi$$

$$Y_{70}^{10}(z,t) + \int_0^z [k_s(\xi)]^{-1}\{c_4(\xi)[Y_{30}^{10}(\xi,\tau) - Y_{70}^{10}(\xi,\tau)]$$
$$+ c_3(\xi)W_{10}^{10}(\xi,\tau) - c_2(\xi)W_{90}^{10}(\xi,\tau)\}_{\tau=t+\tau_s(z)-\tau_s(\xi)}\,d\xi$$
$$= \int_0^z [k_s(\xi)]^{-1}[r_5^0(\xi)\Phi_5(\xi,\tau)$$
$$+ r_4^0(\xi)\Phi_6(\xi,\tau)]_{\tau=t+\tau_s(z)-\tau_s(\xi)}\,d\xi, \quad (z,t)\in D_3.$$

Now choose an arbitrary point $(z, \tau_p(z))$ lying on the characteristic $\tau = \tau_p(\xi)$, draw the characteristic $\tau + \tau_s(\xi) = \tau_p(z) + \tau_s(z)$ through it and integrate the first of equalities (5.119) from one point of the crossing with the boundary of the domain D to another one.

As a result, the equality involving $\gamma_3(z)$ is obtained. This equality, when solved with respect to $\gamma_3(z)$, is as follows:

$$\gamma_3(z) = -\left[1 + \frac{k_s(z)}{k_p(z)}\right]\Bigg\{\tfrac{1}{2}\gamma_4(\xi_1^*(z,\tau_p(z)))$$
$$+ \int_0^z [k_s(\xi)^{-1}[c_4(\xi)(Y_{30}^{10}(\xi,\tau) - Y_{70}^{10}(\xi,\tau)) + c_2(\xi)W_{10}^{10}(\xi,\tau)$$
$$- c_3(\xi)W_{90}^{10}(\xi,\tau) - r_4^0(\xi)\Phi_5(\xi,\tau) - r_5^0(\xi)\Phi_6(\xi,\tau)]_{\tau=\tau_p(z)+\tau_s(z)-\tau_s(\xi)}\,d\xi\Bigg\}. \tag{5.123}$$

Now demonstrate that using $\gamma_3(z)$ one can find a new combination of the coefficients ρ_1^0, a_1^0, and b_1^0, different from the combinations of r_3^0 and r_5^0. For this purpose use (5.118) and (5.120) and (5.99) and (5.104). From (5.118) and (5.104)

$$\begin{aligned}
\alpha_7(z) &= \sqrt{\left(\frac{2k_p(z)}{k_p(0)}\right)}\frac{c_3(z)}{k_p(z)+k_s(z)}\int_0^z \frac{r_1^0(\xi)}{\rho_0(\xi)k_p(\xi)}\,\mathrm{d}\xi \\
\alpha_9(z) &= \sqrt{\left(\frac{2k_p(z)}{k_p(0)}\right)}\frac{c_2(z)}{k_p(z)-k_s(z)}\int_0^z \frac{r_1^0(\xi)}{\rho_0(\xi)k_p(\xi)}\,\mathrm{d}\xi \\
\alpha_{10}(z) &= -\frac{1}{\sqrt{(k_s(z))}}\int_0^z \frac{r_4^0(\xi)d_7(\xi)}{\rho_0(\xi)\sqrt{(k_s(\xi))}}\,\mathrm{d}\xi.
\end{aligned} \tag{5.124}$$

Excluding the derivative α_7' from (5.120) for γ_3 through the first of equalities (5.124) and using formula (5.99) for d_5 the following is obtained

$$\begin{aligned}
\gamma_3(z) = {} & \frac{1}{\rho_0(z)}\sqrt{\left(\frac{2}{k_p(0)k_p(z)}\right)}\frac{c_3(z)}{k_p(z)+k_s(z)}[r_1^0(z)k_s(z)+r_4^0(z)k_p(z)] \\
& + k_s(z)\frac{\mathrm{d}}{\mathrm{d}z}\left[\sqrt{\left(\frac{2k_p(z)}{k_p(0)}\right)}\frac{c_3(z)}{k_p(z)+k_s(z)}\right]\int_0^z \frac{r_1^0(\xi)}{\rho_0(\xi)k_p(\xi)}\,\mathrm{d}\xi \\
& - c_4(z)[\alpha_7(z)-\alpha_9(z)] - c_2(z)\alpha_1(z) + c_3(z)\alpha_3(z) - \frac{1}{\rho_0(z)}r_5^0(z)d_6(z).
\end{aligned}$$

Since $c_3(z) \neq 0$ (resulting from the theorem condition $\lambda_0 \neq 2\mu_0$), the following is obtained from the above

$$\begin{aligned}
r_1^0(z)k_s(z) + r_4^0(z)k_p(z) = {} & \sqrt{\left(\frac{k_p(0)k_p(z)}{2}\right)}\frac{\rho_0(z)}{c_3(z)}[k_p(z)+k_s(z)] \\
& \times\left\{\gamma_3(z) - k_s(z)\frac{\mathrm{d}}{\mathrm{d}z}\left[\sqrt{\left(\frac{2k_p(z)}{k_p(0)}\right)}\frac{c_3(z)}{k_p(z)+k_s(z)}\right]\right. \\
& \left.\times\int_0^z \frac{r_1^0(\xi)}{\rho_0(\xi)k_p(\xi)}\,\mathrm{d}\xi + c_4(\alpha_7-\alpha_9) + c_2\alpha_1 - c_3\alpha_3 + \frac{1}{\rho_0}r_5^0 d_6\right\}.
\end{aligned} \tag{5.125}$$

Equation (5.125) determines the third linearly independent combination of the coefficients ρ_1^0, a_1^0, and b_1^0. Earlier the equations for r_3^0 and r_5^0 have been obtained. Linear independence of the combinations r_3^0, r_5^0, and $r_1^0k_s + r_4^0k_p$ results from the following considerations. From the easily verified equality $k_sr_1^0 + k_pr_4^0 + k_sr_3^0 + k_pr_5^0 = 2\rho_1^0q_1^2(k_p+k_s)$ one can find ρ_1^0, in which case r_5^0 defines a_1^0, and r_3^0 defines b_1^0. Thus, the functions r_3^0, r_5^0, and $r_1^0k_s + r_4^0k_p$ enable one to determine ρ_1^0, a_1^0, and b_1^0, which means that all the elements of the matrix R^0 are determined.

Note that (5.125) is an integral correlation since each of the functions involved in it—$\gamma_3, \alpha_7, \alpha_9, \alpha_1, \alpha_3$, and r_5^0—are determined as an integral operator from the required functions. The function γ_4 involved in the right-hand part of (5.123) is, as follows from (5.120), an integral operator. The function γ_5, after the derivative α_9' has been excluded from it using the second of equalities (5.124), is also an integral operator.

As a result, a closed system of integral equations of the second kind have been obtained with respect to the required functions, as well as to a series of the additional functions α_k and γ_k. This system exhibits Volterra properties with respect to the variable z. Indeed, consider a set of the domains $D(z_0)$ $z_0 > 0$ (Fig. 5.6). For $(z, t) \in D(z_0)$ each of the functions involved in the system of integral equations is expressed as an integral of the required functions with the upper limit of integration equal to z. The system of integral equations is closed in $D(z_0)$ at any $z_0 > 0$. Therefore the method of successive approximations applied to this system has a factorial velocity of convergence, which is typical of the Volterra equations. Due to uniformity of the system of integral equations, its solution is zero and, hence, $\rho_1^0 = 0, a_1^0 = 0, b_1^0 = 0$, and $R^0 = 0$. The investigation carried out so far is the first step in proving the theorem by the method of mathematical induction and, at the same time, a basis for further studies.

Now suppose that, in line with the method of induction, all the $R^\alpha = 0$, $|\alpha| \leqslant k - 1$. Then at $|\alpha| = k$ one gets the equalities identical to relations (5.92) and (5.93)

$$\left(E\frac{\partial}{\partial t} + K\frac{\partial}{\partial z} + C^0\right)W^{1\alpha} = -R^\alpha \frac{\partial}{\partial t} W^{00}$$

$$W^{1\alpha}|_{t<0} \equiv 0, \qquad (TW^{1\alpha})_j|_{z=0} = 0, \quad j = 7, 8, 9$$

$$(TW^{1\alpha})_j|_{z=0} = 0, \quad j = 1, 3 \tag{5.126}$$

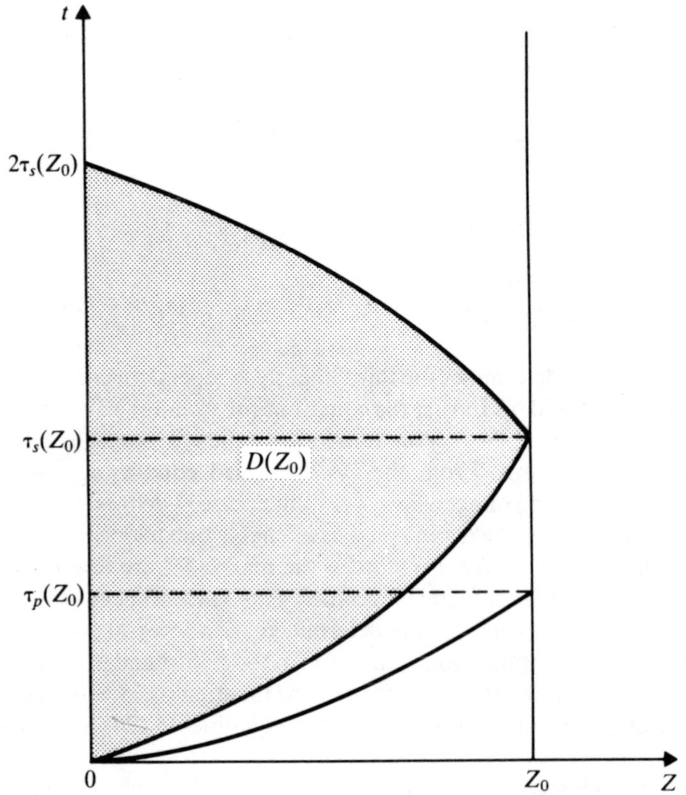

Figure 5.6

$$\left(E\frac{\partial}{\partial t}+K\frac{\partial}{\partial z}+C^0\right)Y^{1\alpha}+C^1W^{1\alpha}=-R^\alpha\frac{\partial}{\partial t}Y^{00}$$

$$Y^{1\alpha}|_{t<0}\equiv 0,\qquad (TY^{1\alpha})_j|_{z=0}=0,\quad j=7,8,9$$
$$(TY^{1\alpha})_{z=0}=0,\quad j=1. \tag{5.127}$$

The above equalities differ from (5.92) and (5.93) only in notation. In this case the roles of R^0, W^{10} and Y^{10} are performed by the functions R^α, $W^{1\alpha}$, and $Y^{1\alpha}$, respectively. Therefore, one can apply the conclusions from problem (5.92) and (5.93) to (5.126) and (5.127). It means that $R^\alpha=0, |\alpha|=k$, and thus, at any $\alpha, \rho_1^\alpha=a_1^\alpha=b_1^\alpha=0$. The theorem is proved.

5.6. Inverse problem for a system of Maxwell equations

Processes of electromagnetic oscillation propagation are described by a system of Maxwell equations

$$\operatorname{rot} H=\varepsilon\frac{\partial E}{\partial t}+\sigma E+j,\qquad \operatorname{rot} E=-\mu\frac{\partial H}{\partial t}. \tag{5.128}$$

Here $E=(E_1, E_2, E_3)$ and $H=(H_1, H_2, H_3)$ are the electric and magnetic field strengths; ε and μ are the dielectric and magnetic permeabilities of the medium; σ is the conductivity of the medium; and $j=(j_1, j_2, j_3)$ is the density of the external electric current. In nonhomogeneous media the parameters ε, μ, and σ are functions of the point $x=(x_1, x_2, x_3)$. These parameters are very important characteristics of media and in geophysics are of the same importance as the medium density and the Lamé elastic parameters. The problem of determining these parameters as functions of the point x is the basic one in electrical geophysical prospecting.

There exist two basic variations of geophysical prospecting by electric means—based on constant and alternating current, the latter being most informative. A mathematical foundation of the method of geophysical prospecting based on alternating current was laid down by Tikhnov [3, 67, 278]. Later the method has been developed by a number of authors (see [221, 268, 279–294]). In this section one of the possible formulations of the inverse problems on determining the electromagnetic parameters ε, μ, and σ under the assumption that they are the functions of only one coordinate x_3 will be investigated.

Consider the medium model widely used in modern geophysics. Let the medium consist of two semi-spaces $x_3\geqslant 0$ and $x_3\leqslant 0$, in either of which the parameters ε, μ, and σ are the smooth functions of the variable x_3 but can suffer a finite discontinuity when going over through the plane $x_3=0$. The parameters ε, μ, and σ are known in the domain $D^-=\{x{:}x_3\leqslant 0\}$ and unknown in the domain $D^+=\{x{:}x_3\geqslant 0\}$. This is the formulation arrived at when considering the problem of determining ε, μ, and σ in the Earth's crust. In this case the Earth's surface is locally replaced by a tangential plane $x_3=0$ (the axis x_3 is directed to the inside of the Earth), the domain D^- simulates the Earth's atmosphere, the electromagnetic parameter of which can be considered known. For applications it would be reasonable in this case to consider the components of the electromagnetic field to be set on the media interface. A full mathematical formulation of the problem to be discussed will be given later.

Let the density of the external electric current be as follows:

$$j = j^0 \delta(x - x^0, t), \qquad j^0 = (1, 0, 0). \tag{5.129}$$

Consider the problem of excitation of an electromagnetic field, initially non-existing, by a pulse current (5.129). In other words, the problem of finding a solution of system (5.128) under the conditions

$$E|_{t<0} \equiv 0, \qquad H|_{t<0} \equiv 0. \tag{5.130}$$

On the media interface the condition of continuity of the tangential components of the field

$$\begin{aligned} E_1^- |_{x_3=0} &= E_1^+ |_{x_3=0}, \qquad & E_2^- |_{x_3=0} &= E_2^+ |_{x_3=0} \\ H_1^- |_{x_3=0} &= H_1^+ |_{x_3=0}, \qquad & H_2^- |_{x_3=0} &= H_2^+ |_{x_3=0}. \end{aligned} \tag{5.131}$$

must be fulfilled. Here the plus sign denotes the limiting values of the components of the vector E, H from the domain D^+, and the minus sign those from the domain D^-.

For a fixed point let $x^0 = (x_1^0, x_2^0, x_3^0)$, $x^0 \in D^-$, the component E_1 of the solution of problem (5.128)–(5.131) be known on the plane $x_3 = 0$

$$E_1^+ |_{x_3=0} = f(x_1, x_2, t) \tag{5.132}$$

the task is to find ε, μ, and σ in the domain D^+.

Note, that the formulation of the problem is obviously over-specified since for defining the three functions by one variable the function of three variables is set. This over-specification facilitates the investigation of the inverse problem. At the same time, as will be shown below, for the inverse problem to be uniquely solved it is sufficient to set a Fourier transform of the function $f(x_1, x_2, t)$ with respect to the variables x_1, x_2 at three different values of the transformation parameter $\nu = (\nu_1, \nu_2)$. Thus, from (5.132) one can derive information on the dimensionality of the required functions and that of the information used coincide.

Prior to investigating the formulated above inverse problem study in detail the properties of the solution to problem (5.128)–(5.131). Henceforth assume that $\varepsilon > 0$, $\mu > 0$, μ, $\varepsilon \in C^3(-\infty, 0) \cap C^3(0, \infty)$, and $\sigma \in C^2(-\infty, 0) \cap C^2(0, \infty)$.

Under the condition (5.129) the system of equations (5.128) can be written as a symmetrical hyperbolic system

$$\left(A \frac{\partial}{\partial t} + \sum_{j=1}^{3} B_j \frac{\partial}{\partial x_j} + B_4 \right) U = F \tag{5.133}$$

where $F = F^0 \delta(x - x^0, t)$

$$U = \begin{Vmatrix} E_1 \\ E_2 \\ E_3 \\ H_1 \\ H_2 \\ H_3 \end{Vmatrix}; \qquad F^0 = \begin{Vmatrix} -1 \\ 0 \\ 0 \\ 0 \\ 0 \\ 0 \end{Vmatrix}; \qquad A = \begin{Vmatrix} \varepsilon I_3 & 0 \\ 0 & \mu I_3 \end{Vmatrix}$$

$$B_j = \begin{Vmatrix} 0 & p_j \\ p_j^* & 0 \end{Vmatrix}, \quad j = 1,2,3; \qquad B_4 = \begin{Vmatrix} \sigma I_3 & 0 \\ 0 & 0 \end{Vmatrix}$$

$$p_1 = \begin{Vmatrix} 0 & 0 & 0 \\ 0 & 0 & 1 \\ 0 & -1 & 0 \end{Vmatrix}; \qquad p_2 = \begin{Vmatrix} 0 & 0 & -1 \\ 0 & 0 & 0 \\ 1 & 0 & 0 \end{Vmatrix}; \qquad p_3 = \begin{Vmatrix} 0 & 1 & 0 \\ -1 & 0 & 0 \\ 0 & 0 & 0 \end{Vmatrix}$$

I_3 is a unit matrix of the third order; p_j^*, $j = 1,2,3$ are the matrices transposed with respect to p_j.

The components of the vector U will be later denoted by U_i, $i = 1,2,\ldots,6$. Conditions (5.130) and (5.131) can be written as follows:

$$U|_{t<0} \equiv 0 \tag{5.134}$$

$$B_3 U^-|_{x_3=0} = B_3 U^+|_{x_3=0}. \tag{5.135}$$

Note that the uniform equation (5.133) is equivalent to the uniform equation

$$\left(A\frac{\partial}{\partial t} + \sum_{j=1}^{3} B_j \frac{\partial}{\partial x_j} + B_4 \right) U = 0$$

considered separately in the domains $x_3 > x_3^0$, $x_3 < x_3^0$, and the condition of sewing at $x_3 = x_3^0$

$$B_3 U|_{x_3 = x_3^0 + 0} - B_3 U|_{x_3 = x_3^0 - 0} = F^0 \delta(x_1 - x_1^0, x_2 - x_2^0, t). \tag{5.136}$$

It is the presentation formula that will henceforth be used in the investigation. For simplification of the calculations assume that the point x^0 belongs to the plane $x_3 = 0$, in which case under a solution to problem (5.128)–(5.131) a limit of the solutions of the corresponding problems is understood, when $x_3^0 \to -0$. Then the limiting solution satisfies the uniform equation

$$\left(A\frac{\partial}{\partial t} + \sum_{j=1}^{3} B_j \frac{\partial}{\partial x_j} + B_4 \right) U = 0, \quad x_3 \neq 0 \tag{5.137}$$

condition (5.134) and the condition on the boundary of media discontinuity

$$B_3(U^+ - U^-)_{x_3=0} = F^0 \delta(x_1 - x_1^0, x_2 - x_2^0, t) \tag{5.138}$$

resulting from comparing conditions (5.135) and (5.136) at $x_3^0 \to -0$.

Now investigate the properties of the solutions of problem (5.137), (5.134), and (5.138) using the Fourier transform with respect to the variables x_1, x_2

$$\tilde{U}(v, x_3, t) = \int_{R^2} U(x,t) \exp\left[i \sum_{j=1}^{2} v_j (x_j - x_j^0) \right] \mathrm{d}x_1 \, \mathrm{d}x_2.$$

In terms of the Fourier transform the initial problem is reduced to the form

$$\left[A\frac{\partial}{\partial t} + B_3 \frac{\partial}{\partial x_3} + B_4 - i(v_1 B_1 + v_2 B_2) \right] \tilde{U} = 0$$

$$\tilde{U}|_{t<0} \equiv 0, \qquad B_3(\tilde{U}^+ - \tilde{U}^-)_{x_3=0} = F^0 \delta(t). \tag{5.139}$$

Now bring the system of equations over to a classical form. To do this introduce the matrix

$$T = \left\| \begin{matrix} q & 0 & q & 0 & 0 & 0 \\ 0 & q & 0 & q & 0 & 0 \\ 0 & 0 & 0 & 0 & 1 & 0 \\ 0 & 1/q & 0 & -1/q & 0 & 0 \\ -1/q & 0 & 1/q & 0 & 0 & 0 \\ 0 & 0 & 0 & 0 & 0 & 1 \end{matrix} \right\| \tag{5.140}$$

$$q = [\mu(x_3)/\varepsilon(x_3)]^{1/4}$$

and change the required function

$$\tilde{U} = TV \tag{5.141}$$

and the independent variable x_3 by the variable

$$z = \int_0^{x_3} \sqrt{(\varepsilon(\xi)\mu(\xi))}\, \mathrm{d}\xi, \quad x_3 = g(z). \tag{5.142}$$

Then the following problem arises for the function V

$$\left(I_6 \frac{\partial}{\partial t} + K \frac{\partial}{\partial z} + C \right) V = 0$$

$$V|_{t<0} \equiv 0, \qquad B_3[(TV)^+_{z=0} - (TV)^-_{z=0}] = F^0 \delta(t). \tag{5.143}$$

Here I_6 is a six-order unit matrix; K is a diagonal matrix

$$K = \mathrm{diag}(-1, -1, 1, 1, 0, 0)$$

C is the matrix of the following structure

$$C = \left\| \begin{matrix} c_1 & 0 & c_1 - c_2 & 0 & -iv_1c_3 & iv_2c_4 \\ 0 & c_1 & 0 & c_1 - c_2 & -iv_2c_3 & -iv_1c_4 \\ c_1 + c_2 & 0 & c_1 & 0 & iv_1c_3 & iv_2c_4 \\ 0 & c_1 + c_2 & 0 & c_1 & iv_2c_3 & -iv_1c_4 \\ -2iv_1c_4 & -2iv_2c_4 & 2iv_1c_4 & 2iv_2c_4 & 2c_1 & 0 \\ 2iv_2c_3 & -2iv_1c_3 & 2iv_2c_3 & -2iv_1c_3 & 0 & 0 \end{matrix} \right\|$$

$$c_1 = \frac{\sigma(g(z))}{2\varepsilon(g(z))}; \qquad c_2 = \frac{1}{4} \frac{\mathrm{d}}{\mathrm{d}z} \ln \frac{\mu(g(z))}{\varepsilon(g(z))}$$

$$c_3 = [\varepsilon(g(z))\mu^3(g(z))]^{-1/4}/2 \qquad c_4 = [\varepsilon^3(g(z))\mu(g(z))]^{-1/4}/2.$$

Introduce the new vector function

$$W = \left\| \begin{matrix} W_1 \\ W_2 \\ W_3 \\ W_4 \end{matrix} \right\|, \quad W_i = V_i, \qquad i = 1, 2, 3, 4. \tag{5.144}$$

The expressions for the last two components of system (5.143) can be written

$$V_5(v,z,t) = 2ic_4(z)\int_0^t \{v_1[W_1(v,z,\tau) - W_3(v,z,\tau)]$$
$$+ v_2[W_2(v,z,\tau) - W_4(v,z,\tau)]\}\exp[-2c_1(z)(t-\tau)]\,\mathrm{d}\tau$$
$$V_6(v,z,t) = 2ic_3(z)\int_0^t \{v_1[W_2(v,z,\tau) + W_4(v,z,\tau)]$$
$$- v_2[W_1(v,z,\tau) + W_3(v,z,\tau)]\}\,\mathrm{d}\tau.$$

Using these formulas V_5 and V_6 can be excluded from the first four components of equality (5.143) and the system of equations is obtained

$$\left[I_4\frac{\partial}{\partial t} + K_0\frac{\partial}{\partial z} + H_0(z)\right]W(v,z,t) + \int_0^t H_1(v,z,t-\tau)W(v,z,\tau)\,\mathrm{d}\tau = 0 \qquad (5.145)$$

where I_4 is a unit matrix

$$K_0 = \mathrm{diag}(-1,-1,1,1)$$
$$H_0(z) = \begin{Vmatrix} c_1 & 0 & c_1 - c_2 & 0 \\ 0 & c_1 & 0 & c_1 - c_2 \\ c_1 + c_2 & 0 & c_1 & 0 \\ 0 & c_1 + c_2 & 0 & c_1 \end{Vmatrix}$$
$$H_1(v,z,t) = c_5(z)\begin{Vmatrix} P_1(v,z,t) & P_2(v,z,t) \\ P_2(v,z,t) & P_1(v,z,t) \end{Vmatrix}$$
$$P_1(v,z,t) = \begin{Vmatrix} v_1^2 e^{-2c_1 t} + v_2^2 & v_1 v_2(e^{-2c_1 t} - 1) \\ v_1 v_2(e^{-2c_1 t} - 1) & v_1^2 + v_2^2 e^{-2c_1 t} \end{Vmatrix}$$
$$P_2(v,z,t) = \begin{Vmatrix} v_2^2 - v_1^2 e^{-2c_1 t} & -v_1 v_2(e^{-2c_1 t} + 1) \\ -v_1 v_2(e^{-2c_1 t} + 1) & v_1^2 - v_2^2 e^{-2c_1 t} \end{Vmatrix}$$
$$c_5(z) = 1/2\varepsilon(g(z))\mu(g(z)).$$

System (5.145) must be solved under the initial conditions

$$W|_{t<0} \equiv 0 \qquad (5.146)$$

and the sewing conditions at $z = 0$. The latter conditions are more convenient when used in a factorial form

$$\begin{pmatrix} W_u^- \\ W_l^+ \end{pmatrix}_{z=0} = R\begin{pmatrix} W_u^+ \\ W_l^- \end{pmatrix}_{z=0} + \phi\delta(t). \qquad (5.147)$$

Here

$$W_u = \begin{pmatrix} W_1 \\ W_2 \end{pmatrix}, \qquad W_l = \begin{pmatrix} W_3 \\ W_4 \end{pmatrix}, \qquad \phi = \begin{pmatrix} \phi_u \\ \phi_l \end{pmatrix}$$
$$\phi_u = \begin{pmatrix} \phi_1 \\ \phi_2 \end{pmatrix} = \tfrac{1}{2}q^-(0)(1+r)\begin{pmatrix} 1 \\ 0 \end{pmatrix}$$

$$\phi_l = \begin{pmatrix} \phi_3 \\ \phi_4 \end{pmatrix} = \tfrac{1}{2} q^-(0)\sqrt{(1-r^2)} \begin{pmatrix} 1 \\ 0 \end{pmatrix}$$

$$r = \frac{[q^+(0)]^2 - [q^-(0)]^2}{[q^+(0)]^2 + [q^-(0)]^2}$$

$$R = \begin{Vmatrix} \sqrt{(1-r^2)} I_2 & r I_2 \\ -r I_2 & \sqrt{(1-r^2)} I_2 \end{Vmatrix}, \qquad I_2 = \begin{Vmatrix} 1 & 0 \\ 0 & 1 \end{Vmatrix}.$$

Henceforth the subscript u attached to the four-dimensional vector means that the upper half of the vector, consisting of its first two components, is considered, while the subscript l denotes the lower half of the vector, consisting of the last two components.

Write the solution of problem (5.145)–(5.147) in the following way

$$W(v, z, t) = \theta(z) \sum_{k=-1}^{1} \alpha^k(v, z)\theta_k(t - z)$$
$$+ \theta(-z) \sum_{k=-1}^{1} \beta^k(v, z)\theta_k(t + z) + \omega(v, z, t). \tag{5.148}$$

Here $\theta_{-1}(t) = \delta(t), \theta_0(t) = \theta(t)$, and $\theta_1(t) = t\theta(t)$. Substituting the expression for W from (5.148) into equalities (5.145) and (5.147) and equating the coefficients at θ'_{-1}, θ_{-1}, and θ_0, the relations for determining α^k, β^k are obtained

$$(I_4 - K_0)\alpha^{-1} = 0, \qquad (I_4 - K_0)\alpha^0 + K_0 \frac{d}{dz}\alpha^{-1} + H^0\alpha^{-1} = 0$$

$$(I_4 - K_0)\alpha^1 + K_0 \frac{d}{dz}\alpha^0 + H_0\alpha^0 + H_1(v, z, 0)\alpha^{-1} = 0$$

$$(I_4 + K_0)\beta^{-1} = 0, \qquad (I_4 + K_0)\beta^0 + K_0 \frac{d}{dz}\beta^{-1} + H_0\beta^{-1} = 0$$

$$(I_4 + K_0)\beta^1 + K_0 \frac{d}{dz}\beta^0 + H_0\beta^0 + H_1(v, z, 0)\beta^{-1} = 0 \tag{5.149}$$

$$\begin{pmatrix} \beta_u^{-1} \\ \alpha_l^{-1} \end{pmatrix}_{z=0} = R \begin{pmatrix} \alpha_u^{-1} \\ \beta_l^{-1} \end{pmatrix}_{z=0} + \phi$$

$$\begin{pmatrix} \beta_u^k \\ \alpha_l^k \end{pmatrix}_{z=0} = R \begin{pmatrix} \alpha_u^k \\ \beta_l^k \end{pmatrix}_{z=0}, \quad k = 0, 1$$

and the relations for determining ω

$$\left[I_4 \frac{\partial}{\partial t} + K_0 \frac{\partial}{\partial z} + H_0(z) \right] \omega(v, z, t)$$
$$+ \int_0^t H_1(v, z, t - \tau)\omega(v, z, \tau)\, d\tau + \Psi(v, z, t) = 0 \tag{5.150}$$

$$\omega|_{t<0} \equiv 0, \begin{pmatrix} \omega_u^- \\ \omega_l^+ \end{pmatrix}_{z=0} = R \begin{pmatrix} \omega_u^+ \\ \omega_l^- \end{pmatrix}_{z=0}.$$

In equalities (5.150) Ψ is determined by the following

$$\Psi(v,z,t)=\gamma(v,z)\theta_1(t-z)+\chi(v,z)\theta_1(t+z)+\Psi^0(v,z,t)$$

$$\gamma(v,z)=K_0\frac{\mathrm{d}\alpha^1}{\mathrm{d}z}+H_0\alpha^1+H_1(v,z,0)\alpha^0+\frac{\partial}{\partial t}H_1(v,z,t)|_{t=0}\alpha^{-1} \tag{5.151}$$

$$\chi(v,z)=K_0\frac{\mathrm{d}\beta^1}{\mathrm{d}z}+H_0\beta^1+H_1(v,z,0)\beta^0+\frac{\partial}{\partial t}H_1(v,z,t)|_{t=0}\beta^{-1}$$

$$\begin{aligned}\Psi^0(v,z,t)=\theta(t-z)\Bigg\{\Bigg[&H_1(v,z,t-z)\\&-H_1(v,z,0)-(t-z)\frac{\partial}{\partial t}H_1(v,z,t)|_{t=0}\Bigg]\alpha^{-1}\\&+\int_z^t[(H_1(v,z,t-\tau)-H_1(v,z,0))\alpha^0(v,z)\\&+(\tau-z)H_1(v,z,t-\tau)\alpha^{-1}(v,z)]\,\mathrm{d}\tau\Bigg\}\\&+\theta(t+z)\Bigg\{\Bigg[H_1(v,z,t+z)-H_1(v,z,0)\\&-(t+z)\frac{\partial}{\partial t}H_1(v,z,t)|_{t=0}\Bigg]\beta^{-1}+\int_{-z}^t[(H_1(v,z,t-\tau)\\&-H_1(v,z,0))\beta^0(v,z)+(\tau+z)H_1(v,z,t-\tau)\beta^{-1}(v,z)]\,\mathrm{d}\tau\Bigg\}.\end{aligned} \tag{5.152}$$

Relations (5.149) make it possible to find α^{-1}, α^0, β^{-1}, β^0 and α_u^1, β_l^1, in succession. However, one cannot obtain α_l^1 and β_u^1 through them. Indeed, the first of equations (5.149) demonstrates that $\alpha_u^{-1}=0$, but it cannot be used to find α_l^{-1}, since the third and fourth components of the equation are the equalities $0=0$. The lower half of the vector α^{-1} can be obtained using the following equation of system (5.149), i.e. the third and fourth components of this equation. In this case the following equation for determining α_l^{-1} is obtained

$$\frac{\mathrm{d}}{\mathrm{d}z}\alpha_l^{-1}+(H_0\alpha^{-1})_l=0$$

or in an expanded form, allowing for the fact that $\alpha_u^{-1}=0$

$$\frac{\mathrm{d}}{\mathrm{d}z}\alpha_l^{-1}+c_1\alpha_l^{-1}=0. \tag{5.153}$$

And vice versa, from the first equation of system (5.149) serving to define β^k one can find that $\beta_l^{-1}=0$, but β_u^{-1} is not defined. The function β_u^{-1} can be obtained from the following equation by way of considering its first two components. As a result, this is the equation for β_u^{-1}

$$-\frac{d}{dz}\beta_u^{-1} + c_1\beta_u^{-1} = 0. \tag{5.153'}$$

From the conditions at $z=0$ one can get the Cauchy data for α_l^{-1} and β_u^{-1}

$$\alpha_l^{-1}|_{z=0} = \phi_l, \qquad \beta_u^{-1}|_{z=0} = \phi_u. \tag{5.154}$$

From equalities (5.153) and (5.154) α_l^{-1} and β_u^{-1} can be found uniquely

$$\alpha_l^{-1}(z) = \phi_l \exp\left[-\int_0^z c_1(\xi)\,d\xi\right]$$

$$\beta_u^{-1}(z) = \phi_u \exp\left[\int_0^z c_1(\xi)\,d\xi\right].$$

In an analogous way one finds α_u^0, β_l^0, and then α_l^0, β_u^0 and α_u^1, β_l^1 in succession. The formulas defining them are as follows:

$$\begin{aligned}
\alpha_u^0(z) &= -\tfrac{1}{2}[c_1(z) - c_2(z)]\alpha_l^{-1}(z)\\
\beta_l^0(z) &= -\tfrac{1}{2}[c_1(z) + c_2(z)]\beta_u^{-1}(z)\\
\alpha_l^0(v,z) &= \Big\{-r\alpha_u^0(0) + \sqrt{(1-r^2)}\beta_l^0(0) + \int_0^z [\tfrac{1}{2}(c_1^2(\xi) - c_2^2(\xi))I_2\\
&\quad - c_5(\xi)P_1(v,\xi,0]\phi_l\,d\xi\Big\}\exp\left[-\int_0^z c_1(\xi)\,d\xi\right]\\
\beta_u^0(v,z) &= \Big\{\sqrt{(1-r^2)}\alpha_u^0(0) + r\beta_l^0(0) + \int_0^z [-\tfrac{1}{2}(c_1^2(\xi) - c_2^2(\xi))I_2\\
&\quad + c_5(\xi)P_1(v,\xi,0)]\phi_u\,d\xi\Big\}\exp\left[\int_0^z c_1(\xi)\,d\xi\right] \qquad\qquad 5.155)\\
\alpha_u^1(v,z) &= -\tfrac{1}{2}[c_1(z) - c_2(z)]\alpha_l^0(v,z)\\
&\quad + \{[-\tfrac{1}{4}(c_1'(z) - c_2'(z)) + \tfrac{1}{2}c_1(z)(c_1(z) - c_2(z))]I_2\\
&\quad - \tfrac{1}{2}c_5(z)P_2(v,z,0)\}\alpha_l^{-1}(z),\\
\beta_l^1(v,z) &= -\tfrac{1}{2}[c_1(z) + c_2(z)]\beta_u^0(v,z)\\
&\quad + \{[\tfrac{1}{4}(c_1'(z) + c_2'(z)) + \tfrac{1}{2}c_1(z)(c_1(z) - c_2(z))]I_2\\
&\quad - \tfrac{1}{2}c_5(z)P_2(v,z,0)\}\beta_u^{-1}(z).
\end{aligned}$$

The remaining undetermined α_l^1 and β_u^1 can be found from the conditions

$$\gamma_l(v,z) = 0, \qquad \chi_u(v,z) = 0 \tag{5.156}$$

and the conditions (5.149) at $z=0$. The role played by conditions (5.156) will be clarified later. By way of calculations one gets the following formulas for α_l^1 and β_u^1

$$\alpha_l^1(v,z) = [-r\alpha_u^1(v,0) + \sqrt{(1-r^2)}\beta_l^1(v,0)]\exp\left[-\int_0^z c_1(\xi)\,d\xi\right]$$

$$+\int_0^z [\tfrac{1}{2}(c_1^2(\xi)-c_2^2(\xi))I_2 - c_5(\xi)P_1(v,\xi,0)]\alpha_l^0(v,\xi)$$

$$\times \exp\left[\int_z^{\xi} c_1(\xi_1)\,\mathrm{d}\xi_1\right]\mathrm{d}\xi + \int_0^z \Big\{[\tfrac{1}{4}(c_1'(\xi)-c_2'(\xi))$$

$$-\tfrac{1}{2}c_1(\xi)(c_1(\xi)-c_2(\xi))](c_1(\xi)+c_2(\xi))I_2$$

$$+c_5(\xi)\left[c_1(\xi)P_2(v,\xi,0)-\frac{\partial}{\partial t}P_1(v,\xi,t)|_{t=0}\right]\Big\}\mathrm{d}\xi\,\alpha_l^{-1}(z)$$

$$\beta_u^1(v,r) = [\sqrt{(1-r^2)}\alpha_u^1(v,0) + r\beta_l^1(v,0)]\exp\left[\int_0^z c_1(\xi)\,\mathrm{d}\xi\right] \tag{5.157}$$

$$+\int_0^z [-\tfrac{1}{2}(c_1^2(\xi)-c_2^2(\xi))I_2 + c_5(\xi)P_1(v,\xi,0)]\beta_u^0(v,\xi)$$

$$\times \exp\left[\int_0^z c_1(\xi_1)\,\mathrm{d}\xi_1\right]\mathrm{d}\xi + \int_0^z \Big\{[\tfrac{1}{4}(c_1'(\xi)+c_2'(\xi))$$

$$+\tfrac{1}{2}c_1(\xi)(c_1(\xi)+c_2(\xi))](c_1(\xi)-c_2(\xi))I_2$$

$$+c_5(\xi)\left[-c_1(\xi)P_2(v,\xi,0)+\frac{\partial}{\partial t}P_1(v,\xi,t)|_{t=0}\right]\Big\}\mathrm{d}\xi\,\beta_u^{-1}(z).$$

Now use (5.151) for writing the expressions for $\gamma_u(v,z)$, $\chi_l(v,z)$, with $(\mathrm{d}/\mathrm{d}z)\alpha_u^1$, α_u^1, α_u^0, $(\mathrm{d}/\mathrm{d}z)\beta_l^1$, β_l^1, β_l^0 excluded through the already obtained formulas. The formulas for calculating $\gamma_u(v,z)$, $\chi_l(v,z)$ are as follows:

$$\gamma_u(v,z) = (c_1-c_2)\alpha_l^1 + \{[\tfrac{1}{2}(c_1'-c_2') - c_1(c_1-c_2)]I_2 + c_5P_2|_{t=0}\}\alpha_l^0$$

$$+\Big\{[\tfrac{1}{4}(c_1''-c_2'') - c_1(c_1'-c_2') - \tfrac{1}{2}c_1'(c_1-c_2)$$

$$+\tfrac{1}{4}(c_1-c_2)(5c_1^2-c_2^2)]I_2 + \tfrac{1}{2}c_5'P_2|_{t=0}$$

$$+c_5\left[-c_1P_2-(c_1-c_2)P_1+\frac{\partial}{\partial t}P_2\right]_{t=0}\Big\}\alpha_l^{-1}$$

$$\chi_l(v,z) = (c_1+c_2)\beta_u^1 - \{[\tfrac{1}{2}(c_1'+c_2') + c_1(c_1+c_2)]I_2 \tag{5.158}$$

$$-c_5P_2|_{t=0}\}\beta_u^0 + \Big\{[\tfrac{1}{4}(c_1''+c_2'') + c_1(c_1'+c_2')$$

$$+\tfrac{1}{2}c_1'(c_1+c_2) + \tfrac{1}{4}(c_1+c_2)(5c_1^2-c_2^2)]I_2 - \tfrac{1}{2}c_5'P_2|_{t=0}$$

$$+c_5\left[-c_1P_2-(c_1+c_2)P_1+\frac{\partial}{\partial t}P_2\right]_{t=0}\Big\}\beta_u^{-1}.$$

The above shows all the α^k, β^k, $k=-1,0,1,\gamma,\chi$ to be continuous functions of the coordinate z and polynomials with respect to the parameter v. It should be underlined that α^k and γ are determined for the values $z \geqslant 0$, and β^k and χ for the values $z \leqslant 0$. The function $\Psi^0(v,z,t)$ is continuous with respect to the variables z,t together with the derivative $(\partial/\partial t)\Psi^0$ in the domains $G^+ = \{(z,t)\colon\ z \geqslant 0\}$,

$G^- = \{(z,t): z \leqslant 0\}$ and has a finite discontinuity at $z=0$. Besides, $\Psi^0 = 0$ in the domain $t \leqslant z$ and it is also a polynomial with respect to the parameter v.

Lemma 5.4. The function $\omega(v,t,z)$ which is a solution of problem (5.150) has the following properties:

(a) $\omega(v,z,t) = 0, t \leqslant |z|$;
(b) *$\omega(v,z,t)$ is continuous with respect to the variables z,t together with the derivative $(\partial/\partial t)\omega$ in the domains G^+, G^- and has a finite discontinuity on their common border;*
(c) *the second derivative with respect to time*

$$\frac{\partial^2}{\partial t^2}\omega(v,z,t) = \Omega(v,z,t)$$

is continuous in the domains K^+, K^-: $K^+ = \{(z,t): t \geqslant z \geqslant 0\}$, $K^- = \{(z,t): t \geqslant -z \geqslant 0\}$, and has a finite discontinuity at $z=0$ and $t=|z|$.

The property (a) is fairly obvious. It results from the fact that the function $\Psi(v,z,t)$, which is a nonuniform term of system (5.150), has an analogous property: $\Psi = 0, t \leqslant |z|$. Since the Cauchy data of system (5.150) are zero, and the carrier of the function Ψ is focused inside the characteristic angle $K^+ \cup K^-$, then the solution carrier is also focused inside this angle. It does not, however, mean that $\omega = 0$ at $t = |z|$, but the fact that ω turns to zero together with the derivative with respect to t on the sides of the characteristic angle will follow from the system of integral equations we are going to construct below.

For simplicity denote through $\Phi^\omega(v,z,t)$ several terms of system (5.150), i.e.

$$\Phi^\omega(v,z,t) = H_0(z)\omega(v,z,t) + \int_{|z|}^{t} H_1(v,z,t-\tau)\omega(v,z,\tau)\,d\tau + \Psi(v,z,t).$$

Integrating system (5.150) along the characteristics and using the condition at $z=0$, one gets

$$\omega_u(v,z,t) + \int_{(t+z)/2}^{t} \Phi_u^\omega(v,t+z-\tau,\tau)\,d\tau = 0$$

$$\omega_l(v,z,t) + \int_{t-z}^{t} \Phi_l^\omega(v,\tau-t+z,\tau)\,d\tau - r\int_{(t-z)/2}^{t-z} \Phi_u^\omega(v,t-z-\tau,\tau)\,d\tau$$

$$+\sqrt{(1-r^2)}\int_{(t-z)/2}^{t-z} \Phi_l^\omega(v,\tau-t+z,\tau)\,d\tau = 0, \quad (z,t)\in K^+$$

$$\omega_u(v,z,t) + \int_{t+z}^{t} \Phi_u^\omega(v,t+z-\tau,\tau)\,d\tau \tag{5.159}$$

$$+\sqrt{(1-r^2)}\int_{(t+z)/2}^{t+z} \Phi_u^\omega(v,t+z-\tau,\tau)\,d\tau$$

$$+r\int_{(t+z)/2}^{t+z}\Phi_l^{\omega}(v,\tau-t-z,\tau)\,\mathrm{d}\tau=0$$

$$\omega_l(v,z,t)+\int_{(t-z)/2}^{t}\Phi_l^{\omega}(v,\tau-t+z,\tau)\,\mathrm{d}\tau=0,\quad (z,t)\in K^-.$$

The scheme of integration is shown in Fig. 5.7.

System (5.159) is a system of integral Volterra equations, with the involved in it functions H_0, H_1, and Ψ continuous in the domain $K^+\cup K^-$ everywhere except for the points belonging to the straight line $z=0$. In each of these points there exist their finite limiting values from the domains K^+ and K^- and thus, the solution of system (5.159) also shares this property. By setting $t=|z|$ in system (5.159), one gets

$$\omega_u(v,z,z)=0,\quad z\geqslant 0 \tag{5.160}$$
$$\omega_l(v,z,-z)=0,\quad z\leqslant 0$$

$$\omega_l(v,z,z)+\int_0^{z}\Phi_l^{\omega}(v,\tau,\tau)\,\mathrm{d}\tau=0,\quad z\geqslant 0$$

$$\omega_u(v,z,-z)+\int_0^{-z}\Phi_u^{\omega}(v,-\tau,\tau)\,\mathrm{d}\tau=0,\quad z\leqslant 0. \tag{5.161}$$

Equations (5.161) are mutually independent integral equations for the functions $\omega_l(v,z,z)$ in the domain $z\geqslant 0$ and $\omega_u(v,z,-z)$ in the domain $z\leqslant 0$. Since

$$\Psi_l(v,z,z)=0,\qquad \Psi_u(v,z,-z)=0,\quad z\leqslant 0, \tag{5.162}$$

these equations are uniform. The only possible solutions for them are zero ones

$$\omega_l(v,z,z)=0,\qquad \omega_u(v,z,-z)=0,\quad z\leqslant 0. \tag{5.163}$$

Therefore, the property (a) is proved completely and the property (b) partially. To complete the proofs of the property (b) one must make sure that the derivative $\partial\omega/\partial t=\omega_t$ shares the same property as the function ω itself. It should be recalled that the function ω_t obeys the same system of equalities (5.150) as the function ω, provided the function Ψ is replaced in it by the function Ψ_t. One can make it sure

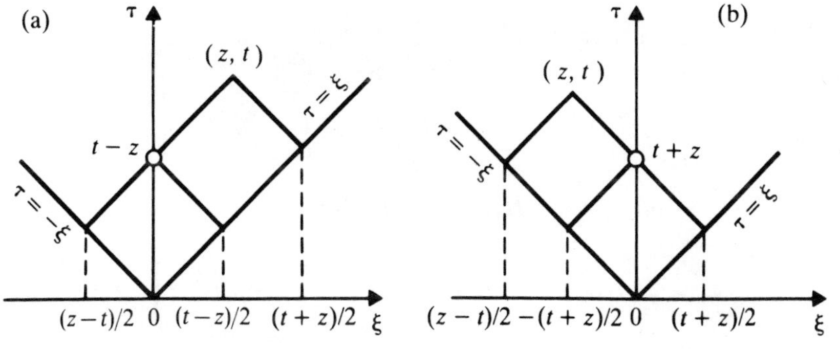

Figure 5.7.

by way of direct integration of the system of equalities (5.150) with respect to the variable t. The only difficulty arising in this case is the differentiation of the integral term. Write it out

$$\frac{\partial}{\partial t}\int_0^t H_1(v,z,t-\tau)\omega(v,z,\tau)\,d\tau$$
$$=\frac{\partial}{\partial t}\int_0^t H_1(v,z,\tau)\omega(v,z,t-\tau)\,d\tau$$
$$=\int_0^t H_1(v,z,\tau)\omega_t(v,z,t-\tau)\,d\tau$$
$$=\int_0^t H_1(v,z,t-\tau)\omega_\tau(v,z,\tau)\,d\tau.$$

In line with the above the integral equations (5.159) are valid for the function ω_t with ω replaced by ω_t and Ψ by Ψ_t. The function Ψ_t is continuous in the domains K^+ and K^- and piecewise-continuous in G^+ and G^- and suffers a finite discontinuity on the lines $z=0$ and $t=|z|$. At the same time, due to fulfilling conditions (5.156), the following equalities, analogous to (5.162) are valid $(\Psi_t)_l|_{t=z}=0$ and $(\Psi_t)_u|_{t=-z}=0, z\leqslant 0$. Therefore, the function ω_t possesses the properties totally identical to those of the function ω; it is continuous in the domains K^+ and K^-, becomes zero at $t=|z|$ and therefore appears continuous in the domains G^+ and G^-.

Now prove the property (c). The function Ω satisfies the equalities, obtained from equalities (5.150) by twice differentiating over the variable t

$$\left[I_4\frac{\partial}{\partial t}+K_0\frac{\partial}{\partial z}+H_0(z)\right]\Omega(v,z,t)+\int_0^t H_1(v,z,t-\tau)\Omega(v,z,\tau)\,d\tau$$
$$+\gamma_1(v,z)\delta(t-z)+\chi(v,z)\delta(t+z)+\Psi^{00}(v,z,t)=0 \tag{5.164}$$
$$\Omega|_{t<0}=0,\qquad \left\|\begin{matrix}\Omega_u^-\\ \Omega_l^+\end{matrix}\right\|=R\left\|\begin{matrix}\Omega_u^+\\ \Omega_l^-\end{matrix}\right\|.$$

Here

$$\Psi^{00}(v,z,t)=\frac{\partial^2}{\partial t^2}\Psi^0(v,z,t)=\theta(t-z)\left[\frac{\partial^2}{\partial t^2}H_1(v,z,t-z)\alpha^{-1}(z)\right.$$
$$\left.+\frac{\partial}{\partial t}H_1(v,z,t-z)\alpha^0(v,z)+H_1(v,z,t-z)\alpha^1(v,z)\right]$$
$$+\theta(t+z)\left[\frac{\partial^2}{\partial t^2}H_1(v,z,t+z)\beta^{-1}(z)\right.$$
$$\left.+\frac{\partial}{\partial t}H_1(v,z,t+z)\beta^0(v,z)+H_1(v,z,t+z)\beta^1(v,z)\right].$$

The function $\Psi^{00}(v,z,t)$ is continuous in the domains K^+ and K^-.

Now demonstrate that the same property is shared by the function $\Omega(v, z, t)$ as well. Denote

$$F^{\Omega}(v, z, t) = H_0(z)\Omega(v, z, t) + \int_{|z|}^{t} H_1(v, z, t-\tau)\Omega(v, z, \tau)\,d\tau + \Psi^{00}(v, z, t).$$

For the function Ω the following system of integral equations, analogous to system (5.159) is valid

$$\begin{aligned}
&\Omega_u(v, z, t) + \int_{(t+z)/2}^{t} F_u^{\Omega}(v, t+z-\tau, \tau)\,d\tau + \tfrac{1}{2}\gamma_u(v, (t+z)/2) = 0\\
&\Omega_l(v, z, t) + \int_{t-z}^{t} F_l^{\Omega}(v, \tau-t+z, \tau)\,d\tau\\
&\quad - r\int_{(t-z)/2}^{t-z} F_u^{\Omega}(v, t-z-\tau, \tau)\,d\tau\\
&\quad + \sqrt{(1-r^2)}\int_{(t-z)/2}^{t-z} F_l^{\Omega}(v, \tau-t+z, \tau)\,d\tau\\
&\quad - \tfrac{1}{2}r\gamma_u(v, (t-z)/2) + \tfrac{1}{2}\sqrt{(1-r^2)}\chi_l(v, (z-t)/2) = 0, \quad (z, t)\in K^+\\
&\Omega_u(v, z, t) + \int_{t+z}^{t} F_u^{\Omega}(v, t+z-\tau, \tau)\,d\tau\\
&\quad + \sqrt{(1-r^2)}\int_{(t+z)/2}^{t+z} F_u^{\Omega}(v, t+z-\tau, \tau)\,d\tau\\
&\quad + r\int_{(t+z)/2}^{t+z} F_l^{\Omega}(v, \tau-t-z, \tau)\,d\tau\\
&\quad + \tfrac{1}{2}\sqrt{(1-r^2)}\gamma_u(v, (t+z)/2) + \tfrac{1}{2}r\chi_l(v, -(t+z)/2) = 0\\
&\Omega_l(v, z, t) + \int_{(t-z)/2}^{t} F_l^{\Omega}(v, \tau-t+z, \tau)\,d\tau\\
&\quad + \tfrac{1}{2}\chi_l(v, (z-t)/2) = 0, \quad (z, t)\in K^-.
\end{aligned} \tag{5.165}$$

Note that in deducing the above equations, extensive use has been made of conditions (5.156) through which α_l^1 and β_u^1 have been determined. The system of equations (5.165) has continuous kernels and nonuniform terms in the domains K^+ and K^- and thus its solution $\Omega(v, z, t)$ is continuous in these domains. At $z = 0$ the function Ω, generally speaking, suffers a finite discontinuity. Besides, at $t = |z|$ its values are other than zero, and hence, in the domains G^+ and G^- the function Ω is piecewise-continuous.

Therefore, Lemma 5.4 is proved. As a result, the complete characteristics of the function $W(v, z, t)$ properties have been obtained and one can now go over to analysing the inverse problem. Write the data of the inverse problem (5.132) as follows:

$$q^+(0)[W_1^+(v, 0, t) + W_3^+(v, 0, t)] = \tilde{f}(v, t). \tag{5.166}$$

Here $\tilde{f}(v, t)$ is the image of the Fourier function $f(x_1, x_2, t)$.

Using the expression (5.148) and the function ω properties established by Lemma 5.4, it is concluded that for the inverse problem to be solved the function $\tilde{f}(v,t)$ must be of the following form

$$\tilde{f}(v,t) = a\delta(t) + \tilde{f}_0(v,t)\theta(t) \tag{5.167}$$

in which case at every fixed v the function $\tilde{f}_0(v,t)$ is a twice differentiable function of the variable t. Besides

$$\begin{aligned} a &= q^+(0)[\alpha_1^{-1}(0) + \alpha_3^{-1}(0)] \\ \tilde{f}_0(v,0) &= q^+(0)[\alpha_1^0(v,0) + \alpha_3^0(v,0)] \\ \frac{\partial}{\partial t}\tilde{f}_0(v,t)|_{t=0} &= q^+(0)[\alpha_1^1(v,0) + \alpha_3^1(v,0)]. \end{aligned} \tag{5.168}$$

Theorem 5.7. Setting the function $\tilde{f}(v,t)$ for three different values of the parameter $v = v^k = (v_{1k}, v_{2k})$, $k = 1,2,3$, satisfying the condition

$$\begin{vmatrix} 1 & v_{11}^2 & v_{21}^2 \\ 1 & v_{12}^2 & v_{22}^2 \\ 1 & v_{13}^2 & v_{23}^2 \end{vmatrix} \neq 0 \tag{5.169}$$

defines the functions $\varepsilon(x_3)$, $\mu(x_3)$, and $\sigma(x_3)$ uniquely.

It should be recalled that in a geometrical sense conditions (5.169) mean that on the plane of the variables v_1^2 and v_2^2 the three points v^k, $k = 1,2,3$ must not belong to the same straight line.

It shall be proved below that setting the functions $\tilde{f}(v^k,t)$ for $k = 1, 2, 3$ under condition (5.169) enables one to define the three coefficients $c_1(z)$, $c_2(z)$, and $c_5(z)$ in a unique way, and besides, the value of $q^+(0)$. The required electromagnetic parameters ε, μ, and σ can be found through them, using

$$\begin{aligned} \varepsilon(g(z)) &= \frac{1}{\sqrt{(2q^+(0)c_5(z))}}\exp\left[-2\int_0^z c_2(\xi)\,d\xi\right] \\ \mu(g(z)) &= \sqrt{\left(\frac{q^+(0)}{2c_5(z)}\right)}\exp\left[2\int_0^z c_2(\xi)\,d\xi\right] \\ \sigma(g(z)) &= 2c_1(z)\varepsilon(g(z)), \qquad g(z) = \int_0^z \sqrt{(2c_5(\xi))}\,d\xi. \end{aligned} \tag{5.170}$$

Now prove the theorem using, first of all, equalities (5.168). In this case the numbers a, $\tilde{f}_0(v^k,0)$, $(\partial/\partial t)\tilde{f}_0(v^k,t)|_{t=0}$, $k = 1, 2, 3$, can be considered set.

The first of equalities (5.168) can be written as follows:

$$a = q^+(0)\alpha_3^{-1}(0) = q^+(0)\phi_3 = \frac{[q^-(0)q^+(0)]^2}{[q^-(0)]^2 + [q^+(0)]^2}.$$

From which one gets

$$q^+(0) = \left(\frac{a[q^-(0)]^2}{[q^-(0)]^2 - a}\right)^{1/2}$$

and there arises a necessary condition for the inverse problem solvability $0 < a$

$< [q^-(0)]^2$. Henceforth no attention will be paid to the solvability conditions of the type since when proving the uniqueness theorem it is natural to assume that the problem considered has a solution.

As soon as $q^+(0)$ is obtained, the quantity r becomes known. Then the second of equalities (5.168) at $v = v^k$ can be written, using (5.155) and (5.154), as follows:

$$\begin{aligned}\tilde{f}_0(v^k, 0) &= q^+(0)[(1+r)\alpha_1^0(v^k, 0) + \sqrt{(1-r^2)}\beta_3^0(v^k, 0)]\\ &= -\tfrac{1}{2}q^+(0)\{(1+r)[c_1^+(0) - c_2^+(0)]\phi_3\\ &\quad + \sqrt{(1-r^2)}[c_1^-(0) + c_2^-(0)]\phi_1\}.\end{aligned}$$

Since $\phi_3 \neq 0$, $1 + r \neq 0$, the following combination from the above is obtained: $c_1^+(0) - c_2^+(0)$. Therefore, the values of $\alpha^0(v, 0)$ and $\beta^0(v, 0)$ become known at any v.

The last of equalities (5.168) result in the relations

$$\begin{aligned}\frac{\partial}{\partial t}\tilde{f}_0(v^k, t)|_{t=0} &= q^+(0)[(1+r)\alpha_1^1(v^k, 0) + \sqrt{(1-r^2)}\beta_3^1(v^k, 0)]\\ &= q^+(0)(1+r)\{-\tfrac{1}{2}[c_1^+(0) - c_2^+(0)]\\ &\quad \times \alpha_1^0(v^k, 0) + [-\tfrac{1}{4}(c_1'(0) - c_2'(0))^+ + \tfrac{1}{2}c_1^+(0)\\ &\quad \times (c_1^+(0) - c_2^+(0)) - \tfrac{1}{2}c_5^+(0)(v_{2k}^2 - v_{1k}^2)]\phi_3\}\\ &\quad + q^+(0)\sqrt{(1-r^2)}\beta_3^1(v^k, 0), \quad k = 1, 2, 3.\end{aligned}$$

Since the values of $\beta_3^1(v^k, 0)$ are expressed through $\beta_3^0(v^k, 0)$ and the values of the coefficients c_1, c_2, c_1', c_2', c_5 at $z = -0$, then they are known. Therefore, due to condition (5.169), from the above equalities one can get two quantities

$$c_5^+(0), \qquad \tfrac{1}{4}(c_1'(0) - c_2'(0))^+ - \tfrac{1}{2}c^+(0)(c_1^+(0) - c_2^+(0)).$$

Therefore one can also find $\alpha^1(v, 0)$ and $\beta^1(v, 0)$.

Note, that the equalities defining β^{-1}, β^0, and β^1, involve the values of $q^+(0)$, $\alpha_u^0(0)$, and $\alpha_u^1(v, 0)$. It has been demonstrated that these values can be obtained; therefore, all the terms of the equalities mentioned are known and one can find β^{-1}, β^0, β^1, and χ in succession through the formulas written above. Include the remaining unknown functions α^{-1}, α^0, α^1, and γ_u, into the number of the functions being determined alongside with the coefficients c_1, c_2, and c_5. Show that using the functions $\tilde{f}_0(v^k, t)$ at $k = 1, 2, 3$ one can construct the equations for determining c_1, c_2, and c_3. In this case it is convenient to introduce the additional unknown functions

$$\lambda(z) = c_1(z) - c_2(z), \qquad \rho(z) = \tfrac{1}{4}\lambda'(z) - \tfrac{1}{2}c_1(z)\lambda(z). \tag{5.171}$$

Note, that in line with the above proved $\lambda^+(0)$ and $\rho^+(0)$ are known.

From equalities (5.166) and (5.167) one has

$$\begin{aligned}\frac{\partial^2}{\partial t^2}\tilde{f}_0(v^k, t) &= q^+(0)[\Omega_1^+(v^k, 0, t) + \Omega_3^+(v^k, 0, t)]\\ &= q^+(0)[(1+r)\Omega_1^+(v^k, 0, t) + \sqrt{(1-r^2)}\Omega_3^-(v^k, 0, t)],\\ &\qquad\qquad t \geqslant 0, \quad k = 1, 2, 3.\end{aligned}$$

Substituting the expressions for the function Ω components from equations (5.165) into the right-hand part of these equalities

$$\frac{\partial^2}{\partial t^2}\tilde{f}_0(v^k,t) = -q^+(0)\Big\{(1+r)\Big[\tfrac{1}{2}\gamma_1(v^k,t/2) + \int_{t/2}^{t} F_1^\Omega(v^k,t-\tau,\tau)\,\mathrm{d}\tau\Big] + \sqrt{(1-r^2)}\Big[\tfrac{1}{2}\chi_3(v^k,-t/2) + \int_{t/2}^{t} F_3^\Omega(v^k,\tau-t,\tau)\,\mathrm{d}\tau\Big]\Big\}, \quad t \geqslant 0, \quad k=1,2,3.$$

Solving these equalities with respect to $\gamma_1(v^k,t/2)$, the following equations for determining $\gamma_1(v^k,t/2)$ are obtained

$$\gamma_1(v^k,t/2) = -\frac{2}{q^+(0)(1+r)}\frac{\partial^2}{\partial t^2}\tilde{f}_0(v^k,t) + 2\int_{t/2}^{t} F_1^\Omega(v^k,t-\tau,\tau)\,\mathrm{d}\tau + \sqrt{\left(\frac{1-r}{1+r}\right)}\Big[\chi_3(v^k,-t/2) + 2\int_{t/2}^{t} F_3^\Omega(v^k,\tau-t,\tau)\,\mathrm{d}\tau\Big],$$

$$t \geqslant 0, \quad k=1,2,3. \qquad (5.172)$$

Through the functions $\gamma_1(v^k,z)$ at $k=1,2,3$ one can construct the equations for λ, ρ, c_1, c_2, c_5. Indeed, from equality (5.158) one gets

$$\gamma_1(v^k,z) = \gamma_{11}(z) + v_{1k}^2\gamma_{12}(z) + v_{2k}^2\gamma_{13}(z) + \gamma_{14}(v^k,z), \quad k=1,2,3,$$

$$\gamma_{11}(z) = \rho'(z) - \tfrac{1}{2}\rho(z)c_1(z) + \tfrac{1}{4}\lambda^2(z)[c_1(z)+c_2(z)] \qquad (5.173)$$

$$\gamma_{12}(z) = \tfrac{1}{2}c_5'(z) + 3c_5(z)c_1(z) - c_5(z)\lambda(z)$$

$$\gamma_{13}(z) = \tfrac{1}{2}c_5'(z) - c_5(z)c_1(z) - c_5(z)\lambda(z) \qquad (5.174)$$

$$\gamma_{14}(v^k,z) = [2\rho(z) + c_5(z)(v_{2k}^2 - v_{1k}^2)]\alpha_3^0(v^k,z) - 2v_{1k}v_{2k}c_5(z)\alpha_4^0(v^k,z) + \lambda(z)\alpha_3^1(v^k,z), \quad k=1,2,3. \qquad (5.175)$$

Due to condition (5.169), equalities (5.173) can be solved with respect to $\gamma_{1k}(z)$, $k=1,2,3$

$$\gamma_{1k}(z) = \sum_{j=1}^{3} d_{kj}[\gamma_1(v^j,z) - \gamma_{14}(v^j,z)], \quad k=1,2,3. \qquad (5.176)$$

Here d_{kj} are the numerical coefficients depending only on v^1, v^2, and v^3.

From equalities (5.174) and (5.171) one gets

$$\rho(z) = \rho(0) + \int_0^z [\gamma_{11}(\xi) + \tfrac{1}{2}\rho(\xi)c_1(\xi) - \tfrac{1}{4}\lambda^2(\xi)(c_1(\xi)+c_2(\xi))]\,\mathrm{d}\xi$$

$$\lambda(z) = \lambda(0) + \int_0^z [4\rho(\xi) + 2\lambda(\xi)c_1(\xi)]\,\mathrm{d}\xi \qquad (5.177)$$

$$c_5(z) = c_5(0) + 2\int_0^z [\gamma_{12}(\xi) - 3c_5(\xi)c_1(\xi) + c_5(\xi)\lambda(\xi)]\,\mathrm{d}\xi$$

$$c_1(z) = [\gamma_{12}(z) - \gamma_{13}(z)]/4c_5(z), \qquad c_2(z) = c_1(z) - \lambda(z).$$

Note, that $c_5(z) > 0$, since $\varepsilon > 0$, $\mu > 0$. Now make some comments on the last of the equalities obtained: the quantities $\rho(0)$, $\lambda(0)$, and $c_5(0)$ are known, and thus, $\rho(z)$, $\lambda(z)$, and $c_5(z)$ are also known to the accuracy of the integral terms. The functions α_3^0, α_4^0, and α_3^1, involved in equalities (5.175), and hence, $\gamma_{14}(v^k, z)$ are known to the same accuracy as well. As a result, $c_1(z)$ and $c_2(z)$ are defined within the same accuracy.

With the integral terms allowed for, equalities (5.172) and (5.175)–(5.177) are a nonlinear system of integral equations for determining $\gamma_1(v^k, z)$, $\gamma_{14}(v^k, z)$, $\gamma_{1k}(z)$, ρ, λ, c_1, c_2, c_3. To obtain a closed system, the above equalities should be supplemented with equations (5.165) for $\Omega(v^k, z, t)$, equalities (5.154), (5.155), and (5.157) for $\alpha^{-1}(z)$, $\alpha^0(v^k, z)$, $\alpha^1(v^k, z)$ and equality (5.158) for the component $\gamma_2(v^k, z)$. As a result, a closed system of equations is obtained that can be considered at any finite $T > 0$ in the domain $K_T = \{(z, t): |z| \leqslant t \leqslant T - |z|\}$. At small T the principle of contracted self-mapping can be applied to the system of equations in the space of the functions continuous throughout K_T everywhere except for the straight line $z = 0$. At the same time under the *a priori* supposition that $c_5(z) > 0$ for this system valid is the theorem of uniqueness at any finite T, and therefore $c_1(z)$, $c_2(z)$, and $c_5(z)$ can be obtained through this system in a unique way. In this case formulas (5.170) make it possible to define uniquely ε, μ, and σ as functions of the variable x_3. Thus Theorem 5.7 is proved.

By way of conclusion note that in the case when the functions ε, μ, and σ can be expressed as $\varepsilon = \varepsilon_0(x_3) + \varepsilon_1(x)$, $\mu = \mu_0(x_3) + \mu_1(x)$, and $\sigma = \sigma_0(x_3) + \sigma_1(x)$, where ε_0, μ_0, and σ_0 are known functions and ε_1, μ_1, and σ_1 are small and finite in the domain D^+, one can consider a linearized formulation of the problem of determining ε_1, μ_1, and σ_1. Its investigation can be easily carried out by the scheme uniting in itself the basic elements of this and preceding sections, provided one assumes that the point $x^0 = (x_1^0, x_2^0, -0)$ runs through the whole set of points of the plane $x_3 = 0$, and that the information given has the form

$$E_1^+|_{x_3=0} = f(x_1, x_2, t, x_1^0, x_2^0).$$

This information uniquely defines ε_1, μ_1, and σ_1 (see [268]).

Chapter 6

Inverse problems for parabolic and elliptical type second-order equations

In determining thermophysical and filtration parameters of a medium and density of distribution of heat sources there arise inverse problems for parabolic type equations. This problem has been investigated by a great number of researchers; mainly during the 1970s. Here, first of all, the works by Beznoshchenko [295–301], Iskenderov [302–304] (also with Budak [305–307]), Klibanov [181, 308–314], Cannon [315–321], should be recalled and it should be noted that they have greatly influenced the creation of the theory of inverse problems for parabolic type equations. Numerous formulations of the inverse problems in this field have been studied in [246, 322–342].

Inverse problems for elliptical type equations arise in the case when the required parameters of a medium are determined from observing the established physical processes that are either oscillating or thermal. Interest has been stimulated by the creation of sources of excitation of the waves of a vibrational type. At the same time it should be noted that the theory of inverse problems for elliptical type equations has been developed to a lesser extent than that for hyperbolic and parabolic type equations. Unfortunately, there is a comparatively limited amount of literature devoted to this field [40, 343–349].

As will be shown in Section 6.3, a fairly wide class of inverse problems for second-order parabolic and elliptical type equations is reduced to inverse problems for hyperbolic equations, which can be investigated by the methods discussed in Chapter 3. At the same time there exist a great number of important problems that cannot be reduced to those studied.

In this chapter I will first consider some special formulations of the inverse problems in order to illustrate the problems arising as a result of their investigation, followed by the discussion of the methods developed for studying the inverse problems for equations of parabolic and elliptical types.

6.1. Problem of determining the density of heat sources

This problem was considered by Vasil'ev [290]. The presentation suggested below is a slight modification of that given by Romanov [43, see also 40].

For the equation

$$u_t = \Delta u + \phi(t) f(x, y) \tag{6.1}$$

the problem of determining the function $f(x, y)$ will be considered in the domain

$y \geqslant 0$ of the space of the variables $x, y, x = (x_1, \ldots, x_n)$, if it is known that under the conditions

$$u|_{t=0} = 0, \qquad (u_y + hu)_{y=0} = 0 \tag{6.2}$$

the solution of (6.1) assumes at the points of the plane $y = 0$ the given values

$$u|_{y=0} = F(x, t). \tag{6.3}$$

In this case $\phi(t)$ is considered a given function, h is a given finite number. In (6.1) Δ is the Laplace operator with respect to the variables x, y.

The problem discussed can be treated like that of determining the density of heat sources operating in the semispace $y \geqslant 0$. It is the function $f(x, y)$ that defines their density. This problem has a very definite physical sense: if the function $\phi(t) = e^{-\lambda_0 t}$, it is associated with the problem of determining the density of radioactive heat sources by the thermal radiation on the Earth's surface. In this case λ_0 is the half-life of a radioactive element.

Theorem 6.1. Let $\phi(t) \not\equiv 0$ and increases at $t \to \infty$ not faster than $Ce^{\alpha t}$, where α is a fixed constant, $f(x, y)$ is a limited function, having at every fixed y a finite Fourier transform

$$f(v, y) = \int_{R^n} f(x, y) \exp[i(v, x)] \, dx, \quad v = (v_1, \ldots, v_n)$$

depending on y in a continuous way. In this case the function $f(x, y)$ is uniquely defined by setting the function $F(x, t)$.

In fulfilling the conditions of the theorem, the solution to (6.1) and (6.2) increases with respect to the variable t not faster than $C\, e^{\alpha t}$. Therefore, one can apply the Fourier–Laplace transform

$$\tilde{u}(v, y, p) = \int_{R^n} dx \int_0^\infty u(x, y, t) \exp[i(v, x) - pt] \, dt, \quad \operatorname{Re} p > \alpha.$$

In this case (6.1)–(6.3) are reduced to the form

$$\tilde{u}_{yy} - (p + |v|^2)\tilde{u} + \tilde{\phi}(p)\tilde{f}(v, y) = 0, \qquad (\tilde{u}_y + h\tilde{u})_{y=0} = 0 \tag{6.4}$$

$$\tilde{u}|_{y=0} = \tilde{F}(v, p). \tag{6.5}$$

Here $\tilde{\phi}(p)$ is the Laplace transform of the function $\phi(t)$

$$\tilde{\phi}(p) = \int_0^\infty \phi(t) \exp(-pt) \, dt.$$

At every fixed value of the parameters p, v, (6.4) is a common differential equation. A limited solution of problem (6.4) can be easily constructed using the Green's function $G(y, \eta, \lambda)$, $\lambda = \sqrt{(p + |v|^2)}$ for this problem. With respect to the variable y this function obeys the equation $G_{yy} - \lambda^2 G = \delta(y - \eta)$, the condition $(G_y + hG)_{y=0} = 0$ and is continuous and limited on the segment $[0, \infty]$. At $y = \eta$ the derivative of this function G_y has a finite jump equal to unity:

$G_y|_{y=\eta+0} - G_y|_{y=\eta-0} = 1$. One can easily prove that the function $G(y, \eta, \lambda)$ is as follows:

$$G(y,\eta,\lambda) = \frac{1}{h-\lambda}\begin{cases}\left(\cosh \lambda y - h\dfrac{\sinh \lambda y}{\lambda}\right)e^{-\lambda\eta}, & 0 \leqslant y \leqslant \eta\\[2ex] \left(\cosh \lambda\eta - h\dfrac{\sinh \lambda\eta}{\lambda}\right)e^{-\lambda y}, & y \geqslant \eta.\end{cases} \tag{6.6}$$

Through the Green's function the solution of problem (6.4) is written in the following form

$$\tilde{u}(v, y, p) = \tilde{\phi}(p)\int_0^\infty G(y, \eta, \sqrt{(p+|v|^2)})\tilde{f}(v,\eta)\,d\eta.$$

Setting here $y=0$, the equation for $\tilde{f}(v,\eta)$ is obtained

$$\tilde{F}(v,p) = \frac{\tilde{\phi}(p)}{h-\sqrt{(p+|v|^2)}}\int_0^\infty (e^{-\eta\sqrt{(p+|v|^2)}}\tilde{f}(v,\eta)\,d\eta, \quad \operatorname{Re} p > \alpha. \tag{6.7}$$

In this equation v functions as a parameter. The function $\tilde{\phi}(p)$ being analytical in the domain $\operatorname{Re} p > \alpha$, it can become zero only at isolated points. Therefore, (6.7) can be expressed in the following way

$$\int_0^\infty e^{-\lambda y}\tilde{f}(v,y)\,dy = \Phi(v,\lambda) \tag{6.8}$$

where the function

$$\Phi(v,\lambda) = (h-\lambda)\frac{\tilde{F}(v,\lambda^2-|v|^2)}{\tilde{\phi}(\lambda^2-|v|^2)}$$

is known in the domain $G(v)$

$$G(v) = \{\lambda : \operatorname{Re}(\lambda^2 - |v|^2) > \alpha, \operatorname{Re}\lambda > 0\},$$

which can be presented in the complex plane $\lambda = \sigma + i\tau$ as $\sigma > \sqrt{(\alpha + |v|^2 + \tau^2)}$.

The function $\Phi(v,\lambda)$, as is seen from (6.8), at every fixed v is a Laplace transform with respect to the variable y of the function $\tilde{f}(v,y)$. But the function $\tilde{f}(v,y)$ is uniquely determined by the Laplace transform values within the domain $G(v)$, for instance, it can be found by the formula (see [350])

$$\tilde{f}(v,y) = \lim_{n\to\infty}\left\{\left[\frac{(-1)^n}{n}s^{n+1}\frac{\partial^n}{\partial s^n}\Phi(v,s)\right]_{s=n/y}\right\}. \tag{6.9}$$

There exists another way of constructing $\tilde{f}(v,y)$, associated with the analytical extension of the function $\Phi(v,\lambda)$ into the semiplane $\operatorname{Re}\lambda > 0$ and the use of a common formula of inversion. Since $\tilde{f}(v,y)$, $v \in R^n$, uniquely defines $f(x,y)$, then the theorem is valid.

6.2. Problem of determining diffusion coefficients

In the domain $y \geqslant 0$ of the space of the variables x, y, $x = (x_1, \ldots, x_n)$ consider the diffusion equation

$$a\frac{\partial}{\partial t}u = \Delta u + \delta(x - x^0, y, t) \tag{6.10}$$

with the initial and boundary conditions

$$u|_{t<0} = 0, \qquad (u_y + hu)_{y=0} = 0. \tag{6.11}$$

In (6.10) $a = a(x, y)$ is a positive and limited from above and below with positive constants function; x^0 is an arbitrary point of the space R^n functioning as a parameter, the coefficient a being directly associated with the diffusion coefficient.

Transform problem (6.10) and (6.11) in terms of Laplace, denoting through $v(x, y, p, x^0)$ the Laplace image of the function $u(x, y, t, x^0)$

$$\Delta v + pav + \delta(x - x^0, y) = 0, \qquad (v_y + hv)_{y=0} = 0. \tag{6.12}$$

Now consider the problem of determining the coefficient $a(x, y)$ by the information

$$\left.\frac{\partial}{\partial p}v\right|_{y=p=0} = f(x, x^0) \tag{6.13}$$

where x, x^0 are arbitrary points of R^n.

This inverse problem can be equally referred to as that for both the parabolic (6.10) and the elliptical (6.12) equations.

Theorem 6.2. Let $a(x, y)$ be expressed in the domain $y \geqslant 0$ in the following way

$$a(x, y) = a_0 + a_1(x, y) \tag{6.14}$$

where a_0 is a positive constant and at every fixed y the function $a_1(x, y)$ has a finite Fourier transform continuously depending on y. In this case the function $a(x, y)$ is uniquely determined by setting the information (6.13).

Express the solution to problem (6.12) as a series with respect to the parameter p

$$v(x, y, p, x^0) = \sum_{n=0}^{\infty} p^n v_n(x, y, x^0). \tag{6.15}$$

Then

$$\left.\frac{\partial}{\partial p}v\right|_{p=0} = v_1(x, y, x^0)$$

and hence

$$f(x, x^0) = v_1(x, 0, x^0). \tag{6.16}$$

Substituting series (6.15) into equalities (6.12) and equating the coefficients at the zero and first powers of p, one gets the equation for finding v_0 and v_1

$$\Delta v_0 + \delta(x - x^0, y) = 0, \qquad [(v_0)_y + hv_0]_{y=0} = 0 \tag{6.17}$$

$$\Delta v_1 + av_0 = 0, \qquad [(v_1)_y + hv_1]_{y=0} = 0. \tag{6.18}$$

These equations, as well as (6.12) should be supplemented with the condition of the solution decrease at $|x| + y \to \infty$, if $n > 1$, or with that of its boundedness, if $n = 1$.

The solution to problem (6.17) can be written as $v_0(x, y, x^0) = w_0(x - x^0, y)$. If the difference $x - x^0$ is denoted through ξ, then the functions $w_0(\xi, y)$ and $w_1(\xi, y, x^0) = v_1(x^0 + \xi, y, x^0)$ in equalities (6.17) and (6.18) assume the form

$$\Delta w_0 + \delta(\xi, y) = 0, \qquad [(w_0)_y + hw_0]_{y=0} = 0 \tag{6.19}$$

$$\Delta w_1 + a(x^0 + \xi, y)w_0(\xi, y) = 0, \qquad [(w_1)_y + hw_1]_{y=0} = 0. \tag{6.20}$$

In these equalities Δ is the Laplace operator with respect to the variables ξ, y.

Apply to equalities (6.19) the Fourier transform with respect to the variable ξ, and to equalities (6.20)—that with respect to the variables ξ, x^0

$$\tilde{w}_1(v, y, \kappa) = \int_{R^{2n}} w_1(\xi, y, x^0) \exp\{i(v, \xi) + (\kappa, x^0)]\} \, d\xi \, dx^0.$$

Then for the Fourier images $\tilde{w}_0(v, y)$ and $\tilde{w}_1(v, y, \kappa)$ of the functions w_0, w_1, one gets the equalities

$$(\tilde{w}_0)_{yy} - |v|^2 \tilde{w} + \delta(y) = 0, [(\tilde{w}_0)_y + h\tilde{w}_0]_{y=0} = 0 \tag{6.21}$$

$$(\tilde{w}_1)_{yy} - |v|^2 \tilde{w}_1 + \tilde{a}(\kappa, y)\tilde{w}_0(v - \kappa, y) = 0$$
$$[(\tilde{w}_1)_y + h\tilde{w}_1]_{y=0} = 0. \tag{6.22}$$

Here $\tilde{a}(\kappa, y)$ is the Fourier image of the function $a(x, y)$. The solutions to problems (6.21) and (6.22) can be easily constructed using the Green's function (6.66), where $\lambda = |v|$. In this case

$$\tilde{w}_0(v, y) = -G(y, 0, |v|) = \frac{1}{|v| - h} e^{-|v|y}$$

$$\tilde{w}_1(v, y, \kappa) = \int_0^\infty G(y, \eta, |v|) G(\eta, 0, |v - \kappa|) \tilde{a}(\kappa, \eta) \, d\eta.$$

setting here $y = 0$ and using (6.6), one has

$$\tilde{w}_1(v, 0, \kappa) = \frac{1}{(h - |v|)(h - |v - \kappa|)} \int_0^\infty \exp\{-[|v| + |v - \kappa|]y\} \tilde{a}(\kappa, y) \, dy \tag{6.23}$$

The left-hand part of the above equality can be expressed through the data (6.16) of the inverse problem

$$\tilde{w}_1(v, 0, \kappa) = \int_{R^{2n}} f(\xi + x^0, x^0) \exp\{i[(v, \xi) + (\kappa, x^0)]\} \, d\xi \, dx^0 = \tilde{f}(v, \kappa).$$

Hence, (6.23) can be written as an integral equation for determining $\tilde{a}(\kappa, y)$

$$\int_0^\infty \exp\{-[|v| + |v - \kappa|]y\} \tilde{a}(\kappa, y) \, dy = \phi(v, \kappa)$$

$$\phi(v, \kappa) = (h - |v|)(h - |v - \kappa|) \tilde{f}(v, \kappa). \tag{6.24}$$

From expression (6.14) one gets

$$\tilde{a}(\kappa, y) = a_0\delta(\kappa) + \tilde{a}_1(\kappa, y).$$

Therefore, $\phi(\nu, \kappa)$ is a generalized function of the type

$$\phi(\nu, \kappa) = c(\nu)\delta(\kappa) + \phi_0(\nu, \kappa)$$

where $c = a_0/2|\nu|$ and $\phi_0(\nu, \kappa)$ is a regular function.

The singular part of the function $\phi(\nu, \kappa)$ defines the parameter a_0, while with the regular part one can easily determine $\tilde{a}_1(\kappa, y)$. Indeed, at every fixed κ the equation

$$\int_0^\infty \exp[-(|\nu| + |\nu - \kappa|)y]\tilde{a}_1(\kappa, y)\,\mathrm{d}y = \phi_0(\nu, \kappa) \tag{6.25}$$

is a Laplace transform of the function $\tilde{a}_1(\kappa, y)$ with respect to the variable y, with the transform parameter $s = |\nu| + |\nu - \kappa|$. Since the parameter of this transform runs through all the positive semi-axis, then $\tilde{a}_1(\kappa, y)$ is defined uniquely. In order to construct the solution of (6.25) use can be made of the inversion formula (6.9). Therefore the theorem is proved.

6.3. Relations among inverse problems for parabolic, elliptical, and hyperbolic type equations

The fact that one can establish a one-to-one correspondence between the solutions of the problems of mathematical physics describing absolutely different physical processes under definite conditions was noted long ago. The simplest case of deducing such a correspondence refers to the equations with constant coefficients when using the Fourier and Laplace integral transforms. Reznitskaya [333] was the first to point out the usefulness of relations between the solutions of equations of various types when considering inverse problems. In studying the inverse problems for equations of parabolic or elliptical types it proves possible to go over to investigating some equivalent problems for equations of a hyperbolic type, a reverse procedure being sometimes also useful. Here is the essence of the method shown on the concrete examples of formulating inverse problems.

In the domain $D \subset R^n$, limited by the surface S, consider a mixed problem for the equation of a parabolic type

$$u_t = Lu + f(x, t), \qquad u|_{t=0} = \phi(x), \qquad \left.\frac{\partial u}{\partial n}\right|_S = \psi(x, t). \tag{6.26}$$

Here L is a uniformly elliptical operator with continuous coefficients depending only on the variable x; n is a conormal to S. Correlate problem (6.26) and the problem for the equation of a hyperbolic type

$$\tilde{u}_{tt} = L\tilde{u} + \tilde{f}(x, t), \qquad \tilde{u}|_{t=0} = \phi(x), \qquad \tilde{u}_t|_{t=0} = 0, \qquad \left.\frac{\partial \tilde{u}}{\partial n}\right|_S = \tilde{\psi}(x, t). \tag{6.27}$$

Let the functions $\tilde{f}$ and $\tilde{\psi}$, that are the smooth functions of their arguments, growing at $t \to \infty$ not faster than $C\,\mathrm{e}^{\alpha t}$, be related to the functions f and ψ through the expressions

$$f(x,t)=\int_0^\infty \tilde{f}(x,t)G(t,\tau)\,d\tau, \qquad \psi(x,t)=\int_0^\infty \tilde{\psi}(x,t)G(t,\tau)\,d\tau \tag{6.28}$$

where

$$G(t,\tau)=\frac{1}{\sqrt{(\pi t)}}e^{-\tau^2/4t}$$

is the solution of the equation of heat conductivity $G_t = G_{\tau\tau}$. Then

$$u(x,t)=\int_0^\infty \tilde{u}(x,\tau)G(t,\tau)\,d\tau. \tag{6.29}$$

Indeed

$$u_t=\int_0^\infty \tilde{u}(x,\tau)G_t(t,\tau)\,d\tau=\int_0^\infty \tilde{u}(x,\tau)G_{\tau\tau}(t,\tau)\,d\tau$$

$$=(\tilde{u}G_\tau-\tilde{u}_\tau G)|_{\tau=0}^{\tau=\infty}+\int_0^\infty \tilde{u}_{\tau\tau}(x,\tau)G(t,\tau)\,d\tau=\int_0^\infty \tilde{u}_{\tau\tau}(x,\tau)G(t,\tau)\,d\tau$$

$$Lu=\int_0^\infty G(t,\tau)L\tilde{u}(x,\tau)\,d\tau.$$

Therefore for $t>0$

$$u_t-Lu-f=\int_0^\infty (\tilde{u}_{\tau\tau}-L\tilde{u}-\tilde{f})G(t,\tau)\,d\tau=0.$$

On the other hand

$$u|_{t=0}=\lim_{t\to 0}\int_0^\infty \tilde{u}(x,\tau)G(t,\tau)\,d\tau$$

$$=\lim_{t\to 0}\frac{2}{\sqrt{\pi}}\int_0^\infty \tilde{u}(x,2\sqrt{(t)}\tau)e^{-\tau^2}\,d\tau=\tilde{u}(x,0)=\phi(x)$$

$$\left.\frac{\partial u}{\partial n}\right|_S=\int_0^\infty \left.\frac{\partial \tilde{u}}{\partial n}\right|_S G(t,\tau)\,d\tau=\int_0^\infty \tilde{\psi}(x,\tau)G(t,\tau)\,d\tau=\psi(x,t).$$

Thus, (6.29) provides the solution to problem (6.26) through the solution to problem (6.27) provided the data of these problems are related by (6.28). The solution to problem (6.26) is known to be unique under fairly general assumptions on the operator L and the surface S, and to be set by (6.29). It should be recalled that equality (6.29) is inverse at every fixed x, since it can be expressed as a Laplace transform with the transformation parameter $p=1/4t$

$$u(x,1/4p)=\sqrt{\left(\frac{p}{\pi}\right)}\int_0^\infty e^{-zp}\tilde{u}(x,\sqrt{z})\frac{dz}{\sqrt{z}}. \tag{6.30}$$

Therefore, (6.29) establishes a one-to-one correspondence between the solutions of problems (6.26) and (6.27).

Assume now that for (6.26) the inverse problem [to find $f(x,t)$ or a certain coefficient involved in the operator L] is considered by the solution of problem (6.26) given at the point S

$$u|_S = g(x,t), \quad t \geqslant 0. \tag{6.31}$$

Bring to correspondence with the function $g(x,t)$ the function $\tilde{g}(x,t)$, $x \in S$, as a solution to the equation

$$g(x,t) = \int_0^\infty \tilde{g}(x,\tau) G(t,\tau)\, \mathrm{d}\tau, \quad x \in S, \quad t \geqslant 0. \tag{6.32}$$

For $x \in S$ the function $\tilde{g}(x,t)$ is found uniquely. Assume

$$\tilde{u}|_s = \tilde{g}(x,t), \quad t \geqslant 0. \tag{6.33}$$

Then the inverse problem (6.26) and (6.31) is equivalent to the corresponding inverse problem (6.27) and (6.33).

One more point should be recalled. It is known from mathematical physics that for the solution of problem (6.27) under fairly general assumptions on the operator L coefficients the following estimate is valid $|\tilde{u}(x,t)| \leqslant C\,\mathrm{e}^{\alpha t}$. Equation (6.29) and the equivalent formula (6.30) demonstrate that in the domain $t > 0$ the function $u(x,t)$ is an analytical function of the argument t. Therefore, it is sufficient to set the function $g(x,t)$ in (6.31) in any small vicinity of the value $t = 0$, for instance, within the interval $0 \leqslant t \leqslant \delta$, $\delta > 0$.

Due to the above established relations between the inverse problems, one can transfer a number of the results obtained Sections 6.2–6.4 for hyperbolic equations onto parabolic equations. It is possible in all the cases when the set where the inverse problem data are given is a cylinder with its elements parallel to the axis t.

In particular, the simplest inverse problem for the equation of a parabolic type on determining the coefficient $q(x)$ involved in the equation

$$u_t = u_{xx} - q(x)u \tag{6.34}$$

by the information on the Cauchy problem solution

$$u|_{t=0} = \delta(x) \tag{6.35}$$

set at $x = 0$

$$u|_{x=0} = f_1(t), \qquad u_x|_{x=0} = f_2(t) \tag{6.36}$$

is reduced to the following inverse problem

$$\begin{gathered} \tilde{u}_{tt} = \tilde{u}_{xx} - q(x)\tilde{u} \\ \tilde{u}|_{t=0} = \delta(x), \qquad \tilde{u}_t|_{t=0} = 0 \\ \tilde{u}|_{x=0} = \tilde{f}_1(t), \qquad \tilde{u}_x|_{x=0} = \tilde{f}_2(t). \end{gathered} \tag{6.37}$$

Here the functions $\tilde{f}_k(t)$ are related to $f_k(t)$ through

$$f_k(t) = \int_0^\infty G(t,\tau)\tilde{f}_k(\tau)\, \mathrm{d}\tau, \quad k = 1, 2.$$

Problem (6.37) is obtained from problem (2.47)–(2.49) by differentiating it once over the variable t. It means that in the inverse problem (6.34)–(6.36) the coefficient $q(x) \in C(-\infty, \infty)$ is uniquely determined by setting the functions $f_1(t)$, $f_2(t)$ on the intercept $[0, \delta]$, where δ is an arbitrary positive number.

Now go over to the equations of an elliptical type. Let the domain D_0 of the space of the variables $x, y (x \in R^n)$ have the form of a semi-infinite cylinder with the elements parallel to the axis y

$$D_0 = D \times R^+, \quad x \in D, \qquad R^+ = \{y: y \geqslant 0\}$$

Consider in the domain D_0 the following equation

$$u_{yy} + Lu + f(x, y) = 0 \tag{6.38}$$

where L is a uniformly elliptical operator with respect to the variables $x_1, \ldots, x_n$ with the y-independent coefficients. Denote the boundary of the domain D by S and formulate for (6.38) the problem of finding the solution of (6.38) that on the boundary of the domain D satisfies the conditions

$$u|_{y=0} = \phi(x), \qquad \left.\frac{\partial u}{\partial n}\right|_{\Gamma} = \psi(x, y), \qquad \Gamma = S \times R^+ \tag{6.39}$$

and the condition of decreasing the solution at $y \to \infty$.

Alongside with (6.38) and (6.39) consider (6.27) where $\tilde{f}$ and $\tilde{\psi}$ are related to f and ψ by

$$f(x, y) = \int_0^\infty \tilde{f}(x, t) H(y, t)\, dt$$

$$\psi(x, y) = \int_0^\infty \tilde{\psi}(x, t) H(y, t)\, dt \tag{6.40}$$

Here

$$H(y, t) = \frac{2}{\pi} \frac{y}{y^2 + t^2}$$

is the solution of the Laplace equation

$$H_{tt} + H_{yy} = 0.$$

For the integrals (6.40) to exist assume that $\tilde{f}$ and $\tilde{\psi}$ are limited, making an analogous assumption on the solution of problem (6.27).

In this case a unique solution of problem (6.38) and (6.39) is given by

$$u(x, y) = \int_0^\infty \tilde{u}(x, t) H(y, t)\, dt. \tag{6.41}$$

Now prove the validity of the formula under the assumption that all the operations carried out below have a sense.

One has

$$u_{yy} = \int_0^\infty H_{yy}(y,t)\tilde{u}(x,t)\,dt = -\int_0^\infty H_{tt}(x,t)\tilde{u}(x,t)\,dt$$

$$= (-H_t\tilde{u} + H\tilde{u}_t)|_{t=0}^{t=\infty} - \int_0^\infty H(y,t)\tilde{u}_{tt}(x,t)\,dt$$

$$= -\int_0^\infty H(y,t)\tilde{u}_{tt}(x,t)\,dt$$

$$Lu = \int_0^\infty H(y,t)L\tilde{u}(x,t)\,dt.$$

Therefore

$$u_{yy} + Lu + f(x,t) = \int_0^\infty (-\tilde{u}_{tt} + L\tilde{u} + \tilde{f})H(y,t)\,dt = 0.$$

At the same time

$$\left.\frac{\partial u}{\partial n}\right|_\Gamma = \int_0^\infty \left.\frac{\partial \tilde{u}}{\partial n}\right|_S H(y,t)\,dy = \int_0^\infty \tilde{\psi}(x,t)H(y,t)\,dt = \psi(x,y)$$

$$u|_{y=0} = \lim_{y\to 0}\int_0^\infty \tilde{u}(x,t)H(y,t)\,dt$$

$$= \lim_{y\to 0}\frac{2}{\pi}\int_0^\infty \tilde{u}(x,y\tau)\frac{d\tau}{1+\tau^2} = \tilde{u}(x,0) = \phi(x).$$

Thus, the function $u(x, y)$ determined by (6.41) is indeed the solution of (6.38) and (6.39).

At any fixed x equality (6.41) is uniquely inversed as an integral transform with the Cauchy kernel. Therefore, if for an elliptical equation we consider the problem on determining the coefficient involved in the operator L by the information on the solution of (6.38) and (6.39) of the type

$$u|_\Gamma = g(x,y) \tag{6.42}$$

then the inverse problem (6.38)–(6.42) can be reduced to the equivalent inverse problem (6.27) and (6.33). In this case the functions g and $\tilde{g}$ are correlated by the inverse agreement

$$g(x,y) = \int_0^\infty \tilde{g}(x,t)H(y,t)\,dt, \quad x\in S, y\in R^+.$$

In line with this, a whole number of the inverse problems for (6.26) can be investigated on the basis of their reducing to the problems studied in Chapters 4–6.

From the above it follows that the transition from an elliptical equation to a hyperbolic one is possible only if the two following conditions are met: (1) the domain where an equation is considered must be a cylindrical one with its elements parallel to the axis y and (2) the equation coefficients must be y-independent. These limitations are quite serious since in the applied problems the variables x and y often enjoy equal rights

Consider one more class of inverse problems for the equations of an elliptical type which can be studied by way of reducing them to the inverse problems for hyperbolic equations on a concrete example. For the equation

$$\Delta u + \omega^2 c(x) u = f(x), \quad x \in R^n \tag{6.43}$$

consider the problem of determining the function $f(x)$ or the coefficients $c(x) \geqslant c_0 > 0$ by the following information. For a series of frequencies $\omega_k (k = 1, 2, \ldots)$ one knows the solution of (6.43) satisfying the conditions of radiation at the points of a certain surface S

$$u(x, \omega_k) = g(x, \omega_k), \quad x \in S, \quad k = 1, 2, \ldots. \tag{6.44}$$

Equation (6.43) arises if one considers the established harmonic oscillations described by the wave equation $c(x) v_{tt} = \Delta v + f(x) e^{i\omega t}$. Expressing the function v as $v = u(x) e^{i\omega t}$, one gets (6.43), which, in the case when $c = 1$, is a Helmholtz equation. The inverse problems on determining $f(x)$ for Helmholtz equations have been studied in [348, 349].

Consider the following Cauchy problem

$$c(x) w_{tt} = \Delta w + f(x) \delta(t), \quad w|_{t<0} \equiv 0. \tag{6.45}$$

Under the condition of limitedness of $c(x)$ and $f(x)$ the solution of this problem allows the estimate $|w| \leqslant C e^{\alpha t}$. Therefore, there exists the Laplace transform $\tilde{w}(x, p)$ of the function $w(x, t)$ at $\operatorname{Re} p > \alpha$, which is an analytical function of the variable p. Under certain conditions [for instance, $c = 1$, $f(x)$ is a finite function] the function $\tilde{w}(x, p)$ allows an analytical extension onto the imaginary axis $p = i\omega$. In this case $\tilde{w}(x, i\omega) = u(x, \omega)$. Hence,

$$\tilde{w}(x, i\omega_k) = g(x, \omega_k), \quad k = 1, 2, \ldots, x \in S.$$

If $\omega_k \to \omega_0$, where ω_0 is an internal point of the domain of analyticity of the function $\tilde{w}(x, i\omega)$, then by the values of $\tilde{w}(x, i\omega_k)$ one can find $\tilde{w}(x, p)$ throughout the whole domain of analyticity and hence calculate

$$w|_S = \tilde{g}(x, t). \tag{6.46}$$

As a result, one comes to the inverse problem (6.45) and (6.46) equivalent to problem (6.43) and (6.44).

6.4. On specific formulations of inverse problems where the coefficient to be determined is independent of one of the variables

The inverse problems with the coefficient to be determined independent of one of the variables involved in the differential equation have been considered throughout this book for the equations describing nonstationary processes with the coefficients to be determined independent of time. A characteristic feature of all the problems considered so far is the fact that the set on which an additional information is given does not coincide with the domain of determining the required coefficient. There exists, however, a wide range of physically justified formulations of inverse problems for all the basic equations of mathematical physics where the domain of defining the required coefficient and that of the given information on the solution of the differential equation coincide. Basic results in this field have been obtained by Iskenderov and Beznoshchenko and refer to parabolic type equations. In the formulation of the inverse problems studied by Iskenderov the unknown coefficients are involved in the additional condition on the domain boundary, while in the formulations by Beznoshchenko the additional condition is set on the plane orthogonal to the variable the coefficients are independent of. It should be recalled that the method employed by Beznoshchenko is based on obtaining an additional problem where the required coefficient is involved in the additional condition.

Here are some examples of such formulations. Let D be the domain of the space of the variables x, y, $x=(x_1,\dots\dots,x_n)$ with the smooth boundary Γ, n be the direction of the normal to Γ, $\Omega = D\times[0,T]$, $S=\Gamma\times[0,T]$, $T>0$ be a given number. The task is to find the coefficient $a=a(x,t)>0$ provided one knows that the solution of the boundary problem

$$\begin{aligned} &u_t - a\Delta u = h(x,y,t), \quad (x,y,t)\in\Omega \\ &u|_{t=0}=\phi(x,y), \qquad u|_S = f(x,y,t) \end{aligned} \tag{6.47}$$

satisfy at the points of S the condition

$$a\frac{\partial u}{\partial n}\bigg|_S = g(x,y,t). \tag{6.48}$$

Here h, ϕ, f, and g are the given functions. The left-hand part of (6.48) in a physical sense is a heat flux through the surface Γ and is a measurable quantity. The required coefficient, as is seen from this formulation, is involved in both the differential equation and the additional condition, and it is determined, naturally, only at the points of the domain S_0 coinciding with the projection of S onto the space of the variables x, t. Problem (6.47) and (6.48) has been studied by Iskenderov [351], who established uniqueness of its solution and considered the problem for a quasi-linear equation, when $a=a(x,u)$, within this approach to formulation.

Here is another example of the problems considered by Beznoshchenko [299]. In the space of the variables $x, t, x=(x_1,\dots,x_n)$ consider a Cauchy problem for the simplest parabolic type equation

$$u_t = \Delta u + qu, \qquad u|_{t=0}=\phi(x). \tag{6.49}$$

Let $q=q(x_1,\dots,x_{n-1},t)$ and the following notation $\bar{x}=(x_1,\dots,x_{n-1})$ be introduced. The task is to find $q(\bar{x},t)$ if one knows the solution of problem (6.49) at the points of the plane $x_n=0$

$$u|_{x_n=0} = f(\bar{x},t). \tag{6.50}$$

The scheme of studying this problem is as follows. Assume that q, f, and ϕ are sufficiently smooth limited functions and the condition of correspondence $\phi(\bar{x},0)=f(\bar{x},0)$ is met. Introduce the additional function $w=(\partial^2/\partial x_n^2)u$, which satisfies the conditions

$$w_t=\Delta w+qw, \qquad w|_{t=0}=\phi_{x_nx_n} \tag{6.51}$$

$$w|_{x_n=0}=f_t-\sum_{k=1}^{n-1}\frac{\partial^2}{\partial x_k^2}f-q\phi|_{x_n=0}. \tag{6.52}$$

In the case when $\phi|_{x_n=0}\neq 0$, relation (6.52) can be solved with respect to the function to be determined

$$q(\bar{x},t)=\left[f_t-\sum_{k=1}^{n-1}\frac{\partial^2}{\partial x_k^2}f-w|_{x_n=0}\right]\bigg/\phi(\bar{x},0). \tag{6.53}$$

Replace the differential problem (6.51) with the equivalent integral equation

$$w(x,t)=\int_{R^n}G(x,t,\xi,0)\phi_{\xi_n\xi_n}\,\mathrm{d}\xi+\int_0^t\mathrm{d}\tau\int_{R_n}G(x,t,\xi,\tau)q(\bar{\xi},\tau)\,\mathrm{d}\xi. \tag{6.54}$$

Here $\xi=(\xi_1,\ldots,\xi_n)$, $\bar{\xi}=(\xi_1,\ldots,\xi_{n-1})$, $G(x,t,\xi,\tau)$ is the fundamental solution of the equation of heat conductivity

$$G(x,t,\xi,\tau)=(2\sqrt{(\pi(t-\tau))})^{-n}\exp[-|x-\xi|^2/4(t-\tau)].$$

Using equality (6.53) one gets a new integral equation

$$q(\bar{x},t)=\frac{1}{\phi(\bar{x},0)}\left[f_t-\sum_{k=1}^{n-1}\frac{\partial^2}{\partial x_k^2}f-\int_{R^n}G(\bar{x},0,t,\xi,0)\phi_{\xi_n\xi_n}\,\mathrm{d}\xi\right.$$
$$\left.-\int_0^t\mathrm{d}\tau\int_{R^n}G(\bar{x},0,t,\xi,\tau)q(\bar{\xi},\tau)w(\xi,\tau)\,\mathrm{d}\xi\right]. \tag{6.55}$$

The system of equations (6.54) and (6.55) is a closed system of nonlinear integral equations with respect to the functions w, q, solvable at small t. On the basis of these equations one can also get an estimate of the conditional stability of the solution (see [299]).

More complex problems for linear and quasilinear equations can be studied in an analogous way.

References

1. Kostrov, B. V. The inverse problem of the theory of seismic focus (in Russian). *Izv. AN SSSR (seriya fizika zemli) 9*, 18–29 (1968).
2. Novikov, P. S. On uniqueness of the inverse problem of the theory of potential (in Russian). *Dokl. AN SSSR 18*, 165–168 (1938).
3. Tikhonov, A. N. On uniqueness of the solution of a problem of electrosurveying (in Russian). *Dokl. AN SSSR 69*, 797–800 (1949).
4. Levitan, B. M. *Operators of a Generalized Shift and Some of their Applications* (in Russian). Fizmatgiz, Moscow (1962).
5. Levitan, B. M. and Sargasyan, I. S. *Introduction to the Spectral Theory* (in Russian). Nauka, Moscow (1970).
6. Marchenko, V. A. *Sturm–Liouville Operators and their Applications.* Naukova Dymka, Kiev (1978).
7. Agranovich, Z. S. and Marchenko, V. A. *The Inverse Problem for the Theory of Dissipation* (in Russian). Kharkov University Press (1960).
8. Sabatier, P. C. and Chadan, K. *Inverse Problems in the Quantum Theory of Dissipation* (in Russian). Mir, Moscow (1980).
9. Lax, P. and Phillips, R. *Theory of Dissipation* (in Russian). Mir, Moscow (1971).
10. Faddeev, L. D. The inverse problem of the quantum theory of dissipation. In: *Modern Problems of Mathematics*, Vol. 3 (in Russian). VINITI, Moscow, 93–180 (1974).
11. Alekseev, A. S. and Megrabov, A. G. Direct and inverse problems of the plane wave dissipation on non-homogeneous transition layers. In: *Mathematical Problems of Geophysics* (in Russian). V. Ts. SO AN SSSR, Novosibirsk, issue 3, pp. 8–36 (1972).
12. Alekseev, A. S. and Megrabov, A. G. Inverse problems for a string with the condition of an inclined derivative at one end and inverse problems of the plane wave dissipation on non-homogeneous layers (in Russian). *Dokl. Akad. Nauk SSSR, 219*, 308–310 (1974).
13. Alekseev, A. S. and Megrabov, A. G. Inverse problems of the spherical wave dissipation on non-homogeneous layers. In: *Mathematical Problems of Geophysics* (in Russian). V. Ts. SO AN SSSR, Novosibirsk, issue 4, pp. 8–29 (1973).
14. Megrabov, A. G. A method of restoration of the density and velocity in a non-homogeneous layer as functions of depth by a set of the plane waves reflected from the layer at various angles. In: *Mathematical Problems of Geophysics* (in Russian). V. Ts. SO AN SSSR, Novosibirsk, issue 5, part 2, pp. 78–107 (1974).
15. Megrabov, A. G. Direct and inverse problems of plane wave dissipation in non-homogeneous vransition layers (elliptical case). In: *Mathematical Problems of Geophysics* (in Russian). V. Ts. SO AN SSSR, Novosibirsk, issue 3, pp. 113–123 (1972).
16. Megrabov, A. G. Inverse problems for the elliptical equation in a band related to the problem plain wave dissipation in non-homogeneous layers. In: *Mathematical*

Problems of Geophysics (in Russian). V. Ts. SO AN SSSR, Novosibirsk, issue 4, pp. 103–115 (1973).
17. Megrabov, A. G. Inverse problems for the elliptical equations in a semi-plane and a band and inverse problems of plane wave dissipation in non-homogeneous layers (in Russian). *Dokl. AN SSSR 220*, 315–317 (1975).
18. Megrabov, A. G. Inverse problems for the equations of a mixed type. In: *Mathematical Problems of Geophysics* (in Russian). V. Ts. SO AN SSSR, Novosibirsk, issue 6, part 1, pp. 122–144 (1975).
19. Megrabov, A. G. Inverse problems for the hyperbolic and elliptical equations with data on a denumerable set of the solutions and inverse problems of plane wave dissipation. In: *Mathematical Problems of Geophysics* (in Russian). V. Ts. SO AN SSSR, Novosibirsk, issue 4, pp. 116–130 (1973).
20. Megrabov, A. G. Inverse problems of plane wave dissipation in non-homogeneous layers with a free or fixed boundary (hyperbolic case). In: *Mathematical Problems of Geophysics* (in Russian). V. Ts. SO AN SSSR, Novosibirsk, issue 4, pp. 84–102 (1973).
21. Megrabov, A. G. Some inverse problems for the equation of a mixed type (in Russian). *Dokl. AN SSSR 234*, pp. 305–307 (1977).
22. Bukhgeim, A. L. The inverse problem of dissipation in the Kirchhoff's approximation (in Russian). *Dokl. Akad. Nauk SSSR 254*, 1292–1294 (1980).
23. Bukhgeim, A. L. The inverse problem of dissipation in the Kirchhoff's approximation. In: *Uniqueness, Stability and Methods of Solving Inverse and Incorrect Problems* (in Russian). V. Ts. SO AN SSSR, Novosibirsk, pp. 17–27 (1980).
24. Isakov, V. M. On reconstructing a domain by dissipation data (in Russian). *Dokl. AN SSSR 230*, 520–522 (1976).
25. Sobolev, S. L. *Some Applications of Functional Analysis in Mathematical Physics* (in Russian). SO AN SSSR, Novosibirsk (1962).
26. Sobolev, S. L. *Equations of Mathematical Physics* (in Russian). Nauka, Moscow, (1966).
27. Anikonov, D. S. On an inverse problem for the transport equation (in Russian). *SMZh 16*, 432–439 (1975).
28. Anikonov, D. S. On inverse problems for the transport equation (in Russian). *Diff. Uravneniya 10*, 7–17 (1974).
29. Anikonov, D. S. On uniqueness of determining the coefficient and the right-hand part of the transport equation (in Russian). *Diff. Uravneniya 11*, 8–18 (1975).
30. Bellman, R., Kashef, B. and Vasudaran, R. The inverse problem of estimating heart parameters from cardiograms. *Math. Biosci. 19*, 221–230 (1974).
31. Calligani, J. *Identification Problems in Electrocardiology*. Vol. 32. Birkhaus, Basel, pp. 32–48 (1976).
32. Glasko, V. B. On uniqueness of determining the structure of the Earth's core by the surface Rayleigh waves (in Russian). *ZhVM i MF 11*, 1498–1509 (1971).
33. Backus, G. and Gilbert, F. Uniqueness in the inversion of inaccurate gross Earth data. *Phil. Trans R. Soc., London, 266A*, 123–192 (1970).
34. Keller, J. B. Inverse problems. *Am. Math. Monthly 83*, 107–118 (1976).
35. Anikonov, Yu. E. *Some Methods of Investigating Multi-dimensional Inverse Problems for Differential Equations* (in Russian). Nauka, Novosibirsk (1978).
36. Gerver, M. L. *The Inverse Problem for a One-dimensional Wave Equation with an Unknown Source of Oscillation* (in Russian). Nauka, Moscow (1974).
37. Goncharskij, A. V., Cherepashchuk, A. M. and Yagola A. G. *Numerical Methods of Solving the Inverse Problems of Astrophysics* (in Russian). Nauka, Moscow (1978).

38. Imanaliev, M. I. *Methods of Solving Non-linear Inverse Problems and their Application* (in Russian). ILIM Press, Frunze (1977).
39. Lavrent'ev, M. M. and Klibanov, M. V. On one inverse problem for the equation of a hyperbolic type (in Russian). *Diff. Uravneniya 11*, 1647–1651 (1975).
40. Lavrent'ev, M. M., Romanov, V. G. and Vasil'ev, V. G. *Multi-dimensional Problems for Differential Equations* (in Russian). Nauka, Novosibirsk (1969).
41. Lavrent'ev, M. M., Romanov, V. G. and Shishatskij, S. P. *Incorrect Problems of Mathematical Physics and Analysis* (in Russian). Nauka, Moscow (1980).
42. Romanov, V. G. *Some Inverse Problems for Equations of a Hyperbolic Type* (in Russian). Nauka, Novosibirsk (1972).
43. Romanov, V. G. *Inverse Problems for Differential Equations* (in Russian). Novosibirsk University Press, Novosibirsk (1973).
44. *Mathematical Problems of Geophysics*. V. Ts. SO AN SSSR, Novosibirsk, issues 1–6 (1969–1975).
45. *Inverse Problems for Differential Equations* (*Proceedings of the All-Union Symposium*) 1–6 February 1971. (in Russian). V. Ts. SO AN SSSR, Novosibirsk (1972).
46. Sabatier, P. S. (Ed.). *Applied Inverse Problems*. Lecture Notes in Physics, 85., Springer, Berlin (1978).
47. Anger, G. (Ed.). *Inverse and Improperly Posed Problems in Differential Equations*. Akademie Berlin (1979). (*Proceedings of the Conference on Mathematical and Numerical Methods*, Hall/Saale, GDR, 29 May– 2 June, 1979).
48. Hadamard, G. *The Cauchy Problem for Linear Equations with Partial Derivatives of a Hyperbolic Type* (in Russian). Nauka, Moscow (1978).
49. Godunov, S. K. *Equations of Mathematical Physics* (in Russian). Nauka, Moscow (1971).
50. Petrovskij, I. G. *Lectures on Equations with Partial Derivatives* (in Russian). Fizmatgiz, Moscow (1961).
51. Mikhlin, S. G. *A Course in Mathematical Physics* (in Russian). Nauka, Moscow (1968).
52. Ivanov, V. K., Vasin, V. V. and Tanana, V. P. *Theory of Linear Incorrect Problems and its Applications* (in Russian). Nauka, Moscow (1978).
53. Tikhonov, A. N. and Arsenin, V. Ya. *Methods of Solving Incorrect Problems* (in Russian). Nauka, Moscow (1974).
54. Tikhonov, A. N. On stability of inverse problems (in Russian). *Dokl. AN SSSR 39*, 195–198 (1943).
55. Lavrent'ev, M. M. On the Cauchy problem for the Laplace equation (in Russian). *Izv. AN SSSR* (*seriya matem.*) *20*, 819–842 (1956).
56. Arsenin, V. Ya. *On Methods of Solving the Incorrectly Set Problems* (in Russian). MIFI, Moscow (1973).
57. Ivanov, V. K. On linear incorrect problems (in Russian). *Dokl. AN SSSR 145*, 270–272 (1962).
58. Ivanov, V. K. On improperly posed problems (in Russian). *Mat. zb. 61*, 211–223 (1963).
59. Lavrent'ev, M. M. On some incorrect problems of mathematical physics. SO AN SSSR, Novosibirsk (1962).
60. Lavrent'ev, M. M. *Conventionally Correct Problems for Differential Equations* (in Russian). Novosibirsk University Press, Novosibirsk (1973).
61. Lattes, R. and Lions, J. L. *Method of Quasi-inversion and its Applications* (in Russian). Mir, Moscow (1970).

62. Morozov, V. A. Linear and non-linear incorrect problems. In: *Mathematical Analysis* (in Russian). VINITI, Moscow, pp. 129–178 (1973).
63. Morozov, V. A. *Methods of Solving Unstable Problems* (in Russian). Moscow University Press, Moscow (1967).
64. Morozov, V. A. *Regular Methods of Solving Improperly Posed Problems.* Moscow University Press, Moscow (1974).
65. Strakhov, V. N. Analytical extension of two-dimensional potential fields and its application to solving the inverse problem of magnetic and gravitational surveying, parts 1–3 (in Russian). *Izv. AN SSSR (seriya geofiz.) 3–4*,9, 307–316, 336–447, 491–507 (1962).
66. Strakhov, V. N. Theory of an approximate solution of linear incorrect problems in a Hilbert space and its use in surveying geophysics, parts 1 and 2 (in Russian). *Izv. AN SSSR (seriya fizika zemli) 8*, 30–53; *9*, 64–96 (1969).
67. Tikhonov, A. N. On mathematical basis of the theory of electromagnetic probing (in Russian). *ZhVM i MF 5*, 545–548 (1965).
68. Tikhonov, A. N. On non-linear equations of the first kind (in Russian). *Dokl. AN SSSR 161*, 1023–1026 (1965).
69. Tikhonov, A. N. On regularity of improperly posed problems (in Russian). *Dokl. AN SSSR 153*, 49–52 (1963).
70. Tikhonov, A. N. On solution of improperly posed problems and the method of regularization (in Russian). *Dokl. AN SSSR 151*, 501–504 (1963).
71. Tikhonov, A. N. On solving nonlinear integral equations of the first kind (in Russian). *Dokl. AN SSSR 156*, 1296–1299 (1964).
72. Tikhonov, A. N. On stable methods of summation of the Fourier series (in Russian). *Dokl. AN SSSR 156*, 268–271 (1964).
73. John, F. A note on 'improper' problems in partial differential equations. *Commun Pure Appl. Math. 8*, 591–594 (1955).
74. John, F. *Partial Differential Equations.* Springer, New York, (1971).
75. Pucci, C. On the improperly posed Cauchy problems for parabolic equations. *Symposium on Numerical Treatment of Partial Differential Equations with Real Characteristics*, Rome, pp. 140–144 (1959).
76. Kolmogorov, A. N. and Fomin, S. V. *Elements of the Theory of Functions and Functional Analysis* (in Russian). Nauka, Moscow (1968).
77. Vladimirov, V. S. *Generalized Functions in Mathematical Physics* (in Russian). Nauka, Moscow (1976).
78. Gel'fand, I. M. and Shilov, G. E. *Generalized Functions and their Manipulation.* Generalized Functions Series (in Russian). Fizmatgiz, Moscow, issue 1 (1959).
79. Mizohata, S. *Theory of Equations with Partial Derivatives* (in Russian). Mir, Moscow (1977).
80. Titchmarsh, E. S. *Introduction into the Theory of Fourier Integrals*, Vol. 1 (in Russian). OGIZ, Moscow (1948).
81. Parijskij, B. S. The inverse problem for a wave equation with a depth effect. In: *Some Direct and Inverse Problems of Seismology* (*Computing Seismology*, issue 4) (in Russian). Nauka, Moscow, pp. 139–169 (1968).
82. Blagoveshchenskij, A. S. On the local method of solving the non-stationary inverse problem for an inhomogeneous string (in Russian). *Trudy* (*Mat. inst. imeni V. A. Steklova, Leningrad*) *65*, 28–38 (1971).
83. Stepanov, V. N. Uniqueness of the solution of an inverse problem of dissipation. In: *Uniqueness, Stability and Methods of Solving Inverse and Incorrect Problems* (in Russian). V. Ts. SO AN SSSR, Novosibirsk, pp. 77–81 (1980).

84. Smirnov, V. I. *A Course of Higher Mathematics*, Vol. 4 (in Russian). Gostechizdat, Moscow (1957).
85. Gel'fand, I. M. and Levitan, B. M. On determining a spectral equation by its spectral functions (in Russian). *Izv. AN SSSR (seriya matem.) 15*, 309–360 (1951).
86. Krein, M. G. On one method of an effective solution of the inverse boundary problem (in Russian). *Dokl. AN SSSR, 94*, 767–770 (1954).
87. Blagoveshchonskij, A. S. The one-dimensional boundary problem for a hyperbolic equation of the second order. In: *Mathematical Problems of the Theory of Wave Propagation* (in Russian), Vol. 2. Nauka, Leningrad, pp. 85–90 (1969).
88. Romanov, V. G. Theorem of uniqueness of the one-dimensional inverse problem for a wave equation. In: *Mathematical Problems of Geophysics* (in Russian). V. Ts. SO AN SSSR, Novosibirsk, issue 2, pp. 100–142 (1971).
89. Romanov, V. G. The problem of determining the one-dimensional velocity of signal propagation in a semi-space under the condition of oscillation of one of the points of this semi-space. In: *Mathematical Problems of Geophysics* (in Russian). V. Ts. SO AN SSSR, Novosibirsk, issue 3, pp. 164–186 (1972).
90. Yakhno, V. G. *The Inverse Problem for a Wave Equation* (in Russian). V. Ts. SO AN SSSR, Novosibirsk, preprint 206 (1979).
91. Yakhno, V. G. *The Inverse Problem for a Wave Equation (Constructing a Solution of an Inverse Problem)* (in Russian). V. Ts. SO AN SSSR, Novosibirsk, preprint 235 (1980).
92. Yakhno, V. G. The inverse problem for a wave equation. In: *Uniqueness, Stability and Methods of Solving Inverse and Incorrect Problems* (in Russian). V. Ts. SO AN SSSR, Novosibirsk, pp. 126–140 (1980).
93. Yakhno, V. G. The one-dimensional inverse problem for a wave equation (in Russian). *Dokl. AN SSSR 255*, 807–810 (1980).
94. Babitch, V. M. Fundamental solutions of the dynamical equations of the theory of elasticity for a non-homogeneous medium (in Russian). *PMM 25*, 38–45 (1961).
95. Babitch, V. M., Kapilevitch, M. B., Mikhlin, S. G. *et al. Linear Equations of Mathematical Physics* (in Russian). Series SMB, Nauka, Moscow (1964).
96. El'sgol'ts, L. E. *Differential Equations and Variational Calculus* (in Russian). Nauka, Moscow (1965).
97. Petrovskij, I. G. *Lectures on Theory of Regular Differential Equations*. Nauka, Moscow (1964).
98. Romanov, V. G. On uniqueness of determining an isotropic Riemann metric inside a domain through distances between the boundary points (in Russian). *Dokl. AN SSSR 218*, 295–297 (1974).
99. Romanov, V. G. On uniqueness of the solution of the inverse kinematic problem in a circle within a class of velocities close to constant ones. In: *Mathematical Problems of Geophysics* (in Russian). V. Ts. SO AN SSSR, Novosibirsk, issue 5, part 2, pp. 108–143 (1974).
100. Gerver, M. L. and Markushevich, V. M. Determining the velocity of the seismic wave propagation by the hodograph. In: *Methods and Problems of the Analysis of Seismic Observations* (In Russian). Nauka, Moscow, pp. 3–51 (1967).
101. Gerver, M. L. and Markushevich, V. M. Studying ambiguity when determining the velocity of the seismic wave propagation by the hodograph (in Russian). *Dokl. AN SSSR 163*, 1377–1380 (1965).
102. Geiko, V. S. On the conditions of uniqueness of determining the velocity of a seismic wave distribution outside wave guides by the hodograph (in Russian). *Dokl. AN SSSR 253*, 74–77 (1980).

103. Markushevich, V. M. On stability of the inverse problems of geometrical seismology (in Russian). *Dokl. AN SSSR 178*, 94–97 (1968).
104. Markushevich, V. M. and Reznikov, E. L. Determination of the velocity cross-section by the hodograph with the lower velocity boundary given. In: *Computing and Statistical Methods of Interpreting Seismic Data* (*Computing Seismology*, issue 6) (in Russian). Nauka, Moscow, pp. 160–198 (1973).
105. Markushevich, V. M. and Reznikov E. L. Solution of the inverse problem of geometrical seismology at the velocity limited *a priori* (in Russian). *Dokl. AN SSSR 214*, 803–806 (1974).
106. Gerver, M. L. and Markushevich, V. M. On characteristic properties of seismic hodographs (in Russian). *Dokl. AN SSSR, 175*, 334–337 (1967).
107. Gerver, M. L. and Markushevich, V. M. Properties of the hodograph of a surface source. In: *Some Direct and Inverse Problems of Seismology* (*Computing Seismology*, issue 4) (in Russian). Nauka, Moscow, pp. 15–63 (1968).
108. Bukhgeim, A. L. On one algorithm of solving the inverse kinematic problem of seismology. In: *Numerical Methods in Seismic Investigations* (in Russian). Nauka, Novosibirsk, pp. 152–155 (1983).
109. Bonchkovskij, V. F. *Internal Structure of the Earth* (in Russian). Izd. AN SSSR, Moscow (1953).
110. Bullen, K. E. *Introduction to Theoretical Seismology* (in Russian). Mir, Moscow (1966).
111. Jeffreys, G. *The Earth, Its Origin, History and Structure* (in Russian). Izd. Inostr. Lit. Moscow (1960).
112. Johnson, L. R. Array measurements of *P*-velocities in the upper mantle. *J. Geophys. Res. 72*, 92–101 (1967).
113. Keilis-Borok, V. I. Seismology and logic. In: *Some Direct and Inverse Problems of Seismology* (*Computing Seismology*, issue 4), (in Russian). Nauka, Moscow, pp. 317–350 (1968).
114. Kuzin, I. P. Velocities of the elastic waves in the focal zone of Kamchatka (in Russian). *Izv. AN SSSR* (*seriya fizika zemli*) *12*, 25–39 (1972).
115. Savarenskij, E. F. *Studying the Earth's Internal Structure by Seismic Data* (in Russian). Moscow (1963).
116. Tarakanov, R. Z. and Levyi, N. B. An asthenospheric model of the upper Earth's mantle by seismic data (in Russian). *Dokl. AN SSSR 176*, 571–574 (1967).
117. Valyus, V. P., Keilis-Borok, V. I. and Levshin, A. L. Determination of the velocity cross-section of the upper mantle of Europe (in Russian). *Dokl. AN SSSR 185*, 564–567 (1969).
118. Valyus, V. P., Keilis-Borok, V. I. and Levshin, A. L. Determination of the velocities of the transverse waves of the upper mantle of Europe. In: *Algorithms of Interpreting Seismic Data* (in Russian). Nauka, Moscow, issue 5, pp. 214–238 (1971).
119. Gel'fand, I. M., Graev, M. I. and Vilenkin, N. Ya. *Integral Geometry and the Related Problems of the Theory of Representation.* Generalized Functions Series (in Russian). Fizmatgiz, Moscow, issue 5 (1962).
120. Romanov, V. G. On restoration of a function through the integrals with respect to a set of curves (in Russian). *SMZh 8*, 1206–1208 (1967).
121. Müntz, G. *Integral Equations*, Vol. 1 (in Russian). Gostekhizdat, Moscow (1934).
122. Petrovskij, I. G. *Lectures on the Theory of Integral Equations* (in Russian). Nauka, Moscow (1965).

123. Alekseev, A. S., Lavrent'ev, M. M., Mukhometov, R. G., Nersesov, I. L. and Romanov, V. G. A numerical method of determining the structure of the Earth's upper mantle. In: *Mathematical Problems of Geophysics* (in Russian). V. Ts. SO AN SSSR, Novosibirsk, issue 2, pp. 143–165 (1971).
124. Alekseev, A. S., Lavrent'ev, M. M., Mukhometov, R. G. and Romanov, V. G. A numerical method of solving the three-dimensional inverse kinematic problem of seismology. In: *Mathematical Problems of Geophysics* (in Russian). V. Ts. SO AN SSSR, Novosibirsk, issue 1, pp. 179–201 (1969).
125. Romanov, V. G. *Inverse Problems for Differential Equations. The Inverse Kinematic Problem of Seismology* (in Russian). Novosibirsk University Press, Novosibirsk (1978).
126. Medvedev, S. N. The problem of integral geometry for a set of curves in a circular ring (in Russian). *ZhVM i MF 20*, 531–538 (1980).
127. Romanov, V. G. A one-dimensional inverse problem for a telegraph equation (in Russian). *Diff. Uravneniya 4*, 87–101 (1968).
128. Romanov, V. G. On one formulation of the inverse problem for a generalized wave equation (in Russian). *Dokl. AN SSSR 181*, 554–557 (1968).
129. Mukhometov, R. G. On the problem of integral geometry. In: *Mathematical Problems of Geophysics* (in Russian). V. Ts. SO AN SSSR, Novosibirsk, issue 6, part 2, pp. 212–242 (1975).
130. Mukhometov, R. G. On the problem of integral geometry on the geodesic lines of the Riemann metric. In: *Conventionally Correct Problems and Problems of Geophysics* (in Russian), V. Ts. SO AN SSSR, Novosibirsk, pp. 86–110 (1979).
131. Mukhometov, R. G. *On the Problem of Restoring the Anisotropic Riemann Metric in an* n*-Dimensional Domain* (in Russian). V. Ts. SO AN SSSR, Novosibirsk, preprint 136 (1978).
132. Mukhometov, R. G. The inverse kinematic problem of seismology on a plane. In: *Mathematical Problems of Geophysics* (in Russian). V. Ts. SO AN SSSR, Novosibirsk, issue 6, part 2, pp. 243–252 (1975).
133. Mukhometov, R. G. The problem of restoring the two-dimensional Riemann metric and integral geometry (in Russian). *Dokl. AN SSSR 232*, 32–35 (1977).
134. Romanov, V. G. Integral geometry on the geodesic lines of an isotropic Riemann metric (in Russian). *Dokl. AN SSSR 241*, 290–293 (1978).
135. Bernshtejn, I. N. and Gerver, M. L. On the problem of integral geometry for a set of geodesic lines and on the inverse kinematic problem of seismology (in Russian). *Dokl. Akad. Nauk SSSR 243*, 302–305 (1978).
136. Bernshtejn, I. N. and Gerver, M. L. The condition of metrics discernibleness by the hodographs. In: *Methods and Algorithms of Interpreting Seismological Data* (in Russian). Nauka, Moscow, pp. 50–73 (1980).
137. Bukhgeim, A. L. Normal solvability of certain special operator equations of the first kind (a sufficient condition). In: *Mathematical Problems of Geophysics* (in Russian), V. Ts. SO AN SSSR, issue 6, part 1, pp. 42–54 (1975).
138. Anikonov, Yu. E. On geometrical methods of inverse problem investigation. In: *Mathematical Problems of Geophysics* (in Russian). V. Ts. SO AN SSSR, Novosibirsk, issue 2, pp. 7–53 (1971).
139. Anikonov, Yu. E. On one of the applications of geometry methods. In: *Mathematical Problems of Geophysics* (in Russian). V. Ts. SO AN SSSR, Novosibirsk, issue 3, pp. 63–76 (1972).

140. Anikonov, Yu. E. On one of the classes of operator equations (in Russian). *SMZh 13*, 1383–1386 (1972).
141. Anikonov, Yu. E. On one of the problems of determining the Riemann metric (in Russian). *Dokl. Akad. Nauk SSSR 204*, 1287–1288 (1972).
142. Anikonov, Yu. E. On operator equations of the first kind (in Russian). *Dokl. Akad. Nauk SSSR 207*, 257–258 (1972).
143. Anikonov, Yu. E. On quasi-monotone operators. In: *Mathematical Problems of Geophysics* (in Russian). V. Ts. SO AN SSSR, Novosibirsk, issue 3, pp. 86–99 (1972).
144. Romanov, V. G. On one class of uniqueness of the inverse kinematic problem solution. In: *Mathematical Problems of Geophysics* (in Russian). V. Ts. SO AN SSSR, Novosibirsk, issue 4, pp. 147–164 (1973).
145. Markushevich, V. M. and Reznikov, E. L. Solution of the inverse problem of geometrical seismology for horizontally inhomogeneous media. II. Certain sufficient conditions of the solution uniqueness for the case when the velocity depends on two variables. In: *Investigation of Seismicity and the Earth's Models* (*Computing Seismology*, issue 9) (in Russian). Nauka, Moscow, pp. 92–133 (1976).
146. Mukhometov, R. G. and Romanov, V. G. On the problem of finding the isotropic Riemann metric in an *n*-dimensional space (in Russian). *Dokl. AN SSSR 243*, 41–44 (1978).
147. Beil'kin, G. Ya. Stability and uniqueness of the solution of the inverse problem of seismology for a multi-dimensional case. In: *Boundary Problems of Mathematical Physics and the Co-problems of the Theory of Functions*. 11 (in Russian), Nauka, Leningrad, pp. 3–6 (1979). (*Zap. nauch. seminarov LOMI*, Vol. 84).
148. Mukhometov, R. G. On one problem of restoring the Riemann metric (in Russian). *SMZh 22*, 119–135 (1981).
149. Markushevich, V. M. Characteristic properties of the hodograph of a depth source. In: *Some Direct and Inverse Problems of Seismology* (*Computing Seismology*, issue 4) (in Russian). Nauka. Moscow, pp. 64–77 (1968).
150. Anikonov, Yu. E. Some remarks on the theory of inverse problems. In: *Incorrect Mathematical Problems and Problems of Geophysics* (in Russian). V. Ts. SO AN SSSR, Novosibirsk, pp. 16–23 (1976).
151. Anikonov, Yu. E. Uniqueness theorem for the multi-dimensional inverse kinematic problem of seismology. In: *Mathematical Problems of Geophysics* (in Russian). V. Ts. SO AN SSSR, Novosibirsk, issue 5, part 2, pp. 18–29 (1975).
152. Anikonov, Yu. E., Pivovarova, N. B. and Slavina, L. B. The three-dimensional velocity field of the Kamchatka focal zone. In: *Mathematical Problems of Geophysics* (in Russian). V. Ts. SO AN SSSR, Novosibirsk, issue 5, part 1, pp. 92–117 (1974).
153. Gol'din, S. V. One inverse kinematic problem of seismology of reflected waves (in Russian). *Dokl. AN SSSR 233*, 64–67 (1977).
154. Gol'din S. V. and Suvorov, V. D. An analytical extention of the reflected waves hodograph (in Russian). *Dokl. AN SSSR 222*, 825–828 (1975).
155. Courant, R. *Equations with Partial Derivatives* (in Russian). Mir, Moscow (1964).
156. Vladimirov, V. S. *Equations of Mathematical Physics* (in Russian). Nauka, Moscow (1967).
157. Babitch, V. M. Fundamental solutions of the hyperbolic equations with variable coefficients (in Russian). *Matem. Sb. 52*, 709–738 (1960).
158. Romanov, V. G. Differential properties of the fundamental solution of an equation of the second kind of a hyperbolic type. In: *Incorrect Mathematical Problems and*

Problems of Geophysics (in Russian). V. Ts. SO AN SSSR, Novosibirsk, pp. 110–121 (1979).
159. Ladyzhenskaya, O. A. *Boundary Problems of Mathematical Physics* (in Russian). Nauka, Moscow (1973).
160. Stepanov, V. N. Some geometrical problems in the inverse problem of dissipation in the Kirchhoff's approach. In: *Approximation Methods of Solutions and Problems of Inverse Problem Correctness* (in Russian). V. Ts. SO AN SSSR, Novosibirsk, pp. 113–123 (1981).
161. Anikonov, Yu. E. and Romanov, V. G. On uniqueness of determining the first-order form by its integrals with respect to geodesic lines. In: *Some Mathematical Problems and Problems of Geophysics* (in Russian). V. Ts. SO AN SSSR, Novosibirsk, pp. 22–27 (1979).
162. Romanov, V. G. On restoration of a function through the integrals with respect to the ellipsoids of rotation with one of their focuses motionless (in Russian). *Dokl. AN SSSR 173*, 766–769 (1967).
163. John, F. *Plane Waves and Spherical Means When Applied to Differential Equations with Partial Derivatives* (in Russian). Izd. Inostr. Lit., Moscow (1958).
164. Romanov, V. G. A problem of integral geometry and a linearized inverse problem for a differential equation (in Russian). *SMZh 10*, 1364–1374 (1969).
165. Romanov, V. G. On one theorem of uniqueness for the problem of integral geometry on a set of curves. In: *Mathematical Problems of Geophysics* (in Russian). V. Ts. SO AN SSSR, issue 4, pp. 140–146 (1973).
166. Romanov, V. G. On some classes of uniqueness of the solution of problems of integral geometry (in Russian). *Mat. Zametki 16*, 657–668 (1974).
167. Romanov, V. G. On some classes of uniqueness of the solution of the Volterra operator equations of the first kind (in Russian). *Funkts. Analiz i ego Prilozh. 1*, 81–82 (1975).
168. Romanov, V. G. The Volterra operator equations of the first kind. Classes of uniqueness. In: *Some Problems of Computing and Applied Mathematica* (in Russian). Nauka, Novosibirsk, pp. 123–135 (1975).
169. Lavrent'ev, M. M. Inverse problems and special operator equations of the first kind. In: *International Congress of Mathematicians*, Nice, 1970 (in Russian). Nauka, Moscow, pp. 130–136 (1972).
170. Bukhgeim, A. L. On one class of the Volterra operator equations of the first kind (in Russian). *Funkts. Analiz i ego Prilozh. 1*, 1–9 (1972).
171. Bukhgeim, A. L. The Volterra operator equations in the scales of the Banach spaces (in Russian). *Dokl. Akad. Nauk SSSR 242*, 272–275 (1978).
172. Lavrent'ev, M. M. and Bukhgeim, A. L. On one class of operator equations of the first kind (in Russian). *Funkts. analiz i ego Prilozh. 4*, 44–53 (1973).
173. Lavrent'ev, M. M. and Bukhgeim, A. L. On one class of the problems of integral geometry (in Russian). *Dokl. AN SSSR 211*, 38–39 (1973).
174. Bukhgeim, A. L. and Yakhno, V. G. On two inverse problems for differential equations (in Russian). *Dokl. Akad. Nauk SSSR 229*, 785–786 (1976).
175. Blagoveshchenskij, A. S. On various formulations of the one-dimensional inverse problem for a telegraph equation. In: *Problems of Mathematical Physics* (in Russian). LGU, Leningrad, issue 4, pp. 40–41 (1970).
176. Blagoveshchenskij, A. S. The inverse problem for the wave equation with an unknown source. In: *Problems of Mathematical Physics* (in Russian). LGU, Leningrad, issue 4, pp. 27–39 (1970).

177. Romanov, V. G. A one-dimensional inverse problem for a wave equation (in Russian). *Dokl. AN SSSR 211*, 1083–1084 (1973).
178. Romanov, V. G. Inverse problems for hyperbolic equations and energetic inequalities (in Russian). *Dokl. AN SSSR 242*, 541–544 (1978).
179. Bukhgeim, A. L. Karlemanian estimations for the Volterra operators and uniqueness of inverse problems. In: *Non-classical Problems of Mathematical Geophysics* (in Russian). V. Ts. SO AN SSSR, Novosibirsk, pp. 56–64 (1981).
180. Bukhgeim, A. L. and Klibanov, M. V. Uniqueness as a whole of one class of multidimensional inverse problems (in Russian). *Dokl. Akad. Nauk SSSR 260*, 269–271 (1981).
181. Klibanov, M. V. Uniqueness of solutions 'as a whole' of some multi-dimensional inverse problems. In: *Non-classical Problems of Mathematical Physics* (in Russian). V. Ts. SO AN SSSR, Novosibirsk, pp. 101–114 (1981).
182. Anikonov, Yu. E. On uniqueness of the solution of integral equations of the first kind (in Russian). *Matem. Zametki, 14*, 493–498 (1973).
183. Anikonov, Yu. E. The inverse kinematic problem of seismology and certain aspects of star dynamics (in Russian). *Dokl. Akad. Nauk SSSR 252*, 14–17 (1980).
184. Anikonov, Yu. E. Theorem of uniqueness of the inverse problem solution for a quasilinear hyperbolic equation (in Russian). *Diff. Uravneniya 12*, 2265–2266 (1976).
185. Anikonov, Yu. E. Theorem of uniqueness of the inverse problem solution for a wave equation (in Russian). *Matem. Zametki 19*, 211–214 (1976).
186. Anikonov, Yu. E. and Moskvitin, V. N. On an inverse problem for a system of dynamical equations of the theory of elasticity (in Russian). *Dokl. Akad. Nauk SSSR 253*, 1086–1087 (1980).
187. Alekseev, A. S. Inverse dynamical problems of seismology. In: *Some Methods and Algorithms for Interpreting Geophysical Data* (in Russian). Nauka, Moscow, pp. 9–84 (1967).
188. Alekseev, A. S. Some inverse problems of the theory of wave propagation (in Russian). *Izv. Akad. Nauk SSSR, ser. geofiz. 11*, 1514–1531 (1962).
189. Alekseev, A. S. and Dobrinsky, V. I. Some aspects of practical application of inverse dynamical problems of seismology. In: *Mathematical Problems of Geophysics* (in Russian). V. Ts. SO AN SSSR, Novosibirsk, issue 6, part 2, pp. 7–53 (1975).
190. Alekseev, A. S. and Dobrinsky, V. I., Neprochnov Yu. P. and Semenov G. A. On the problem of practical application of the theory of inverse dynamical problems of seismology (in Russian). Dokl. *Akad. Nauk SSSR 228*, 1053–1056 (1976).
191. Anikonov, D. S. On uniqueness of the solution of inverse problems for equations of mathematical physics (in Russian). *Diff. Uravneniya 15*, 3–9 (1979).
192. Bamberger, A., Chavent, G. and Lailly, P. Une application de la theorie du controle a un probleme inverse de seismique. *Annales de Geophysique 33*, 183–200 (1977).
193. Berezanskij, Yu. M. On the inverse problem of the spectral analysis for the Shrödinger equation (in Russian). *Dokl. Akad. Nauk SSSR 105*, 197–200 (1955).
194. Berezanskij, Yu. M. On the theorem of uniqueness of the inverse problem of the spectral analysis for the Shrödinger equation (in Russian). *Trudy Mosk. Mat. Ob. 7*, 3–51 (1958).
195. Berezanskij, Yu. M. On uniqueness of determining the Shrödinger equation by its spectral function (in Russian). *Dokl. Akad. Nauk SSSR 93*, 591–594 (1953).
196. Blagoveshchenskij, A. S. Inverse problems of the theory of elastic waves propagation (in Russian). *Izv. AN SSSR* (*seriya fiziki Zemli*) *12*, 50–59 (1978).
197. Blagoveshchenskij, A. S. On the quasi-two-dimensional problem for a wave equation (in Russian). *Trudy* (*Mat. inst. imeni V. A. Steklova, Leningrad*) *65*, 57–69 (1971).

198. Blagoveshchenskij, A. S. Some inverse problems of the theory of hyperbolic equations. In: *Non-classical Methods in Geophysics* (in Russian), V. Ts. SO AN SSSR, Novosibirsk, pp. 17–26 (1977).
199. Blagoveshchenskij, A. S. The inverse boundary problem of the theory of wave propagation in an anisotropic medium (in Russian). *Trudy* (*Mat. inst. imeni V. A. Steklova, Leningrad*) *65*, 39–56 (1971).
200. Blagoveshchenskij, A. S. and Kabanikhin, S. I. *On the Inverse Problem of the Theory of Seismic Wave Propagation in a Semi-finite Irregular Wave Guide* (in Russian). V. Ts. SO AN SSSR, Novosibirsk, preprint 224 (1980).
201. Blagoveshchenskij, A. S. and Lavrent'ev, K. K. Inverse problems of determining the boundary condition in the theory of non-stationary wave propagation. In: *Mathematical Problems of the Theory of Wave Propagation* (in Russian). Nauka, Leningrad, pp. 78–84 (1975).
202. Cannon, J. R. and Dunninger, D. R. Determination of an unknown forcing function in a hyperbolic equation from overspecified data. *Ann. Math. Pure ed Appl., Ser. 4*, *85*, 49–62 (1970).
203. Chavent, G. About the stability of the optimal control solution of inverse problems. In: *Inverse and Improperly Posed Problems in Differential Equations*. Akademie, Berlin, pp. 45–58 (1979).
204. Elubaev, S. On one inverse problem for a telegraph equation (in Russian). *Dokl. Akad. Nauk SSSR 189*, 461–463 (1969).
205. Glushkova, E. S. On uniqueness of some inverse problems for a telegraph equation. In: *Mathematical Problems of Geophysics* (in Russian). V. Ts. SO AN SSSR, Novosibirsk, issue 6, part 2, pp. 130–144 (1975).
206. Isakov, V. M. On uniqueness of the solution of some inverse hyperbolic problems (in Russian). *Diff. Uravneniya 10*, 165–167 (1974).
207. Kabanikhin, S. I. A finite-difference inverse problem for an oscillation equation. In: *Problems of Correctness of Problems of Mathematical Physics* (in Russian). V. Ts. SO AN SSSR, Novosibirsk, pp. 57–69 (1977).
208. Kabanikhin, S. I. *An Inverse Problem for an Acoustic Equation* (*Determining a Medium Density*) (in Russian). V. Ts. SO AN SSSR, Novosibirsk, preprint 246 (1980).
209. Kabanikhin, S. I. Application of energy inequalities to one inverse problem for a hyperbolic equation (in Russian). *Diff. Uravneniya 15*, 61–67 (1979).
210. Kabanikhin, S. I. On a formulation of the two-dimensional problem for an oscillation equation. In: *Incorrect Mathematical Problems and Problems of Geophysics* (in Russian). V. Ts. SO AN SSSR, Novosibirsk, pp. 64–73 (1976).
211. Kabanikhin, S. I. On solvability of one dynamical problem of seismology. In: *Conditionally Correct Mathematical Problems and Problems of Geophysics* (in Russian). V. Ts. SO AN SSSR, Novosibirsk, pp. 43–51 (1979).
212. Kabanikhin S. I. On the finite-difference method of determining the coefficients of a hyperbolic equation (in Russian). *ZhVM i MF 19*, 417–425 (1979).
213. Kabanikhin, S. I. *On the Finite-difference Method of Determining the Coefficient of the Equation* $(\partial^2 u/\partial t^2) = (\partial^2 u/\partial x^2) + (\partial^2 u/\partial y^2) + q(x, y)u$ (in Russian). V. Ts. SO AN SSSR, Novosibirsk, preprint 103 (1978).
214. Kabanikhin, S. I. On uniqueness of determining the coefficients of a hyperbolic equation (in Russian). V. Ts. SO AN SSSR, Novosibirsk, preprint 91 (1978).
215. Krueger, R. J. An inverse problem for a dissipative hyperbolic equation with discontinuous coefficients. *Q. Appl. Math. 34*, 129–147 (1976).
216. Kunetz, G. Quelques examples d'analyse d'enregistrements sismiques. *Geophys. Prospecting 11*, 87–96 (1963).

217. Lavrent'ev, M. M. Inverse problems for differential equations. In: *Proceedings of the Symposium: Inverse Problems for Differential Equations. (in Russian). Nauka, Novosibirsk*, pp. 7–13 (1972).
218. Lavrent'ev, M. M. An inverse problem for a wave equation (in Russian). *Dokl. AN SSSR 157*, 520–521 (1964).
219. Lavrent'ev, M. M. On one class of inverse problems for differential equations (in Russian). *Dokl. AN SSSR 160*, 32–35 (1965).
220. Lavrent'ev, M. M. and Klibanov, M. V. On one integral equation of the first kind and an inverse problem for a parabolic equation (in Russian). *Dokl. AN SSSR 221*, 782–783 (1975).
221. Lavrent'ev, M. M. and Reznitskaya, K. G. An inverse problem with an unknown source. In: *Uniqueness, Stability and Methods of Solving Inverse and Incorrect Problems* (in Russian). V. Ts. SO AN SSSR, pp. 53–63 (1980).
222. Lavrent'ev, M. M. and Romanov, V. G. On three linearized inverse problems for hyperbolic equations (in Russian). *Dokl. AN SSSR 171*, 1279–1281 (1966).
223. Romanov, V. G. An abstract inverse problem and its correctness (in Russian). *Funkts. Analiz i ego Prilozh 3*, 67–74 (1973).
224. Romanov, V. G. On correctness of the inverse problems for some nonlinear equations (in Russian). *Funkts. Analiz i ego Prilozh. 3*, 67–70 (1974).
225. Romanov, V. G. On the theorems of uniqueness of one class of inverse problems (in Russian). *Dokl AN SSSR 104*, 1075–1076 (1972).
226. Romanov, V. G. Theorem of uniqueness and stability for a nonlinear operator equation (in Russian). *Dokl. AN SSSR 207*, 1051–1053 (1972).
227. Yakhno, V. G. Iteration method of constructing the solution of one inverse problem. In: *Conventionally Correct Mathematical Problems and Problems of Geophysics* (in Russian). V. Ts. SO AN SSSR, Novosibirsk, pp. 134–150 (1979).
228. Yakhno, V. G. On classes of uniqueness and stability of one formulation of the inverse problem for general hyperbolic equations of the second kind. In: *Incorrect Mathematical Problems and Problems of Geophysics* (in Russian). V. Ts. SO AN SSSR, Novosibirsk, pp. 145–153 (1979).
229. Yakhno, V. G. On one inverse problem for hyperbolic equations (in Russian). *Mat. Zametki 26*, 39–44 (1979).
230. Yakhno, V. G. The inverse problem for a nonlinear equation of a string oscillation. In: *Mathematical Problems of Geophysics* (in Russian). V. Ts. SO AN SSSR, Novosibirsk, issue 4, pp. 179–192 (1973).
231. Yakhno, V. G. Theorem of existence and uniqueness of one one-dimensional inverse problem for a quasi-linear equation of a hyperbolic kind of the second order. In: *Mathematical Problems of Geophysics* (in Russian). V. Ts. SO AN SSSR, Novosibirsk, issue 5, part 2, pp. 173–183 (1974).
232. Yakhno, V. G. Theorem of existence in the 'small' of one multi-dimensional inverse problem. In: *Problems of Correctness of Inverse Problems of Mathematical Physics* (in Russian). V. Ts. SO AN SSSR, Novosibirsk, pp. 143–155 (1982).
233. Yakhno, V. G. Theorem of uniqueness for a quasi-linear hyperbolic equation. In: *Mathematical Problems of Geophysics* (in Russian). V. Ts. SO AN SSSR, Novosibirsk, issue 5, part 1, pp. 73–91 (1974).
234. Yakhno, V. G. Theorem of uniqueness of one inverse problem for a hyperbolic equation (in Russian). *Diff. Uravneniya 13*, 544–551 (1977).
235. Bidajbekov, E. Y. On one of the inverse problems for the quasi-linear equation of a hyperbolic type. In: *Mathematical Problems of Geophysics* (in Russian). V. Ts. SO AN SSSR, Novosibirsk, issue 4, pp. 61–68 (1973).

236. Bidajbekov, E. Y. On uniqueness of determining a certain differential operator. In: *Mathematical Problems of Geophysics* (in Russian). V. Ts. SO AN SSSR, Novosibirsk, issue 5, part 1, pp. 26–37 (1975).
237. Bidajbekov, E. Y. On uniqueness of determining one of the differential operators. In: *Mathematical Problems of Geophysics* (in Russian). V. Ts. SO AN SSSR, Novosibirsk, issue 3, pp. 100–112 (1972).
238. Glushkova, E. S. On one inverse problem. In: *Incorrect Mathematical Problems and Problems of Geophysics* (in Russian). V. Ts. SO AN SSSR, Novosibirsk, pp. 49–59 (1979).
239. Glushkova, E. S. Theorem of existence of the solution of one inverse problem. In: *Problems of Correctness of Inverse Problems of Mathematical Physics* (in Russian). V. Ts. SO AN SSSR, Novosibirsk, pp. 69–74 (1982).
240. Glushkova, E. S. Theorem of uniqueness of one inverse problem for the quasi-linear equation of a hyperbolic type. In: *Incorrect Mathematical Problems and Problems of Geophysics* (in Russian). V. Ts. SO AN SSSR, Novosibirsk, pp. 35–45 (1976).
241. Glushkova, E. S. On uniqueness of determining the coefficients of the quasi-linear equation of a hyperbolic type. In: *Problems of Correctness of Problems of Mathematical Physics* (in Russian). V. Ts. SO AN SSSR, Novosibirsk, pp. 42–56 (1977).
242. Anikonov, Yu. E. On solvability of the problems of integral geometry (in Russian). *Mat. Zb.* issue 101 (143), 2, 271–279 (1976).
243. Bukhgeim, A. L. On analyticity of the solution of special integral equations of the first kind. In: *Mathematical Problems of Geophysics* (in Russian). V. Ts. SO AN SSSR, Novosibirsk, issue 6, part 2, pp. 110–119 (1975).
244. Bukhgeim, A. L. On certain problems of integral geometry (in Russian). *SMZh 13*, 34–42 (1972).
245. Bukhgeim, A. L. On one class of integral equations of the first kind (in Russian). *Dokl. Akad. Nauk SSSR 215*, 15–16 (1974).
246. Bukhgeim, A. L. On one problem of integral geometry. In: *Mathematical Problems of Geophysics* (in Russian). V. Ts. SO AN SSSR, Novosibirsk, issue 4, pp. 69–73 (1973).
247. Bukhgeim, A. L. On solvability of one class of the problems of integral geometry. In: *Incorrect Mathematical Problems and Problems of Geophysics* (in Russian). V. Ts. SO AN SSSR, Novosibirsk, pp. 37–46 (1979).
248. Garipov, R. M. Non-hyperbolic boundary problem for a wave equation (in Russian). *Dokl. AN SSSR 219*, 777–780 (1974).
249. Garipov, R. M. and Kardakov, V. B. The Cauchy problem for the wave equation with non-spatial initial variety (in Russian). *Dokl. AN SSSR 213*, 1047–1050 (1973).
250. Hormander, L. *Linear Differential Operators with Partial Derivatives* (in Russian). Mir, Moscow (1965).
251. Kardakov, V. V. A stability estimate of the solution of a non-hyperbolic boundary problem for a wave equation. In: *Mathematical Problems of Geophysics* (in Russian). V. Ts. SO AN SSSR, Novosibirsk, issue 6, part 2, pp. 157–166 (1975).
252. Kaistrenko, V. M. On the Cauchy problem for a hyperbolic equation of the second type with the data on a time-like surface (in Russian). *SMZh 16*, 395–398 (1975).
253. Shishatskij, S. P. *A priori* estimates in the problem on extending a wave field from a cylindrical time-like surface (in Russian). *Dokl. AN SSSR 213*, 49–50 (1973).
254. Fridrichs, K. O. Symmetric hyperbolic system of linear differential equation. *Commun. Pure Appl. Math.* 7, 345–392 (1954).
255. Belinskij, S. P. On one of the inverse problems for linear symmetrical t-hyperbolic systems with $n+1$ independent variables (in Russian). *Diff. Uravneniya 12*, 15–23 (1976).

256. Belinskij, S. P. On the inverse problem for linear symmetrical t-hyperbolic systems of equations of the first kind. In: *Mathematical Problems of Geophysics* (in Russian). V. Ts. SO AN SSSR, Novosibirsk, issue 6, part 2, pp. 100–109 (1975).
257. Belinskij, S. P. Theorem of uniqueness of one of the inverse problems for hyperbolic systems of the first kind. In: *Incorrect Mathematical Problems and Problems of Geophysics* (in Russian). V. Ts. SO AN SSSR, Novosibirsk, pp. 24–30 (1976).
258. Kabanikhin, S. I. The finite-difference method of determining the coefficients of a hyperbolic system of the first kind. In: *Uniqueness, Stability and Methods of Solving Inverse and Incorrect Problems* (in Russian). V. Ts. SO AN SSSR, Novosibirsk, pp. 36–43 (1980).
259. Nizhnik, L. P. *The Inverse Non-stationary Problem of Dissipation* (in Russian). Naukova Dumka, Kiev (1973).
260. Pukhnachova, T. P. On uniqueness of determining the coefficients of a symmetric hyperbolic system. In: *Uniqueness, Stability and Methods of Solving Inverse and Incorrect Problems* (in Russian). V. Ts. SO AN SSSR, Novosibirsk, pp. 69–76 (1980).
261. Romanov, V. G. Inverse problems of the propagation of seismic and electromagnetic waves. In: *Methods of Solving Incorrect Problems and Their Applications* (in Russian). V. Ts. SO AN SSSR, Novosibirsk, pp. 111–118 (1982).
262. Romanov, V. G. On one formulation of the inverse problem for symmetric hyperbolic systems of the first order (in Russian). *Mat. Zametki 24*, 231–236 (1978).
263. Romanov, V. G. On one inverse problem for weakly correlated hyperbolic systems of the first order. In: *Incorrect Mathematical Problems and Problems of Geophysics* (in Russian). V. Ts. SO AN SSSR, Novosibirsk, pp. 135–148 (1976).
264. Romanov, V. G. On the problem of determining the first part of hyperbolic systems (in Russian). *Diff. Uravneniya 13*, 509–515 (1977).
265. Romanov, V. G. On uniqueness of the solution of one inverse problem for hyperbolic systems of the first order (in Russian). *Mat. Zametki 24*, 359–366 (1978).
266. Romanov, V. G. The problem of determining the coefficients of a linear hyperbolic system (in Russian). *Diff. Uravneniya 14*, 94–103 (1978).
267. Romanov, V. G. and Belinskij, S. P. On the problem of determining the coefficients of a t-hyperbolic system (in Russian). *Mat. Zametki 28*, 525–532 (1980).
268. Romanov, V. G., Kabanikhin, S. I. and Pukhnachova, T. P. On the theory of inverse problems of thermodynamics (in Russian). *Dokl. AN SSSR 266*, 1070–1073 (1982).
269. Romanov, V. G. and Slinyucheva, L. I. The inverse problem for linear hyperbolic systems of the first kind. In: *Mathematical Problems of Geophysics* (in Russian). V. Ts. SO AN SSSR, Novosibirsk, issue 3, pp. 187–215 (1972).
270. Blagoveshchenskij, A. S. On the inverse problem of the theory of seismic waves propagation. In: *Problems of Mathematical Physics* (in Russian). LGU, Leningrad, issue 1, pp. 68–81 (1966).
271. Borodaeva, N. M. On numerical solution of the one-dimensional inverse dynamical problem of seismology. In: *Some Methods and Algorithms of Interpreting Geophysical Data* (in Russian). Nauka, Moscow, pp. 85–91 (1967).
272. Borodaeva, N. M. On the problem of numerical solution of the one-dimensional inverse dynamical problem in the scheme of surveying sea sediments. In: *Mathematical Problems of Geophysics* (in Russian). V. Ts. SO AN SSSR, Novosibirsk, issue 1, pp. 225–234 (1969).
273. Jobert, G. and Brunean, C. Problèmes direct et inverse linéarises pour les hetèrogéneites bidimensionnéles. Cas des ondes plan SH. *C. R. Acad. Sci., Paris 288B*, 313–316 (1979).

274. Jobert, C., Brunean, C. and Cisternas, A. Problème inverse linéaire pour les amplitudes spectrales. Example de la corde vibrante. *C. R. Acad. Sci., Paris 287 B*, 153–155 (1978).

275. Love, L. *Mathematical Theory of Elasticity* (in Russian). ONTI. Moscow (1935).

276. Volkova, E. A. *On One-dimensional Problems for a System of Equations of the Theory of Elasticity of Anisotropic Media* (in Russian). V.T.s. SO AN SSSR, Novosibirsk, preprint 330 (1981).

277. Romanov, V. G. The inverse Lamb problem in a linear approximation. In: *Numerical Methods in Seismic Investigations* (in Russian). Nauka, Novosibirsk, pp. 170–192 (1983).

278. Tikhonov, A. N. On determining the electric characteristics of deep layers of the Earth's core (in Russian). *Dokl. AN SSSR 73*, 295–297 (1950).

279. Barashkov, A. S. On uniqueness of the solution of one inverse problem (in Russian). *ZhVM i MF 13*, 365–372 (1973).

280. Barashkov, A. S. and Dmitriev, V. I. Inverse problems of probing the ionosphere. In: *Computing Methods and Programming* (in Russian). Moscow University Press, Moscow, issue 20, pp. 315–322 (1973).

281. Barthes, V. and Vasseur, G. *An Inverse Problem for Electromagnetic Prospection. Applied Inverse Problems*. Lecture Notes in Physics, 85. Springer, pp. 325–329 (1978).

282. Belinskij, S. P. On one of the formulations of the inverse problem for a system of Maxwell equations. In: *Non-classical Problems of Mathematical Physics* (in Russian). V. Ts. SO AN SSSR, Novosibirsk, pp. 44–55 (1981).

283. Chetaev, D. N. On the solution of the inverse problem of the theory of electromagnetic probings (in Russian). *Izv. AN SSSR (seriya geof.) 12*, 1864–1866 (1959).

284. Dmitriev, V. I. Direct and inverse problems of magnetotellurgical probing of a complex medium (in Russian). *Izv. AN SSSR (seriya fizika zemli) 1*, 64–70 (1970).

285. Dmitriev, V. I. and Il'inskij, A.S., and Sosnikov, A. G. Development of mathematical methods of investigating direct and inverse problems of electrodynamics (in Russian). *UMN 31*, 123–141 (1976).

286. Dmitriev, V. I. and Rudneva, T. L. On the inverse problem of depth magnetotellurgical probing. In: *Computing Methods and Programming* (in Russian). Moscow University Press, Moscow, issue 20, pp. 241–245 (1973).

287. Filatov, V. A. On the inverse problem of magnetic probing (in Russian). *Dokl. AN SSSR 186*, 1315–1317 (1969).

288. Kabanikhin, S. I. and Pukhnacheva, T. P. On determining the conductivity tensor in an anisotropic three-dimensional inhomogeneous medium. In: *Methods of Solving Incorrect Problems and Their Applications* (in Russian). V. Ts. SO AN SSSR, Novosibirsk, pp. 218–222 (1982).

289. Parker, R. L. The inverse problem of electric conductivity in the mantle. *Geophys. J. R. Astron. Soc. 22*, 121–138 (1971).

290. Pukhnachova, T. P. On determining conductivity and electrical permeability in non-homogeneous anisotropic media. In: *Approximation Methods of the Solution and Problems of Inverse Problem Correctness* (in Russian). V. Ts. SO AN SSSR, Novosibirsk, pp. 97–107 (1981).

291. Pukhnachova, T. P. The inverse problem for Maxwell equations in a medium with an anisotropic conductivity (in Russian). *Diff. Uravneniya 18*, 1780–1787 (1982).

292. Vasil'ev, V. G. and Isaev, G. A. On the solution of the one-dimensional inverse problem of the theory of electromagnetic probing (in Russian). *ZhVM i MF 10*, 759–762 (1970).

293. Weidelt, P. The inverse problem of geomagnetic induction. *Z. Geophys. 38*, 257–289 (1972).
294. Weidelt, P. Inversion of two-dimensional conductivity structures. *Phys. Earth Planet. Int. 10*, 282–291 (1975).
295. Beznoshchenko, N. Ya. On determining the coefficient at the lower term of a general parabolic equation (in Russian). *Diff. Uravneniya 12*, 175–176 (1976).
296. Beznoshchenko, N. Ya. On determining the coefficient in a parabolic equation (in Russian). *Diff. Uravneniya 10*, 24–35 (1974).
297. Beznoshchenko, N. Ya. On determining the coefficient q in the equation $u_t - \Delta u + qu = F$ (the case of the first boundary problem in a semi-space) (in Russian). *SMZh 21*, 22–27 (1980).
298. Beznoshchenko, N. Ya. On determining the coefficients at higher derivatives in a parabolic equation (in Russian). *Diff. Uravneniya 11*, 19–26 (1976).
299. Beznoshchenko, N. Ya. On determining the coefficients at lower terms in a parabolic equation (in Russian). *SMZh 16*, 473–482 (1975).
300. Beznoshchenko, N. Ya. On the existence of a solution to the problem of determining the coefficient q in the equation $u_t - \Delta u + qu = F$ (in Russian). *Diff. Uravneniya 15*, 10–17 (1978).
301. Beznoshchenko, N. Ya. and Prilepko A. I. Inverse problems for equations of a parabolic type. In: *Problems of Mathematical Physics and Calculus* (in Russian). Nauka, Moscow, pp. 51–63 (1977).
302. Iskenderov, A. D. On inverse boundary problems with unknown coefficients for some quasi-linear equations (in Russian). *Dokl. AN SSSR 178*, 999–1003 (1968).
303. Iskenderov, A. D. On one inverse problem for quasi-linear parabolic equations (in Russian). *Diff. Uravneniya 10*, 890–898 (1974).
304. Iskenderov, A. D. Some inverse problems for determining filtration and thermophysical characteristics. In: *Non-classical Methods in Geophysics* (in Russian). V. Ts. SO AN SSSR, Novosibirsk, 54–63 (1977).
305. Budak, B. M. and Iskenderov, A. D. Difference method of solving some coefficient boundary problems (in Russian). *Dokl. Akad. Nauk SSSR 171*, 1054–1057 (1966).
306. Budak, B. M. and Iskenderov, A. D. On one class of the boundary problems with unknown coefficients (in Russian). *DAN SSSR 175*, 13–16 (1967).
307. Budak, B. M. and Iskenderov, A. D. On one class of the inverse boundary problems with unknown coefficients (in Russian). *Dokl. Akad. Nauk SSSR 176*, 20–23 (1976).
308. Klibanov, M. V. On inverse problems for one quasi-linear parabolic equation (in Russian). *Dokl. Akad. Nauk SSSR, 245*, 530–532 (1979).
309. Klibanov, M. V. On one class of inverse problems for a parabolic equation and problems of integral geometry (in Russian). *Dokl. AN SSSR 222*, 29–31 (1975).
310. Klibanov, M. V. On one class of operator equations of the first kind (in Russian). *Funkts. Analiz i ego Pril. 11*, 82–83 (1977).
311. Klibanov, M. V. On one problem of integral geometry and an inverse problem for a parabolic equation (in Russian). *SMZh 17*, 75–84 (1976).
312. Klibanov, M. V. On the method of descent in solving some inverse problems. In: *Mathematical Problems of Geophysics* (in Russian). V. Ts. SO AN SSSR, Novosibirsk, issue 5, part 2, pp. 72–77 (1974).
313. Klibanov, M. V. Reconstruction of the function set by the integrals with respect to one set of the ellipsoids of rotation, and an inverse problem for a parabolic equation. In: *Mathematical Problems of Geophysics* (in Russian). V. Ts. SO AN SSSR, Novosibirsk, issue, 6, part 1, pp. 97–116 (1975).

314. Klibanov, M. V. Uniqueness of the solution of one inverse problem for the equation of a parabolic type. In: *Mathematical Problems of Geophysics* (in Russian). V. Ts. SO AN SSSR, Novosibirsk, issue 5, part 2, pp. 59–71 (1974).
315. Cannon, J. R. Determination of an unknown coefficient in a parabolic differential equation. *Duke Math. J. 30*, 313–323 (1963).
316. Cannon, J. R. Determination of certain parameters in heat conduction problems. *J. Math. Anal. Appl. 8*, 188–201 (1964).
317. Cannon, J. R. Determination of an unknown coefficient in the parabolic equation from overspecified boundary data. *J. Math. Anal. Appl. 18*, 112–114 (1967).
318. Cannon, J. R. Determination of an unknown heat source from overspecified boundary data. *SIAM J. Numer. Anal. 5*, 275–286 (1968).
319. Cannon, J. R., Douglas, J. and Jones, B. E. Determination of diffusivity of a isotropic medium. *Int. J. Engng Sci. 1*, 453–460 (1963).
320. Cannon, J. R. and Duchateau, P. An inverse problem for unknown source in a heat equation. *J. Math. Anal. Appl. 75*, 465–485 (1980).
321. Cannon, J. R. and Duchateau, P. Determining unknown coefficients in a nonlinear heat conduction problem. *SIAM J. Appl. Math. 24*, 298–314 (1973).
322. Chavent, G. and Lemonnier, P. Identificated de la non-linearity d'une equation parabolique quasi-lineaire. *Appl. Math. Optim. 1*, 121–162 (1974).
323. Jones, B. F. The determination of a coefficient in parabolic differential equations. Part 1: Existence and uniqueness. *J. Math. Mech. 11*, 907–919 (1962).
324. Jones, B. F. Various methods for finding unknown coefficients in parabolic differential equations. *Commun. Pure Appl. Math. 16*, 33–34 (1963).
325. Langmach, H. On the determination of functional parameters in some parabolic differential equations. Theory of nonlinear operators, *Proceedings of the Fifth International Summer School* (Centre for Institute of Mathematics and Mechanics, Academy of Sciences, GDR, Berlin, 1977). Academie, Berlin, pp. 175–184 (1978).
326. Lavrent'ev, M. M., Noppe, M. G. and Reznitskaya, K. G. *On the Magnetotellurgical Method with a Sharp Variation in the Field* (in Russian). Trudy SNIIGGIMSa, Novosibirsk, issue 215, pp. 31–40 (1975).
327. Lavrent'ev, M. M. and Reznitskaya, K. G. Theorems of uniqueness of certain nonlinear inverse problems for equations of a parabolic type (in Russian). *Dokl. AN SSSR 208*, 531–533 (1973).
328. Lavrent'ev, M. M., Reznitskaya, K. G. and Yakhno, V. G. *One-dimensional Inverse Problems of Mathematical Physics* (in Russian), Nauka, Novosibirsk (1982).
329. Linke, H. P. Über ein inverses Problem für die Wärmeleitungsgleichung. *Wiss. Z. der Techn. Hochsch. Karl-Marx-Stadt 18*, 445–447 (1976).
330. Linke, H. P. Zwei inverse Problems für die Wärmeleitungsgleichung im zweidimensionalen Fall. *Wiss. Z. der Techn. Hochsch. Karl-Marx-Stadt 20*, 705–710 (1978).
331. Muzylev, N. V. Theorems of uniqueness for some inverse problems of heat conductivity (in Russian). *ZhVM i MF 20*, 388–400 (1980).
332. Prilepko, A. I. Inverse problems of the theory of potential (in Russian). *Mat. Zametki 14*, 755–765 (1973).
333. Reznitskaya, K. G. Relations between the Cauchy problem solutions for the equations of various kinds and inverse problems. In: *Mathematical Problems of Geophysics* (in Russian). V. Ts. SO AN SSSR, Novosibirsk, issue 5, part 1, pp. 55–62 (1974).
334. Reznitskaya, K. G. Theorem of existence and uniqueness of one one-dimensional nonlinear inverse problem of the theory of heat conductivity. In: *Mathematical*

Problems of Geophysics (in Russian). V. Ts. SO AN SSSR, Novosibirsk, issue 4, pp. 131–134 (1973).

335. Reznitskaya, K. G. Theorem of uniqueness of some inverse problems for a diffusion equation. In: *Mathematical Problems of Geophysics* (in Russian). V. Ts. SO AN SSSR, Novosibirsk, issue 6, part 1, pp. 154–159 (1975).
336. Romanov, V. G. On one inverse problem for an equation of a parabolic kind (in Russian). *Mat. Zametki 19*, 597–602 (1976).
337. Rundell, W. Determination of an unknown non-homogeneous term in a linear partial differential equation from overspecified boundary data. *Appl. Anal. 10*, 231–242 (1980).
338. Shishko, N. P. On uniqueness of the solution of two inverse problems (in Russian). *SMZh 16*, 399–404 (1975).
339. Tabaldyev, B. P. Restoration of an unknown function in the coefficients of the equation in partial derivatives of a parabolic kind. In: *Investigations with Integro-Differential Equations* (in Russian). II.IM, Frunze, pp. 275–294 (1977).
340. Volkov, V. M. The inverse problem for a quasi-linear equation of a parabolic kind. In: *Studying Correctness of Inverse Problems and some Operator Equations* (in Russian). V. Ts. SO AN SSSR, Novosibirsk, pp. 27–36 (1981).
341. Yakhno, V. G. On classes of uniqueness and stability of one formulation of the inverse problem for parabolic equations of the second order. In: *Incorrect Mathematical Problems and Problems of Geophysics* (in Russian). V. Ts. SO AN SSSR, pp. 154–165 (1979).
342. Yakhno, V. G. On one inverse problem for a system of parabolic equations (in Russian). *Diff. Uravneniya 15*, 566–569 (1979).
343. Anger, C. Some remarks on inverse problems in differential equations (Elliptische Differentialgleichungen). *Congress in Rostock 1977*. Wilhelm University Press, Rostock, pp. 31–51 (1978).
344. Iskenderov, A. D. On one inverse problem for equations of an elliptical type. In: *Inverse Problems for Differential Equations* (in Russian). V. Ts. SO AN SSSR, Novosibirsk, pp. 107–115 (1972).
345. Iskenderov, A. D. The inverse problem of defining the coefficients of a quasi-linear elliptical equation (in Russian). *Izv. AN Azerb. SSR (seriya fiz.-tech. i matem. nauk) 2*, 80–85 (1978).
346. Temirbulatov, S. I. *Inverse Problems for Elliptical Equations* (in Russian). Kazakh University Press, Alma-Ata (1975).
347. Temirbulatov, S. I. On uniqueness of the solution of an inverse problem of induction logging (in Russian). *SMZh 8*, 471–474 (1967).
348. Zapreev, A. S. Theorem of uniqueness of the solution of one plane inverse problem for a Helmholtz equation. In: *Incorrect Mathematical Problems and Problems of Geophysics* (in Russian). V. Ts. SO AN SSSR, Novosibirsk, pp. 46–63 (1976).
349. Zapreev, A. S. and Tsetsokho, V. A. The inverse problem for a Helmholtz equation (in Russian). V. Ts. SO AN SSSR, Novosibirsk, preprint 22 (1976).
350. Wiener, N. and Paley, R. *Fourier Transform in a Complex Domain* (in Russian). Nauka, Moscow (1964).
351. Iskenderov, A. D. Multi-dimensional inverse problems for linear and quasi-linear parabolic equations (in Russian). *Dokl. AN SSSR 225*, 1005–1008 (1975).
352. Anikonov, Yu. E. On uniqueness of the solution of inverse problems. In: *Mathematical Problems of Geophysics* (in Russian). V. Ts. SO AN SSSR, Novosibirsk, issue 1, pp. 26–40 (1969).

353. Bukhgeim, A. L. Necessary conditions of stability of one class of integro-differential equations. In: *Computing Methods and Programming* (in Russian). V. Ts. SO AN SSSR, Novosibirsk, pp. 78–85 (1975).
354. Iskenderov, A. D. Some inverse problems of determining the right-hand parts of differential equations (in Russian). *Izv. AN Azerb. SSR (seriya fiz.-tech. i matem. nauk) 2*, 58–63 (1976).
355. John, F. Continuous dependence on data for solutions of partial differential equations with a prescribed bound. *Commun. Pure Appl. Math. 13*, 551–585 (1960).
356. John, F. Numerical solution of problems which are not well posed in the sense of Hadamard. *Symposium on Numerical Treatment of Partial Differential Equations with Real Characteristics.* Rome, pp. 103–116 (1959).
357. Klibanov, M. V. An inverse problem for a parabolic equation and one problem of integral geometry (in Russian). *SMZh 17*, 564–569 (1976).
358. Lavrent'ev, M. M. On one class of operator equations on a plane. In: *Conditionally Correct Mathematical Problems and Problems of Geophysics* (in Russian). V. Ts. SO AN SSSR, Novosibirsk, pp. 52–57 (1979).
359. Romanov, V. G. Inverse problems for hyperbolic systems. In: *Computing Methods in Mathematical Physics, Geophysics and Optimal Control* (in Russian). Nauka, Novosibirsk, pp. 128–142 (1978).
360. Romanov, V. G. and Yakhno, V. G. On one linearized formulation of the problem of determining a hyperbolic operator (in Russian). *Mat. Zametki 28*, 391–400 (1980).